High Resolution Radar Cross-Section Imaging

For a complete listing of the *Artech House Radar Library*,
turn to the back of this book . . .

High Resolution Radar Cross-Section Imaging

Dean L. Mensa
Pacific Missile Test Center
Point Mugu, California

Artech House
Boston • London

Library of Congress Cataloging-in-Publication Data

Mensa, Dean L.
High resolution radar cross-section imaging / Dean L. Mensa.
p. cm.
Includes bibliographical references and index.
ISBN 0-89006-389-3
1. Synthetic aperture radar. I. Mensa, Dean L. High resolution radar imaging. II. Title.
TK6592.S95M46 1991 90-1140
621.3848--dc20 CIP

British Library Cataloguing in Publication Data

Mensa, Dean L.
High resolution radar cross section imaging. --2d rev.
1. Synthetic aperture radar systems
I. Title
621.38485

ISBN 0-89006-389-3

685 Canton Street
Norwood, MA 02062

International Standard Book Number: 0-89006-389-3
Library of Congress Catalog Card Number: 90-1140

10 9 8 7 6 5 4 3 2 1

To My Family

Contents

Preface

This book is a revision of a previous edition published in 1981 under the title *High Resolution Radar Imaging*. In the ensuing period, high resolution radar imaging, obtained by combining wide signal bandwidths and inverse synthetic aperture methods, has evolved from an experimental method into a working tool widely applied to diagnostic radar cross-section measurements. These techniques are embodied in a number of commercial instrumentation systems which combine microwave equipment, computers, and graphic displays to acquire, process, and present two-dimensional images.

The emphasis throughout the book is on the context in which radar imaging is applied to radar cross-section measurements. Four of the seven chapters have been revised to include new material and more current examples of radar images. Three new chapters have been added to cover: instrumentation and signal processing, tomographic imaging, and maximum entropy methods. The chapter on instrumentation and signal processing provides guidelines for acquiring the data necessary to produce the images. The text is suitable for a short course presented to practicing engineers during a period of three to five days. As in the previous edition, the theoretical limits of imaging performance are derived and related to physical quantities and parameters of the signal-processing algorithms. Mathematics is used to establish the theoretical basis, but is kept to a minimum; whereever possible, we stress physical interpretation of the processes involved.

The experimental results were obtained at the facilities of the Pacific Missile Test Center. I am indebted to my colleagues, K. Vaccaro, A. Bati, and others in the Radar Signature Branch at the center for their excellent applications of computer resources and for many consultations, and to E. Eichblatt for his invaluable help with the illustrations. In addition, I thank the participants in short courses for their questions, comments, suggestions, and ideas. I am most grateful to my wife, Judy, for her support and encouragement, and for the many hours spent in preparing the manuscript.

Chapter 1
Introduction

1.1 THE CONCEPT OF RADAR IMAGING

Radar sensors respond to electromagnetic waves which are scattered when the propagation of incident waves is disturbed by the presence of an object. The physical mechanism can be heuristically described as follows: the incident fields induce currents in the volume bounded by the object to generate scattered fields, subject to constraints imposed by boundary conditions.

To determine the object's scattering properties, we must visualize the object features which stimulate the scattered fields. A quantitative, spatial map of the current density in the volume occupied by the object would reveal the physical mechanisms responsible for the scattered fields. However, because the scattered fields are determined by the integrated effect of the current distribution, the latter cannot be measured directly by the radar. Nevertheless, we strive to obtain, from measurements of scattered fields by a remote sensor, any information leading to the spatial separation of object features responsible for the scattering.

Here, we define a *radar image* as the spatial distribution of reflectivity corresponding to the object. Adhering to this definition, the radar image thus can be considered a collection of reflection coefficients assigned to an array partitioning the object space. This notion is consistent with the IEEE definition [1], which states that an image is "a spatial distribution of a physical property such as radiation, electric charge, conductivity, or reflectivity, mapped from another distribution of either the same or another physical property."

To characterize the object properly, an image must provide a spatial, quantitative description of the object's physical property of interest with fidelity which equals or exceeds the discrimination limits of the eventual observing system. The complete characterization of complex objects may require several images which differ as a function of the viewing angle. Useful optical images display spatial distributions of optical reflectivity with sufficient detail for the object to be recognized by a human observer. Similarly, a radar image presents a spatial distribution of radar reflectivity sufficient to characterize the object illuminated. Although the similarity of radar images to optical images can convey useful information, the quality of the radar image should not be judged by how closely it approximates the optical image, but rather how faithfully the radar image represents the spatial distribution of radar reflectivity.

An essential feature of any high-quality image is resolution, the ability to represent distinctly two closely-spaced elements of the object, with the resolvable separation related inversely to the aperture size of the imaging system. Because the human visual system is equipped with apertures on the order of 10,000 wavelengths (being orders of magnitude greater in terms of wavelength than radar apertures), human observers are accustomed to perceiving images consisting of millions of resolvable elements. As a result, optical replicas of radar images often appear primitive by the standards used to judge optical images. Nevertheless, radar images convey unique information and, in some cases, can display object features not otherwise obtainable. Figure 1.1 shows examples of radar imagery obtained by synthetic-aperture processing (described in [2]). This type of image, generated by processing radar data collected from an airborne platform, provides a two-dimensional map of terrain features.

The principal applications for radar images are:

1. Diagnostic investigations to identify and characterize radar reflectivity components of complex objects.
2. Analytical and physical simulation of radar signatures for determining responses of radar sensors.
3. Object recognition systems which use the image as an identifier unique to a particular object.
4. Nondestructive and noninvasive testing for imaging through media that support radar propagation.

1.2 OBJECTIVES AND ORGANIZATION OF THE TEXT

Although pertinent to each of the preceding applications, the material in this text is directed to radar imaging in which data from controlled measurements are analyzed to diagnose and characterize the scattering mechanisms of complex objects. Specifically, our objective is to describe and analyze techniques for obtaining high-resolution, quantitative radar images by processing data obtained from *radar cross-section* (RCS) ranges.

The book contains chapters treating: fundamental imaging concepts, instrumentation methods and techniques, signal processing required for imaging, tomographic imaging, and maximum entropy techniques. The concepts and techniques

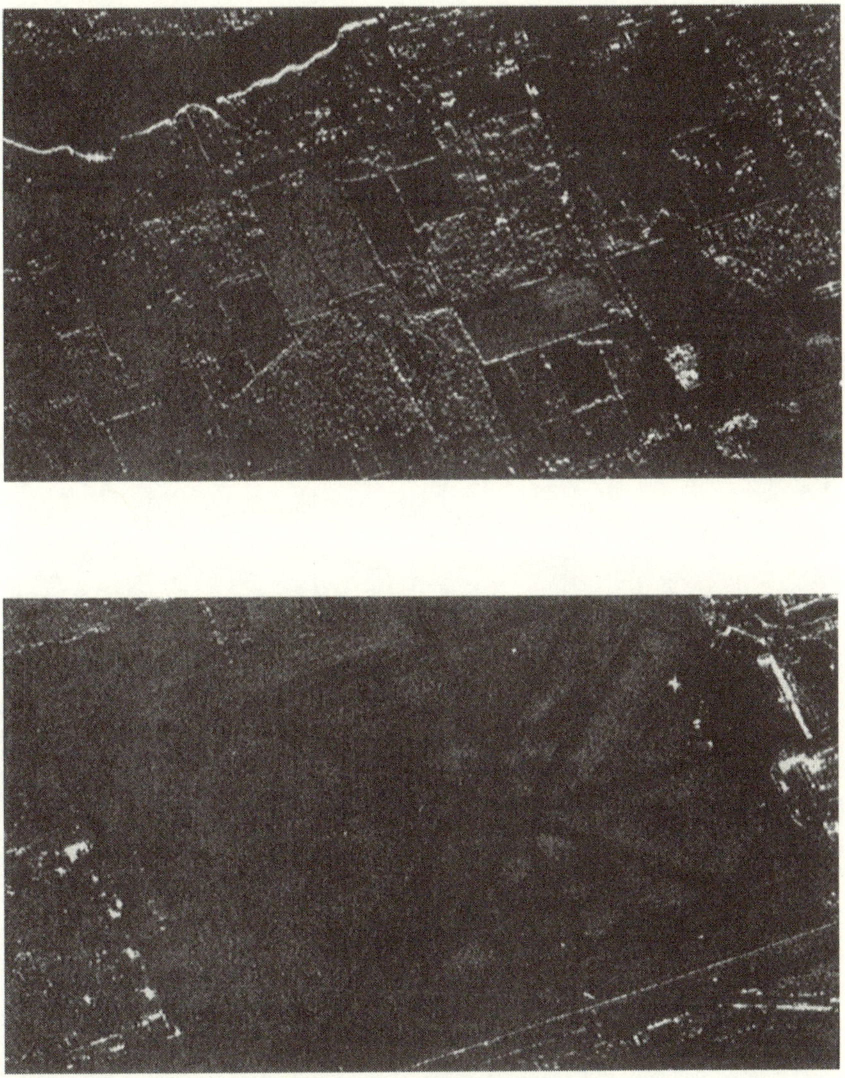

Figure 1.1 Examples of terrain imagery by airborne synthetic aperture processing.

are supported by illustrations and examples of images. The material, directed to practicing engineers or graduate students, can be covered in 20 to 30 hours of formal instruction.

REFERENCES

1. Jay, F., ed., *IEEE Standard Dictionary of Electrical and Electronics Terms,* New York, IEEE, 1984, p. 424.
2. Jensen, H., L.C. Graham, L.J. Porcello, and E.N. Leith, "Side-Looking Airborne Radar," *Scientific American,* October 1977, pp. 84–95.

Chapter 2
Fundamental Imaging Concepts

2.1 DIRECT IMAGING METHODS

This chapter develops the basic techniques applicable to radar imaging. The treatment is limited to one-dimensional resolution to develop a conceptual foundation with minimum mathematical complications; extensions and combinations of the basic processes to achieve multidimensional imaging are considered in subsequent chapters. The emphasis here is on establishing fundamental concepts and applying results reported in the available literature to meet the objectives.

In the most direct form, imaging a three-dimensional object can be accomplished by using a range-gated, short-pulse radar with a pencil-beam antenna, as shown in Figure 2.1. The antenna beam and the range gate are systematically scanned throughout the three-dimensional volume, and the intensity of the signals received is displayed as a function of the spatial coordinates being interrogated. Because the spatial resolution is established by the angular and range resolution of the sensor, the image is obtained directly, without subsequent processing. The example shown in the lower part of Figure 2.2 is a two-dimensional image of a scale model of a boat 4.5 m in length, obtained by raster-scanning the focused spot of a large antenna across the object shown in the upper part of the figure. Because range is not resolved, the image is a projection of the reflectivity distribution along the radar line-of-sight. The image was obtained using a wavelength of 3 mm and an antenna with an aperture diameter of 1 m, focused at a distance of 25 m to produce a circular spot approximately 10 cm in diameter.

The relative simplicity of forming the image, which requires minimal data processing, and the fact that the image can be obtained while the object is stationary constitute the advantages of the direct imaging method. The principal disadvantages of the direct method are:

a. A high degree of spatial resolution requires subnanosecond pulses and large apertures, posing practical limitations.

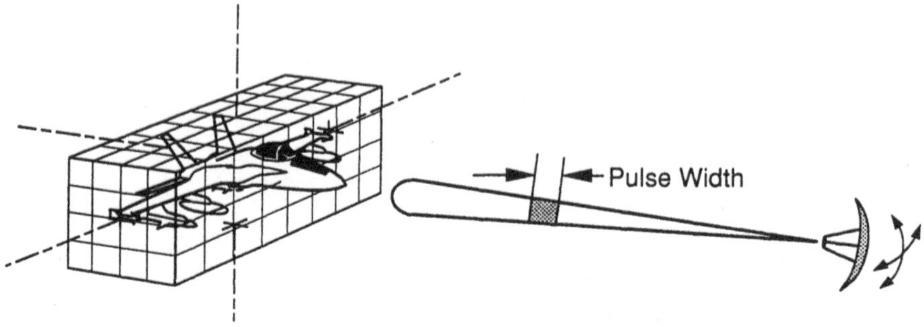

Figure 2.1 Direct imaging by scanning narrow beam of short-pulse radar across target.

Figure 2.2 Microwave image of boat scale model by focused spot scanning.

b. The cross-range resolution, obtained from the antenna beamwidth, degrades as the range increases.

c. The spatially limited irradiation of the object omits interactions and coupling that may occur when the entire object is simultaneously irradiated.

The first two disadvantages can be overcome by imaging at short ranges with millimeter or submillimeter-wave frequencies for which electrically large antennas are feasible, conditions that prevailed for the example of Figure 2.2.

2.2 SYNTHETIC IMAGING METHODS

As supported by material presented in following sections, the fundamental physical properties which provide spatial resolution in any radar system are frequency and spatial diversity.

Techniques of imaging by direct methods utilize frequency and spatial diversities that are embodied in the radar sensor. In the case of a short-pulse radar using a large aperture, for example, the frequency diversity is inherent to the wide-bandwidth pulse spectrum, and the spatial diversity is inherent to the range of angles subtended by the aperture. Spatial resolution can be obtained by synthetic means when results of many observations of the object at different frequencies and angles are coherently combined. The resolutions commensurate with wide frequency bands and large apertures can thus be obtained by combining signals obtained from sequences of narrow band, single-angle observations. The term "synthetic" here refers to the synthesis of short-pulse, narrow-beam radar performance from elemental measurements. These methods are employed in the majority of systems currently applied to high-resolution RCS imaging. The material in the following sections deals exclusively with synthetic methods.

2.2.1 Range Processing

The most basic method of radar imaging involves discrimination on the basis of range. The determination of range is accomplished by measuring the time delay between a distinct feature present in the transmitted waveform and recognizable in the received waveform, and computing range using knowledge of the propagation velocity. The most readily mechanized system for range measurement is the pulsed radar in which the leading and trailing edges of the pulse provide inherent time markers. The round-trip propagation delay τ for a reflector at range R is $2R/c$, where c is the propagation velocity. Thus, a pulse of duration T corresponds to a range increment $\Delta R = cT/2$, and two points separated by greater distances can produce distinguishable pulses that can be resolved.

The resolution of closely-spaced object features can be accomplished by narrowing the transmitted pulse duration and increasing the system bandwidth, B, such that $BT = 1$. A time-bandwidth product approximating unity is used to maximize the sensitivity of pulsed radars in which a carrier is amplitude-modulated by a pulsed waveform. The measurement of range can be treated more generally from a communication theory viewpoint. Two significant aspects of the measurement process are *accuracy* and *resolution;* the former constitutes the ability to provide an unbiased estimate of the absolute range, while the latter is the ability to distinguish closely spaced objects. The standard deviation of a range measurement error in the presence of Gaussian noise is [1,2]:

$$\sigma_R = c(8B^2E/N)^{-1/2} \tag{2.1}$$

where B is the signal bandwidth, and E/N is the radio of signal energy to two-sided noise power spectral density.

Equation (2.1) indicates that an arbitrary degree of range accuracy can be obtained by using signals with wide bandwidth or high energy. In imaging applications, however, the accurate determination of range is of limited significance, while resolution of range is of primary interest. The principal objective for imaging applications, therefore, is to perform quantitative measurements which resolve the range difference between closely-spaced scatterers. The *propagation delay,* from which the range is derived, is determined by comparing the time-shifted transmitted waveform with the received waveform. Because the received waveform is a delayed, attenuated, noisy replica of the transmitted waveform, the time shift required to bring the two waveforms into alignment constitutes a measure of the propagation delay and, therefore, of range.

Using a waveform which is maximally different from its shifted self provides high sensitivity to the degree of alignment, thus allowing resolution of the small differences in time delay. The preceding notion is mathematically expressed by the autocorrelation function:

$$R(\tau) = \int_{-\infty}^{+\infty} s(t)s*(t + \tau)dt \tag{2.2}$$

where $s(t)$ is the signal, τ is the time shift, and the conjugate (denoted by the asterisk) is written for generality to include complex signals. $R(\tau)$ is a real, even function with a maximum at $\tau = 0$; $R(\tau)$ quantitatively describes the degree to which a waveform matches its shifted self, as a function of the shift. The requirement to distinguish between closely spaced scatterers having nearly equal delays dictates an autocorrelation function which falls sharply as τ increases, the ideal $R(\tau)$ being an impulse. Practical autocorrelation functions, having nonzero width, are described by the effective correlation width, $\Delta\tau$.

An alternate description of the signal is the *power spectrum,* or *power spectral density,* which describes the distribution of power per unit bandwidth as a function of frequency. The range of frequencies that contains most of the signal power is the bandwidth, B. $R(\tau)$ and the power spectrum $S(f)$ constitute a Fourier transform pair [3] expressed by

$$S(f) = \int_{-\infty}^{+\infty} R(\tau) \exp(-j2\pi f\tau)\, d\tau \tag{2.3}$$

The uncertainty relations between Fourier transform pairs, namely $B\Delta\tau = 1$ [4], can be invoked to conclude that an autocorrelation function having a narrow width corresponds to a wide signal bandwidth, B. Various statistical measures of $\Delta\tau$ and B can be formulated leading to slightly different constants in the uncertainty relation, which depend on the shape of the autocorrelation function. Nevertheless, the inverse relation is firmly established, and the above criterion is usually accepted to determine the limit of range resolution:

$$\Delta R = c\Delta\tau/2 = c/2B \tag{2.4}$$

Equation (2.4) indicates that the achievable range resolution depends only on the signal bandwidth, and is independent of the transmitted waveform, range, and frequency. Thus, two radars operating at 100 GHz and 3 GHz, each using a signal bandwidth of 1 GHz, allow the same degree of range resolution. Of course, higher frequency systems can provide wider bandwidths, and thus have the potential for greater resolution.

Figure 2.3 summarizes the relations expressed by the preceding equations. Figure 2.4 shows examples of signals with their corresponding autocorrelation functions and power spectra. The signals represent waveforms which would be used to modulate a carrier. Signals with more rapid variations exhibit wider spectra and narrower autocorrelation functions, thus providing higher range resolution.

The general viewpoint described in the preceding paragraphs indicates that signal bandwidth is the fundamental quantity that establishes range resolution. This means that the required range resolution can be obtained when object responses are measured over a band of frequencies. Any radar waveform that supports an extended bandwidth can be used; the specific type of waveform only determines the necessary implementation of the receiver for coherently processing the wideband signal.

The fundamental observation that the spectral extent of the processed signal dictates the achievable resolution is exploited by methods which achieve high range-resolution by synthetic means. In contrast to direct methods, in which all the spectral components of the signal must be present simultaneously, synthetic resolution can be achieved even though spectral components are only present sequentially. In the simplest implementation of high range-resolution by synthetic means, several narrow

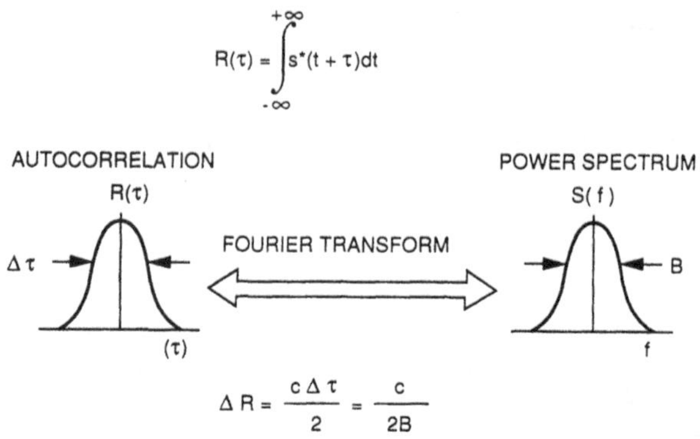

Figure 2.3 Summary of relations for range processing.

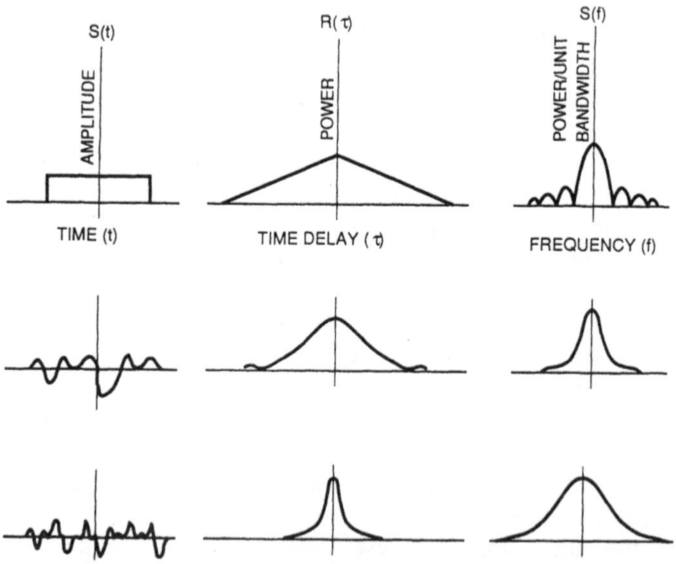

Figure 2.4 Qualitative examples of waveforms, autocorrelation functions, and power spectra.

band measurements are made at discrete frequency increments. Although the individual narrow band responses have insignificant resolution potential, the coherent combination of all the responses provides the resolution allowed by the total bandwidth spanned. This is consistent with the procedure commonly used in the analysis of linear systems, in which the impulse response is obtained from the steady-state frequency response.

A linear system is characterized by $H(f)$, the steady-state response to a sinusoidal input as a function of frequency and $h(t)$, the response to an impulse input. The transfer function and impulse response, represented symbolically in Figure 2.5, constitute a Fourier transform pair expressed by

$$H(f) = \mathcal{F}[h(t)] = \int_{-\infty}^{+\infty} h(t) \exp(-j2\pi ft)\, dt \tag{2.5}$$

$$h(t) = \mathcal{F}^{-1}[H(f)] = \int_{-\infty}^{+\infty} H(f) \exp(+j2\pi ft)\, df \tag{2.6}$$

Equations (2.5) and (2.6) form the underlying theory of linear system analysis; they interrelate frequency and time responses, each sufficient to describe the behavior of the system to arbitrary inputs. The impulse response can be measured experimentally either by observing the response to an input which approximates an impulse, or by performing a Fourier transform on the steady-state response to a sinusoidal input as a function of frequency.

The application of linear system theory to radar backscattering can be made by considering an object, surrounded by an anechoic enclosure, as shown in Figure 2.6, in which the input and output consist of the transmitted and received radar signals, respectively. Linearity is upheld because electromagnetic fields, being linear, obey superposition. The complex amplitude of the output signal, measured relative to the

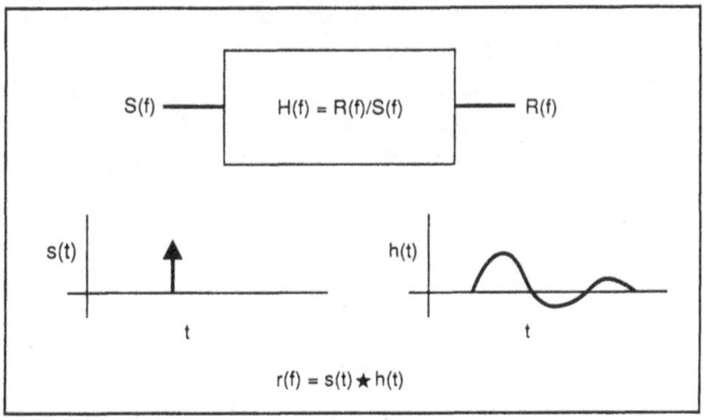

Figure 2.5 Linear system relations.

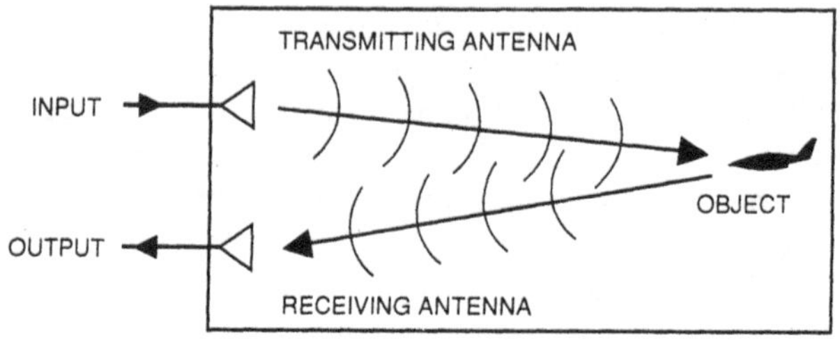

Figure 2.6 Linear system representation of radar target backscattering.

input signal, constitutes a description of the object response, and defines the complex variable $H(f)$. If the latter can be determined over a wide frequency band, the response to very short radar pulses, ultimately approaching impulses, can be determined by computing the Fourier transform of $H(f)$. This allows range (time) discrimination of individual backscattering sources on the object.

As indicated in (2.6), computation of the impulse response requires knowledge of the system response to all frequencies. The inability to determine $H(f)$ for all frequencies poses a limitation, and allows the computation of responses only to input signals with spectra contained in the frequency range for which $H(f)$ is available. Limiting the description of $H(f)$ over the spectral window defined by the range of frequencies B is conceptually equivalent to imposing an ideal bandpass filter in series with the system transfer function. The system transfer function is then represented by the product $H(f)\ W(f)$, where

$$W(f) = \begin{cases} 1, & \text{for } \quad |(f - F)| < B/2 \\ 0, & \text{otherwise} \end{cases} \tag{2.7}$$

The effect of the spectral window can be demonstrated by finding the response of an idealized point scatterer of constant magnitude at range R. The complex amplitude, plotted as a function of frequency in Figure 2.7, exhibits even and odd symmetry about zero frequency for the magnitude and phase, respectively. The band-limited impulse response of the system is obtained by the Fourier transform of the frequency response as follows:

$$\begin{aligned} h(t) &= \mathcal{F}^{-1}[T(f)W(f)] \\ h(t) &= \int_{-F-B/2}^{-F+B/2} \exp(j2\pi ft)\, df + \int_{F-B/2}^{F+B/2} \exp(j2\pi ft)\, df \\ h(t) &= 2B[\sin(\pi Bt)/\pi Bt] \cos(2\pi ft) \end{aligned} \tag{2.8}$$

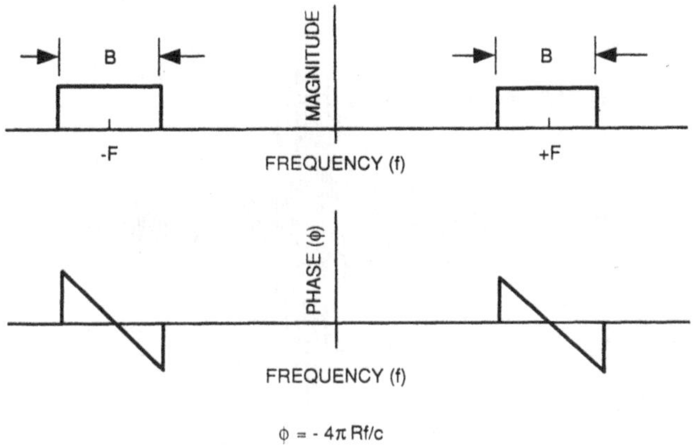

Figure 2.7 Complex amplitude response of point scatterer as a function of frequency.

Equation (2.8), plotted in Figure 2.8, shows the output signal response to an impulse input. The period of rapid oscillations is the reciprocal of the center frequency, F, and the time span between the peak and first null of the envelope is the reciprocal of the bandwidth.

The ability to distinguish two closely-spaced scattering points on an object, therefore, is limited to a differential range corresponding to a time determined by the width of the central lobe of Figure 2.8. Because a time separation $\Delta t = 1/B$ corresponds to a round-trip differential range $c\Delta t/2$, the achievable range resolution is:

$$\Delta R = c/2B \tag{2.9}$$

The preceding analysis indicates that practical constraints of limited bandwidth smear the impulse response of the system, and thus limit range resolution. With band-limited instrumentation systems, therefore, we can expect to recover only the true impulse response of the object convolved with the inverse Fourier transform of the spectral window function.

The effects of increasing the fractional bandwidth, defined as the ratio of bandwidth to center frequency, are illustrated in Figure 2.9, which shows impulse responses corresponding to various fractional bandwidths. As the fractional bandwidth is increased, the central lobe of the impulse response is narrowed; in the limit of an idealized spectral window extending from 0 to $2F$ ($B/F = 2$), the width of the impulse response narrows to a time duration $1/2F$, corresponding to a range extent of $\lambda/4$.

A convenient method of extending the spectral content of a pulsed carrier waveform is to modulate the phase of the carrier during the pulse (*intrapulse modulation*),

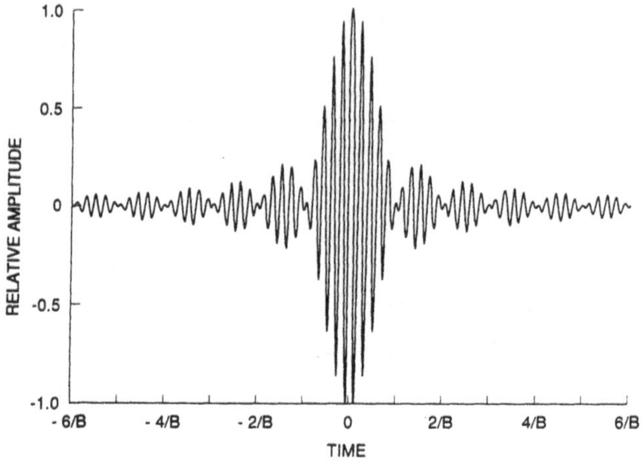

Figure 2.8 Impulse response of rectangular spectral window.

or to alter the frequency of the carrier from pulse to pulse (*interpulse modulation*). These types of waveforms, characterized by large time-bandwidth products $BT \gg 1$, provide both a high degree of resolution by virtue of their large bandwidth and high energy content by virtue of their long duration. This class of waveforms avoids the trade-off between increasing resolution at the cost of sensitivity that is inherent to pulsed waveforms for which $BT = 1$. Aspects of signal energy and sensitivity are addressed in Chapter 3.

2.2.2 Synthetic Aperture Processing

A high degree of resolution in the cross-range direction can be obtained by scanning a focused beam across the object. If the aperture which forms the scanning beam is focused at the object plane, the minimum lateral extent of the focused spot is approximately

$$\Delta = \lambda R/D \tag{2.10}$$

where

λ = wavelength,
D = aperture dimension,
R = observation distance.

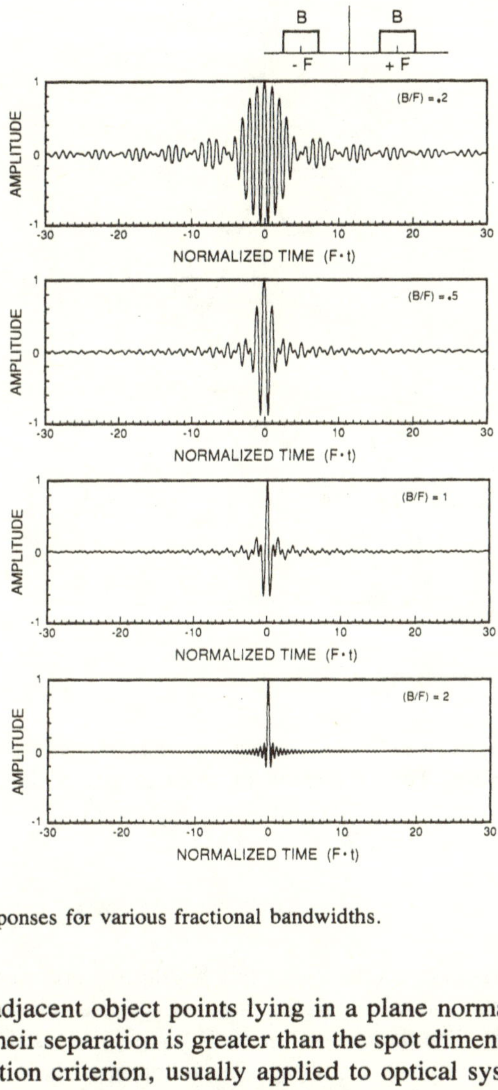

Figure 2.9 Impulse responses for various fractional bandwidths.

Resolution of two adjacent object points lying in a plane normal to the line-of-sight can be expected if their separation is greater than the spot dimension. This constitutes the Rayleigh resolution criterion, usually applied to optical systems [5].

An alternative representation of the Rayleigh resolution criterion, applied in a radar context, can be seen in Figure 2.10. Signals received from distant scatterer 1, located on the x-axis, by elements along the aperture are compared to those corresponding to scatterer 2, displaced by Δ. The signal scattered from 1 is received by aperture elements in phase, while the signal scattered from 2 is received by aperture elements with a varying phase caused by the different round-trip distances.

Two adjacent scatterers can be resolved if their phase histories, observed across the aperture, differ by 360 degrees or more. A differential path length of ± 0.25

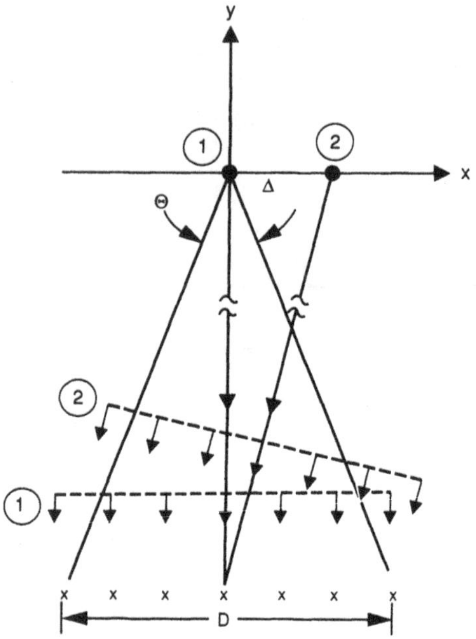

Figure 2.10 Resolution of closely-spaced scatterers.

wavelengths at the aperture extremes causes the required phase variation of 360 degrees across the aperture. The resolution in cross range, Δ, is found as follows:

$$\Delta/R = (\lambda/4)/(D/2)$$

$$\Delta = \lambda R/2D = \lambda/2\,\sin(\Theta) \tag{2.11}$$

Equation (2.11), equivalent to the Rayleigh resolution criterion for radar applications, indicates that the lateral resolution has an inverse relation to the angle subtended by the aperture. The resolutions in (2.10) and (2.11) differ by a factor of two as a consequence of the two-way propagation inherent to radar.

For a fixed wavelength and observation distance, the resolution is improved by increasing the aperture dimension. For example, imaging an object with two-wavelength resolution from a distance of 40 m requires an aperture measuring 10 m, subtending an angle of 14.5 degrees. To obtain the same resolution from a distance of 400 m, the aperture subtending 14.5 degrees would measure 100 m. The large aperture dimension relative to the observation distance places the object in the near field of the antenna. Consequently, the aperture would require focusing to achieve the stated resolution.

The difficulties associated with large physical apertures can be avoided by synthesizing an equivalent aperture which conforms to the surface of the physical aperture. This is accomplished by storing samples of complex field amplitudes obtained with a sensor sequentially stepped through small incremental distances. The stored signals are subsequently summed coherently to produce signals equivalent to those that would be received by the physical aperture. Using this procedure, the equivalent of a large physical aperture is synthesized by summing a sequence of field samples obtained with a small sensor. If the stored data are then processed to achieve a scanning of the synthetic aperture beam across the object, an image can be formed which resolves the object scatterers on the basis of cross range.

To illustrate the equivalence in the directivity of real and synthetic apertures, we compare the responses of a linear real array and a linear synthetic array. The real aperture with uniform illumination can be treated as an array of identical elements, excited simultaneously for transmission, and summed coherently upon reception, as shown in Figure 2.11. The device connected to the array elements performs a summation on reception, and equal power division on transmission. The response of the array to a unit-amplitude point object is the superposition of field

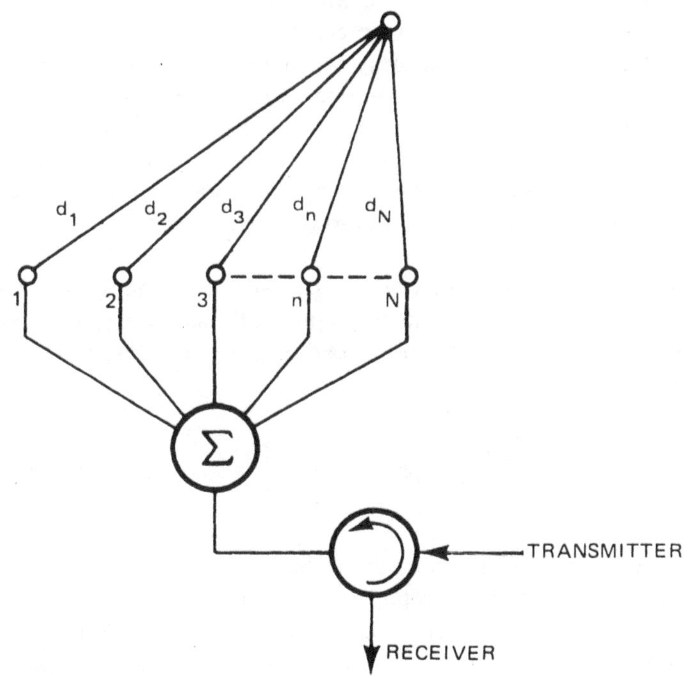

Figure 2.11 Physical aperture modeled as an antenna array.

values received by the individual elements; each receives a contribution from signals transmitted by all elements. The normalized complex field received from a distant point object by the nth element is given by

$$
\begin{aligned}
v_n &= \exp[-j2\pi\lambda^{-1}(d_n + d_1)] + \exp[-j2\pi\lambda^{-1}(d_n + d_2)] + \dots \\
&\quad + \exp[-j2\pi\lambda^{-1}(d_n + d_N)] \\
&= \exp(-j2\pi\lambda^{-1}d_n) \sum_{i=1}^{N} \exp(-j2\pi\lambda^{-1}d_i) \qquad (2.12)
\end{aligned}
$$

The output of the entire array is the sum of signals received by each element:

$$
\begin{aligned}
v_0 &= \sum_{n=1}^{N} v_n = \sum_{n=1}^{N} \exp(-j2\pi\lambda^{-1}d_n) \sum_{i=1}^{N} \exp(-j2\pi\lambda^{-1}d_i) \\
v_0 &= \left[\sum_{n=1}^{N} \exp(-j2\pi\lambda^{-1}d_n) \right]^2
\end{aligned} \qquad (2.13)
$$

The term inside the brackets is the one-way array response; this represents the response of the array if it were used to receive signals from a source located at the distant point. The response of the array when it is used for both transmitting and receiving (two-way) is the square of the one-way response. The cross-products generated in the squaring process represent the coupling terms associated with signals received by one element from transmissions by other elements. Equation (2.13) shows that the two-way response of the array is simply the square of the one-way response, an intuitively satisfying result when the array is viewed as a linear system obeying reciprocity.

The synthetic array is formed by sequentially transmitting and receiving with each individual element and subsequently coherently summing the received signals. The array geometry is identical to the preceding case. The response of the synthetic array will be:

$$
v_0 = \sum_{n=1}^{N} \exp(-j4\pi\lambda^{-1}d_n) = \sum_{n=1}^{N} [\exp(-j2\pi\lambda^{-1}d_n)]^2 \qquad (2.14)
$$

The response of (2.14) differs clearly from that of (2.13). The former represents the square of a sum, while the latter represents the sum of squared terms. To illustrate the difference, the far-field responses of a real and synthetic array are determined. The array consists of N identical elements separated by d, and arranged in a line, as shown in Figure 2.12. The response of the real array is

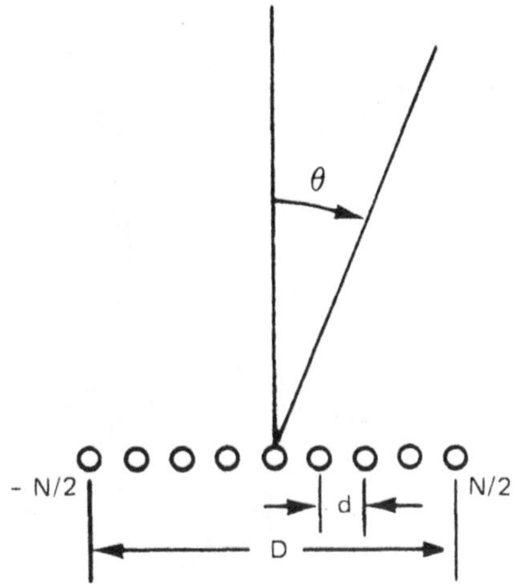

Figure 2.12 Linear array of N identical elements.

$$v(\theta) = \left\{ \sum_{n=-N/2}^{n=+N/2} \exp[j2\pi\lambda^{-1}nd(\sin\theta)] \right\}^2$$

Let $u = \sin\theta$, then

$$v(u) = \left[\sum_{n=-N/2}^{n=+N/2} \exp(j2\pi\lambda^{-1}ndu) \right]^2$$

$$= \left\{ \frac{\sin[\pi du\lambda^{-1}(N+1)]}{\sin[\pi du\lambda^{-1}]} \right\}^2 \tag{2.15}$$

The response of the synthetic array is

$$v(u) = \sum_{n=-N/2}^{n=+N/2} \exp(j4\pi\lambda^{-1}ndu)$$

$$= \frac{\sin[2\pi du\lambda^{-1}(N+1)]}{\sin[2\pi du\lambda^{-1}]} \tag{2.16}$$

The two responses are plotted in normalized form in Figure 2.13. The one-way responses of both real and synthetic arrays are identical; however, the two-way responses differ considerably. This is a consequence of the coupling terms which arise

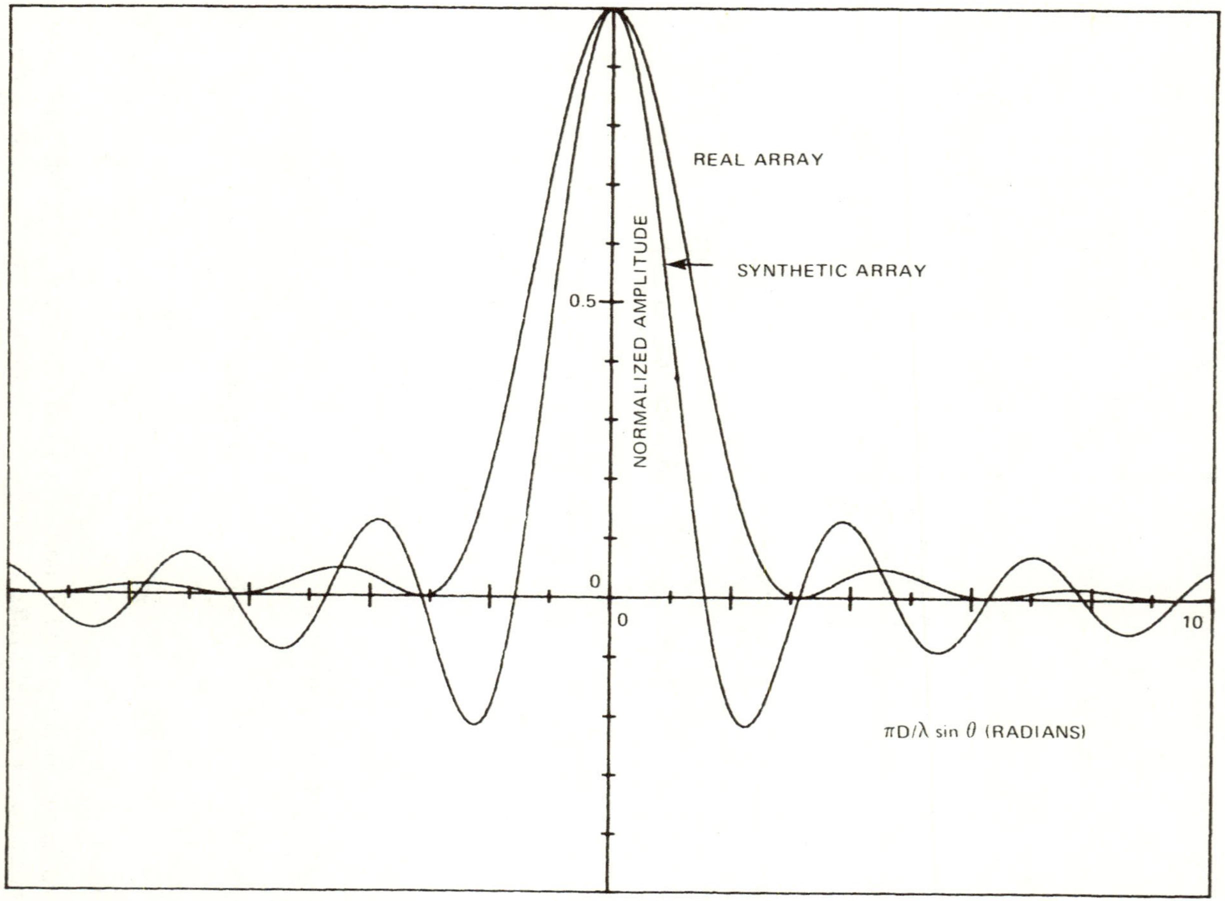

Figure 2.13 Two-way responses of real and synthetic array.

when the array is used for both transmitting and receiving. The two-way response of the synthetic array is equivalent to the one-way response of a real array of twice the length. This feature, of significance in applications requiring maximum achievable resolution, is often overlooked when analogies between synthetic and real apertures are established. Table 2.1 summarizes the beamwidths of one-way and two-way responses of real and synthetic arrays. The tabulated beamwidths are relative to the angular measure in radians, given by the ratio λ/D.

The discussion leading to (2.11) established that a high degree of cross-range resolution can be obtained synthetically by processing signals scattered from the object that are sampled for various viewing angles to provide angular diversity. This process requires relative motion between the radar and the object to produce changes in the relative aspect angle; the resolution is then obtained in the plane containing the angular motion, in a direction normal to the line-of-sight.

Figure 2.14 shows three classical types of *synthetic aperture radar* (SAR): linear SAR, spotlight SAR, and *inverse SAR* (ISAR). Linear SAR, described in [6–8], involves observing stationary objects with a radar moving linearly across the line-of-sight. It has been widely used for mapping terrain features and ground-based objects by airborne radars. Spotlight SAR, described in [9,10], involves observing stationary objects with a radar antenna fixed on the object while the viewing angle is changed. Inverse SAR, described in [11–13], involves a stationary radar viewing objects rotating about an axis normal to the line of sight. Data identical to those obtained from a circular spotlight SAR can be obtained by the equivalent ISAR process, as illustrated in Figures 2.15(a) and 2.15(b), respectively. In Figure 2.15(a), the object described in the object-fixed coordinates (x,y) is viewed from an angular offset θ. In figure 2.15(b), the object has been rotated through the angle θ, and the sensor is collinear with the initial object axis y. The relative aspect angle is identical, and the positions of arbitrary points in the object coordinate frame relative to the radar are identical in each case. The term "inverse" in ISAR refers to the synthetic aperture being formed by the rotation of the object rather than the rotation of the radar.

Table 2.1
Beamwidth of Real and Synthetic Arrays Normalized to (λ/D) radians

	Peak-null Beamwidth	*Null-null Beamwidth*	*Half-power Beamwidth*
One-way real and synthetic array	1	2	0.886
Two-way real array	1	2	0.638
Two-way synthetic array	0.5	1	0.443

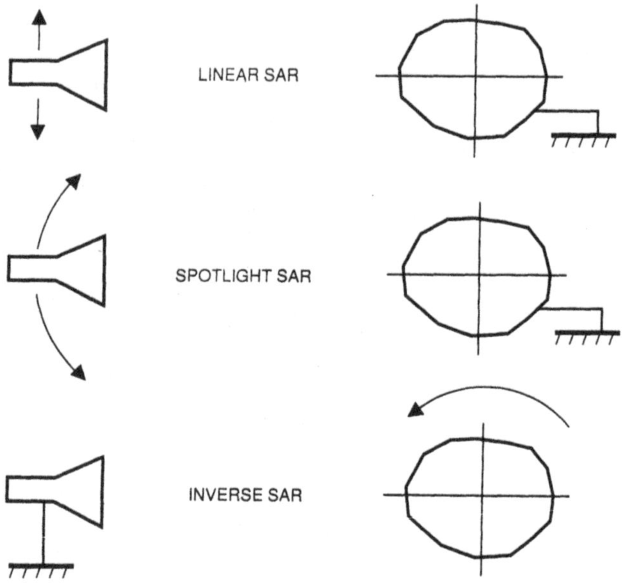

Figure 2.14 Various forms of synthetic aperture radar.

2.2.2.1 Resolution of Synthetic Apertures

Each of the three configurations offers cross-range resolution dictated by the angular extent of the viewing aspect from which the data are obtained.

Resolution of the linear SAR can be derived by considering the idealized geometry of Figure 2.16, in which a point scatterer passes through the angular beamwidth of the real antenna. Because the scatterer is visible while it is traversed by the beam of the real antenna, the maximum length of the synthetic aperture is L. The angular resolution of the synthetic aperture is $\lambda/2L$, corresponding to a cross-range resolution of

$$\Delta = (\lambda R)/(2L) \approx 0.5\lambda/\Theta \tag{2.17}$$

An interesting observation can be made by substituting $\Theta = \lambda/D$, the beamwidth of the real antenna, into (2.17) to obtain:

$$\Delta \approx (\lambda D)/(2\lambda) = D/2 \tag{2.18}$$

Equation (2.18) indicates that the potential resolution of linear SAR equals one-half the size of the real aperture. The apparent paradox that finer resolution is achieved

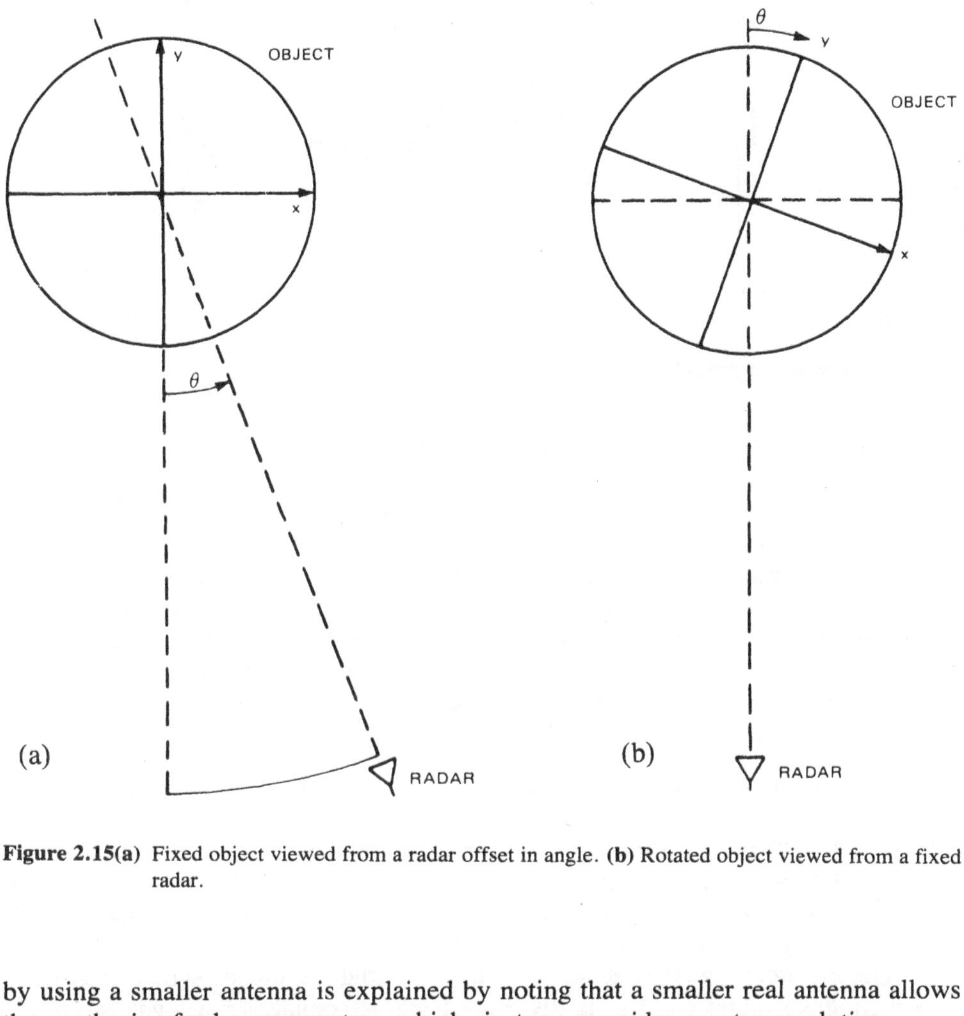

Figure 2.15(a) Fixed object viewed from a radar offset in angle. **(b)** Rotated object viewed from a fixed radar.

by using a smaller antenna is explained by noting that a smaller real antenna allows the synthesis of a longer aperture which, in turn, provides greater resolution.

The resolution of spotlight SAR is obtained by direct application of (2.11). This configuration offers the potential of greater resolution than the linear SAR because the object can be spotlighted over a larger range of observation angles, thus yielding finer resolution.

The resolution of ISAR is given by (2.11), in which the angle Θ is interpreted as the total rotation of the object. The synthetic aperture is formed by recording the output of a single, fixed sensor for a number of object angular positions, and subsequently coherently summing the resulting signals. The process of rotating the object generates an equivalent circular-arc aperture fixed on the object center of rotation, which thus becomes the focal point of the aperture. The ISAR configuration

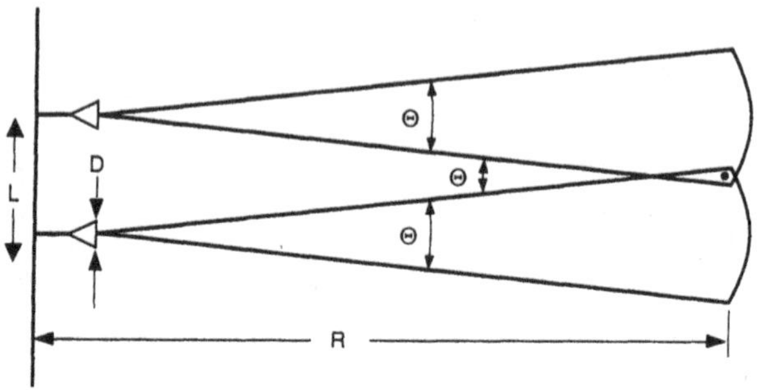

Figure 2.16 Idealized geometry of linear SAR.

conveniently evolves from conventional static measurements performed on RCS ranges, in which rotating objects are uniformly irradiated by stationary radars. Implementation details of systems combining wide band signals and ISAR processing are presented in Chapter 4.

The resolution of both spotlight SAR and ISAR can be improved by increasing the angular extent of the aperture, which can approach 360 degrees in the limit. Under these conditions, the simple relation provided by (2.11) fails to express the achievable resolution because it invoked small-angle approximations. Rigorous derivations of resolution for these cases are presented in Chapter 5.

2.2.2.2 Synthetic Aperture Formation

The circular synthetic aperture is formed by coherently summing signals received by a radar sensor which uniformly irradiates the entire object. Moving the radar along a circular path to vary the viewing angle, as shown in Figure 2.17, is equivalent to rotating the object while the radar is stationary.

The coherent signal summation is the essential feature of synthetic aperture formation; this requires that the signals be available in complex (magnitude and phase) form. Because the sensor is sequentially moved along the aperture to obtain samples of backscattered signals, some form of storage device is required to retain the complex sampled signals for subsequent summing. In early synthetic-aperture processing applied to terrain mapping, the signal information was encoded on film in a manner analogous to a hologram [8]. In RCS applications, the coherent data are sampled, digitized, and stored for subsequent processing by a computer.

The aperture response corresponding to the process in Figure 2.17 is represented by the sum:

$$G = \sum_{l} s(l) \tag{2.19}$$

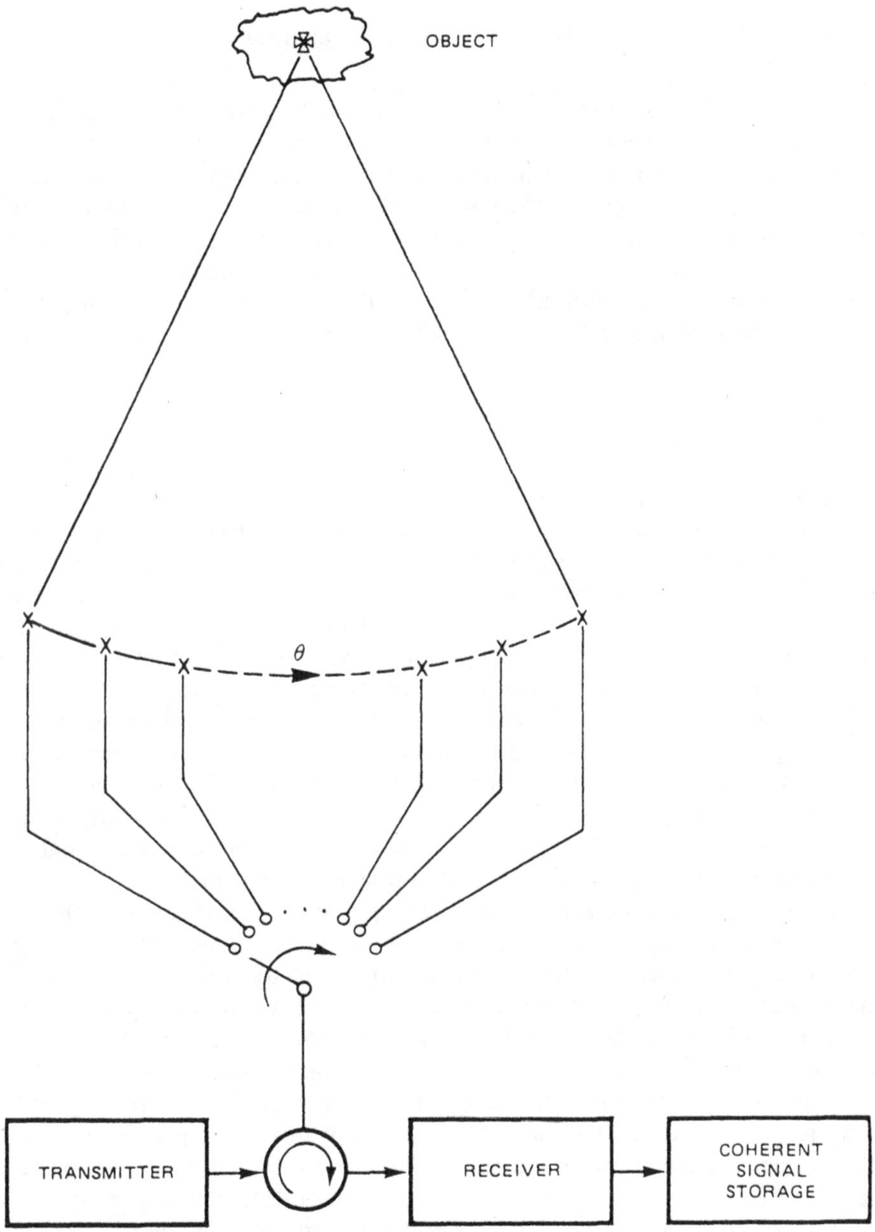

Figure 2.17 Conceptual synthetic aperture forming process.

where

$s(l)$ = the complex signal observed at the angle θ.

$s(l)$ is the complex envelope of the received signal, represented by the magnitude and phase of the receiver output relative to the transmitted signal. The sum of signals received along the aperture is maximum when all signal samples are in-phase. This condition prevails for a point reflector located on the axis of the synthetic aperture. Phases of signals backscattered from points at other locations, however, vary across the aperture, thus reducing the magnitude of the summed output. This process, forming the beam of the synthetic aperture, is analogous to the summation process occurring in array antennas and to the integration process occurring in aperture antennas.

2.2.2.3 Synthetic Aperture Beam Steering

A long synthetic aperture achieves a high degree of cross-range resolution by virtue of its narrow beamwidth. In order to resolve scatterers disposed on the cross-range axis, it is necessary to scan or steer the synthetic aperture beam across this axis. In real antennas, this can be simply accomplished by physical rotation; in phased-array antennas, the steering is accomplished by activating phase shifters connected to the various array elements; in synthetic aperture radar, the steering is accomplished by computational means, as described in the following paragraphs.

Consider the response of the linear SAR to a distant point scatterer on the x-axis of Figure 2.18. The response of the synthetic aperture, obtained by coherent summation of the signal samples occurs for $x = 0$, in which case the signals are received along the aperture essentially in phase. As the scatterer is displaced along the x-axis, the response of the synthetic aperture decreases because the signal samples are received out of phase, and therefore sum to a lesser value.

To steer the synthetic aperture beam to a scatterer located at $x = X$, the received signals must be brought to an in-phase condition prior to the coherent summation. As diagrammed in Figure 2.18, this is accomplished by applying to each received signal a phase correction necessary to force the signals to an in-phase condition. The phase corrections compensate for the unequal round-trip distances between the scatterer and aperture elements. When the phases of signals received from a scatterer at $x = 0$ are identical to those for the scatterer at $x = X$, the aperture is said to be steered to point X. By computing different sets of phase compensations, the beam of the synthetic aperture can be steered to any x coordinate. Because the phase compensations necessary to steer the beam to a particular location depend on the location, aperture coordinate, frequency, and observation distance, the computation of these phase compensations can pose an extensive computational task. Methods for simplifying the computations are addressed in Chapter 4.

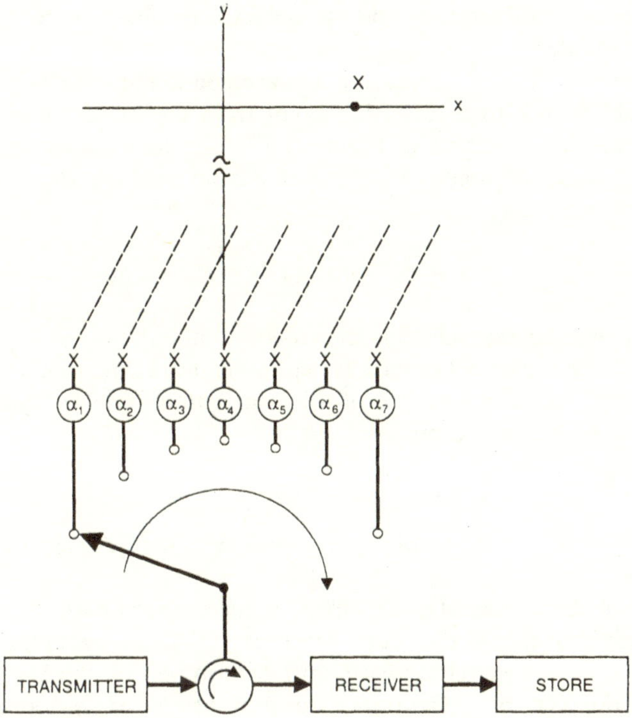

Figure 2.18 Synthetic aperture beam steering.

2.2.3 Doppler Processing

Doppler processing is a widely used method for obtaining spatial resolution of objects irradiated with coherent waves [14–16]. The basis of doppler processing is the observation that the frequency shift of signals reflected from a moving object is directly proportional to the radial component of velocity between the object and a stationary source and receiver. For the special case of a rotating reflector and a stationary source and received, the doppler shift of the reflected signal is proportional to the lateral offset of the reflector measured along an axis normal to the axis of rotation and to the line-of-sight. Doppler processing of signals reflected from a two-dimensional object rotating in a plane, therefore, yields resolution in cross range.

The simple spectral analysis and the fact that doppler processing requires only continuous wave (CW) signals constitute major advantages. Implementations of spectral analyses required for cross-range sorting can use filter banks for analog processing or fast Fourier transform algorithms for digital processing. CW is the preferred mode of operation for coherent sources, and narrow band receivers, used with CW signals, are inherently simple.

The essence of doppler processing is presented in the following statements. Let d be the radial distance of a reflecting point from the center of an object rotating with angular rate Ω, as shown in Figure 2.19. The transmitted signal is given by $e_T(t) = \mathrm{Re}[\exp(j\omega_0 t)]$. At time t, the distance r from a remote ($R_0 >> d$) source and receiver is approximately

$$r = R_0 - d\sin(\Omega t) = R_0 - d\sin\theta \tag{2.20}$$

where Ω is the angular rate and θ is the rotation angle.

Assuming that the object motion is negligible during the signal round-trip travel time, the received signal is a replica of the transmitted signal delayed by the round-trip time $2r/c$, and expressed by

$$\begin{aligned} e_T(t) &\propto \mathrm{Re}\,[\exp(j\omega_0 t - j4\pi r/\lambda)] \\ &= \mathrm{Re}\,\{\exp[j\omega_0 t - j4\pi R_0/\lambda + j2\pi/\lambda 2d\sin(\Omega t)]\} \end{aligned} \tag{2.21}$$

The first term of the exponential argument is the carrier phase; the second term is a constant phase shift due to the round-trip delay associated with the constant range R_0. The third term is a variable phase shift determined by the varying range, and constitutes the doppler shift. The phase variation can be expressed as a function of

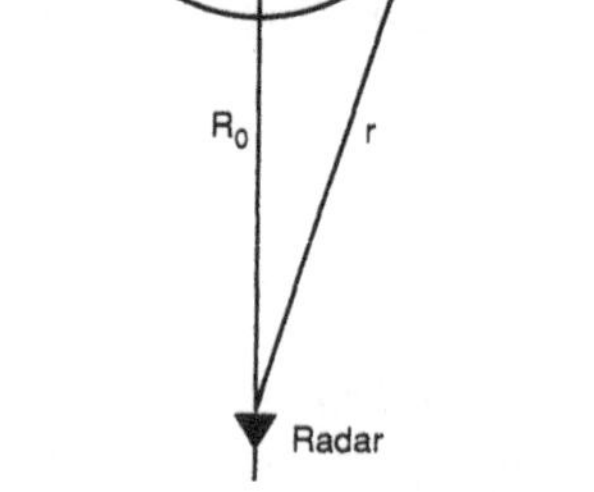

Figure 2.19 Two-dimensional rotating object geometry.

either time or rotation angle; in either case, the effect being described is the variation in received signal phase attributed to variations in the round-trip distance.

The doppler frequency is equal to the temporal phase rate of the complex envelope of the reflected signal.

$$f_t = \frac{1}{2\pi}\frac{d\phi}{dt} = \frac{-2}{\lambda}\frac{dr}{dt} = 2\Omega d/\lambda \cos(\Omega t) = 2\Omega x/\lambda \text{ Hz} \tag{2.22}$$

where ϕ is the phase shift corresponding to the round-trip delay indicated by the last two terms in the exponent of (2.21), and x is cross range.

The phase can also be expressed as a function of the rotation angle $\theta = \Omega t$ in which case the angular rate of change of phase is

$$f_\theta = \frac{1}{2\pi}\frac{d\phi}{d\theta} = \frac{-2}{\lambda}\frac{dr}{d\theta} = 2d/\lambda \cos\theta = 2x/\lambda \text{ cycles/radian} \tag{2.23}$$

The frequencies in (2.22) and (2.23), denoted by subscripts t and θ to indicate time and angle, respectively, are equivalent expressions; the angular and temporal frequencies are proportional to the instantaneous cross-range coordinate, and differ only by the scale factor, Ω. Processing signals reflected from rotating objects recorded as a function of angle yields results identical to those obtained by processing signals recorded as a function of time for fixed rotation rate Ω. Both methods are termed doppler processing in the literature; if a distinction is required, the former process may be termed *temporal doppler processing* and the latter *angular doppler processing*.

The equivalence of (2.22) and (2.23) allows considerable flexibility in practical instrumentations, which can be slanted toward temporal or angular processing. In the former method, the object is rotated at a constant angular rate. On-line processing is accomplished by a bank of contiguous frequency filters, or by analyzing the sampled signal with Fourier transform processors of sufficient speed. Off-line processing can be performed by recording the signal for subsequent analysis; in either case, the signal is usually frequency-translated by ω_0 to retain only the complex envelope. In the latter method, the object is stepped in uniform angular increments, and a series of static measurements is made of the phase and amplitude of the received signal. Provided that the Nyquist sampling criterion is met, a sequence of such measurements is identical to the phase history that would be observed with a continuous rotation.

The direct relation between doppler frequency and cross range allows spectral analysis of the signal to separate contributions of several reflecting points on the basis of cross-range. The frequency resolution for f_t varies inversely with the processing time duration T. Thus, $\Delta f_t \approx 1/T$ and, from (2.22), the cross-range resolution is

$$\Delta x = \frac{\lambda}{2\Omega}\Delta f_t \approx \frac{\lambda}{2\Omega}\frac{1}{T} = \frac{\lambda}{2\Theta} \tag{2.24}$$

where Θ is the angular change during the processing time T.

Equation (2.24) indicates that fine cross-range resolution requires processing over a large interval of both time and angle. However, the instantaneous frequency of the echoes reflected from a particular point may vary considerably over a large processing interval, depending upon the angular position and rate of rotation of the object and the distance of the point from the center of rotation. The variation in the instantaneous frequency imposes limitations on rotational speed, maximum cross-range, and the number of pixels in the final image. To examine how the cross-range resolution is limited by the variation in instantaneous frequency, let the doppler frequency space be divided into small uniform frequency intervals, Δf_t, corresponding to the cross-range resolution cells of width Δx. Assume that the object can be enclosed in a circle of radius R. The movement of a particular reflecting point must be confined to a single resolution cell during the processing interval T. If the rotational velocity, Ω, is constant, the reflecting point will move rapidly through a cross-range resolution cell when Θ is $\pm\pi/2$. Thus, $\Delta x \geqslant \Omega RT$ or $\Delta x \geqslant R\Theta$. Combining this with (2.24) gives:

$$(\Delta x)^2 \geqslant \frac{1}{2}\lambda R \tag{2.25}$$

If the restriction of (2.25) is violated, doppler frequency analysis will produce degraded imagery. To illustrate this degradation, we consider signals from various locations on a rotating object processed over a finite interval. Figure 2.20 shows the locus followed by each point object as it rotates over 0.15 radians; the observation distance, R_0, is 1000λ and the theoretical resolution according to (2.24) is 3.3λ. The phase of the signal received from each object point considered individually is shown as a function of rotation angle in the left column of Figure 2.21; the amplitude is constant and not shown. The Fourier transform of the received signal for each case is shown in the right column plots of Figure 2.21. Because instantaneous frequency is equal to the rate of change of phase, the shape of the doppler spectrum is determined by the phase history. In each case, the peak of the spectral response coincides with the cross-range coordinate of the point at the center of the processing interval; however, the nonlinearity of the phase history causes spectral spreading with degradations which are more pronounced as the phase history deviates from linearity. As the phase nonlinearities increase beyond the values shown, the spectral responses no longer exhibit a single peak, thus producing severe image degradation.

By performing a Fourier transform, multiple doppler frequencies present in the composite signal are separated. Doppler processing is particularly suited to imaging rotating objects due to the relation between frequency and cross range; however,

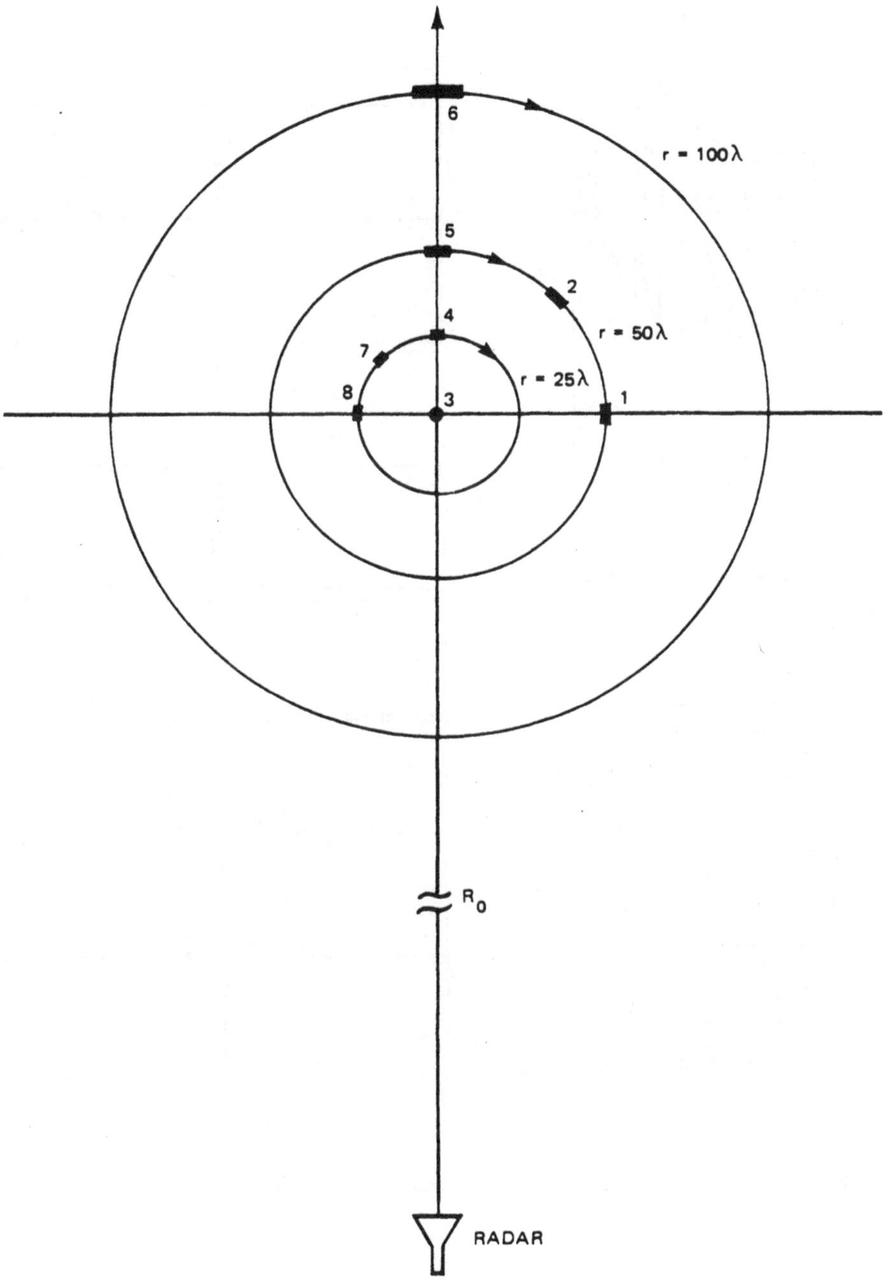

Figure 2.20 Locus of points observed on a rotating field.

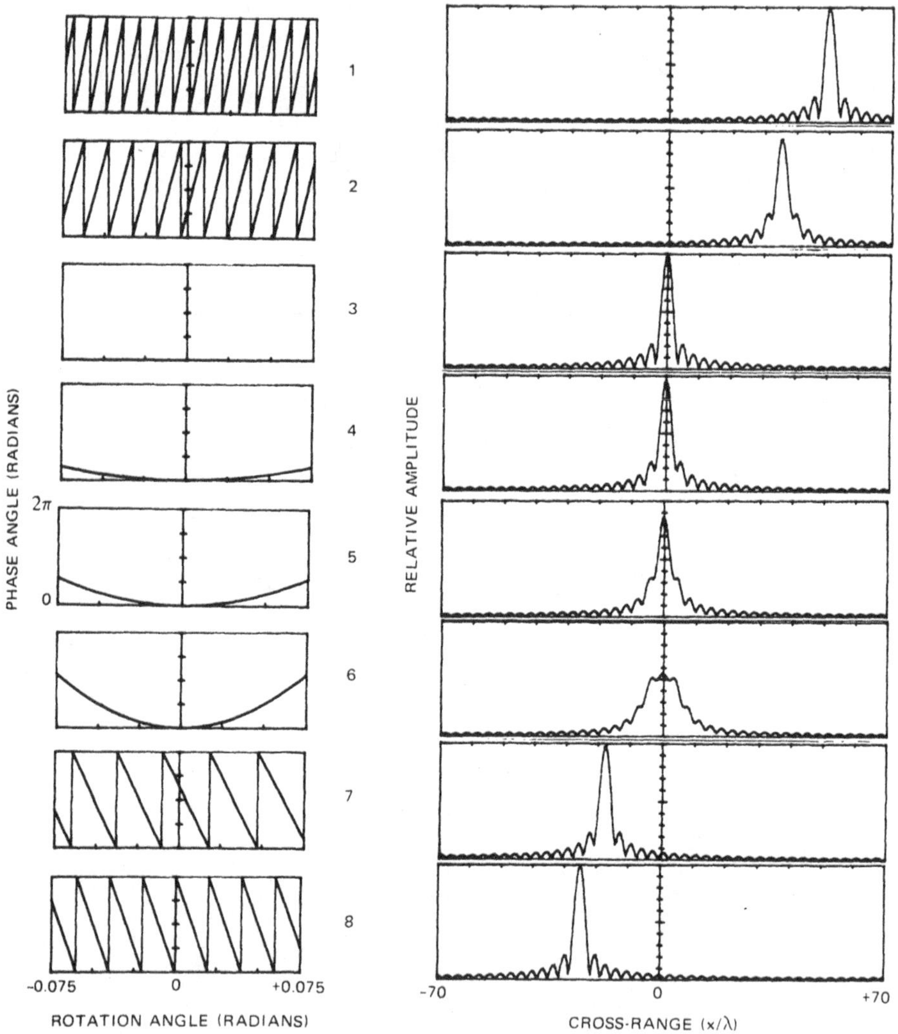

Figure 2.21 Phase of signals received from point objects and corresponding cross-range spectra.

other geometries involving relative radial motion among object elements can be utilized. Imaging a three-dimensional object with doppler processing yields resolution along an axis normal to the axis of rotation and to the radar line-of-sight, as represented in Figure 2.22. Unlike the range-mapping case, in which range profiles could be associated with unique aspect angles, the cross-range profiles obtained by doppler processing require that the object be rotated through a range of angles. Therefore, the cross-range profiles cannot be uniquely associated with a discrete aspect, but only with a range of angles necessary to induce the required relative motion, albeit small.

2.2.4 Holographic Processing

One of the most significant developments in optical imaging has been the concept of holography. In this case, the imaging is performed by recording a pattern of light intensity which is used to reconstruct a replica of the light waves associated with an object. A viewer observing these reconstructed waves perceives an image of the original object. The hologram encodes and stores a record of the waves diffracted from the object. Although microwave imaging concepts can be developed independently of holography, many of the intermediate steps in the recording and reconstruction processes for microwave imaging are directly analogous to holography. If recording microwave data and subsequently reconstructing an image from such data are performed optically, microwave imaging and holography become identical processes. From a review of the literature, it is apparent that the holographic process

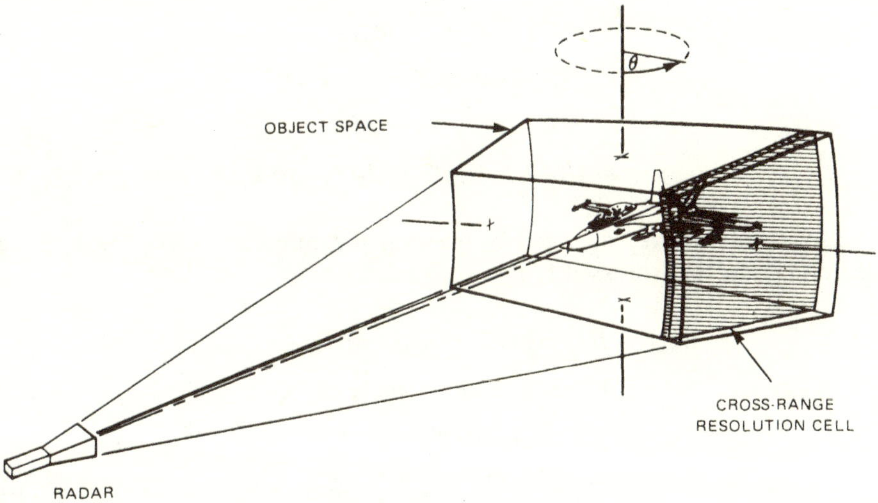

Figure 2.22 Cross-range resolution of three-dimensional object by doppler processing.

applied to coding and image formation from synthetic aperture microwave data was indeed the vehicle by which the two disciplines were joined to achieve significant contributions [17–19].

We will briefly review the concept of optical holography for the purpose of establishing fundamental similarities and differences between optical and microwave imaging. The intent is not to provide a comprehensive summary of optical holography (available in [20–22]), but to consider the holographic process in the context of microwave imaging.

In optical holography, the object image information encoded on the hologram is the intensity pattern of the interference between the object and reference beams. The complete image information is contained in the complex amplitude (magnitude and phase) of the field associated with the object. Because phase sensitive optical storage media are not available, a recording of the interferogram of the object and reference wave is used to store the object complex field information in the form of a pattern of intensity. This elegant encoding scheme allows the subsequent retrieval (reconstruction) of the complex field. The generation of additional images and noise beams is a consequence of the intensity encoding process. The following development summarizes the procedure for recording and reconstructing a holographic image. If $\bar{U}_T = U_O \exp[j\omega t + j\phi_O(x,y)] + U_R \exp[j\omega t + j\phi_R(x,y)]$ is the superposition of complex object and reference waves incident on the hologram or x-y plane, the intensity of the light pattern on the plane is

$$\begin{aligned} I = \bar{U}_T\bar{U}_T^*/2 &= \frac{1}{2} U_O\{\exp[j\omega t + j\phi_O(x,y)] \\ &\quad + U_R \exp[j\omega t + j\phi_R(x,y)]\} \\ &\quad \cdot \{U_O \exp[-j\omega t - j\phi_O(x,y)] \\ &\quad + U_R \exp[-j\omega t - j\phi_R(x,y)]\} \\ &= U_O^2/2 + U_R^2/2 + U_R U_O \cos[\phi_O(x,y) - \phi_R(x,y)] \end{aligned} \tag{2.26}$$

The reconstructed wave, obtained by passing the reference wave through the recorded hologram, is expressed by:

$$\begin{aligned} \bar{U}_{\text{REC}} &= I U_R \exp[j\omega t + j\phi_R(x,y)] \\ &= U_R/2 \exp[j\omega t + j\phi_R(x,y)][U_O^2 + U_R^2] \\ &\quad + U_R^2 U_O/2 \exp[j\omega t + 2j\phi_R(x,y) - j\phi_O(x,y)] \\ &\quad + U_R^2 U_O/2 \exp[j\omega t + j\phi_O(x,y)] \end{aligned} \tag{2.27}$$

The first two terms constitute noise beams associated with undiffracted light from the reference and object waves, the third term represents a real image of the object, and the fourth term is a replica of the original object wave termed the *virtual image*.

The information contained in the holographic record can be obtained by electronic or computer processing of measured values of the object wave field by the system shown in Figure 2.23. The real signals representing the object and reference waves are

$$U_O \cos[\omega t + \phi_O(x,y)] \quad \text{and} \quad U_R \cos[\omega t + \phi_R(x,y)],$$

respectively

The output of the square-law detector is

$$\begin{aligned} V(t,x,y) &= U_O\{\cos[\omega t + \phi_O(x,y)] + U_R \cos[\omega t + \phi_R(x,y)\}^2 \\ &= U_O^2 \cos^2 [\omega t + \phi_O(x,y)] + U_R^2 \cos^2 [\omega t + \phi_R(x,y)] \\ &\quad + 2U_O U_R \cos[\omega t + \phi_O(x,y)] \cos[\omega t + \phi_R(x,y)] \end{aligned} \tag{2.28}$$

Expanding the above expression and removing the high frequency terms yields the output of the low-pass filter.

$$W(t,x,y) = \frac{1}{2} U_O^2 + \frac{1}{2} U_R^2 + U_R U_O \cos[\phi_O(x,y) - \phi_R(x,y)] \tag{2.29}$$

The right-hand side of (2.29) is identical to that of (2.26), indicating that the signal-processing model of Figure 2.23 is equivalent to the process used in optical holography. The squaring and low-pass temporal filtering of real signals yield identical results to the intensity of the summation of complex signal inputs. Image reconstruction using a hologram formed with the recording process of Figure 2.23 will yield twin images and noise beams precisely as in the case of optical holograms.

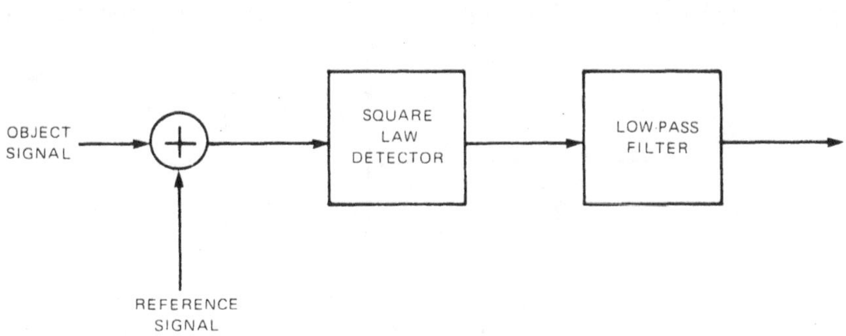

Figure 2.23 Signal processing model for simulation of optical hologram recording.

A second method involving a multiplicative detection scheme is shown in Figure 2.24, and analyzed as follows. The output of the multiplier is

$$\begin{aligned} V(t,x,y) &= U_O \cos[\omega t + \phi_O(x,y)] U_R \cos[\omega t + \phi_R(x,y)] \\ &= \frac{1}{2} U_O U_R \cos[2\omega t + \phi_O(x,y) + \phi_R(x,y)] \\ &\quad + \frac{1}{2} U_O U_R \cos[\phi_O(x,y) - \phi_R(x,y)] \end{aligned} \tag{2.30}$$

Low-pass filtering suppresses the first term to the right of the equality, leaving

$$W(x,y) = \frac{1}{2} U_O U_R \cos[\phi_O(x,y) - \phi_R(x,y)] \tag{2.31}$$

Equation (2.31) is similar to (2.26) and (2.29), except for the omission of the bias terms. Reconstruction of the image by multiplication with the reference yields:

$$\begin{aligned} U_{\text{REC}} &= U_R \cos[\omega t + \phi_R(x,y)] \\ &\quad \cdot \frac{1}{2} U_O U_R \cos[\phi_O(x,y) - \phi_R(x,y)] \\ &= \frac{1}{4} U_O U_R^2 \cos[\omega t + \phi_O(x,y)] \\ &\quad + \frac{1}{4} U_O U_R^2 \cos[\omega t + 2\phi_R(x,y) - \phi_O(x,y)] \end{aligned} \tag{2.32}$$

The first term is a virtual image, constituting a replica of the object wave, and the second represents a displaced real image of the object. The signal processing method

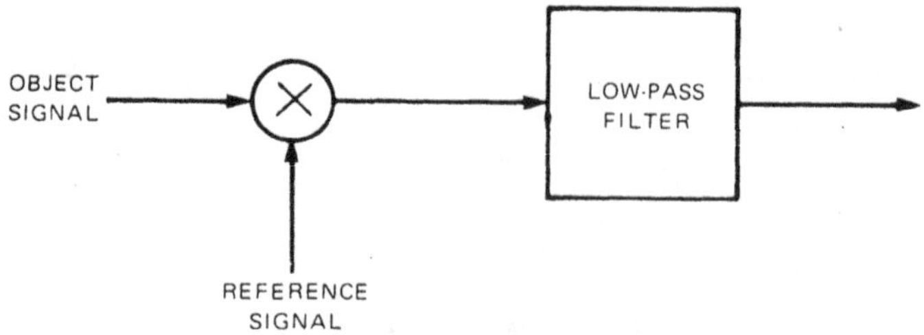

Figure 2.24 Alternate signal processing model for recording image information.

of Figure 2.24 is effective in removing the noise beams analogous to the undiffracted light in the optical hologram, but retains the twin images inherent to the holographic process.

The holographic imaging process is summarized in Figure 2.25. Figure 2.25(a) shows the situation for conventional imaging. The object wave consists of light scattered by the object, which propagates to the aperture of the imaging system and produces an image of the object. The disposition of the illumination source is arbitrary. The formation of the hologram, depicted in Figure 2.25(b), is a record of the intensity of the interferogram of the object and reference waves. The object waves are identical to those in Figure 2.25(a), and the disposition of the reference waves is again arbitrary. The reconstruction of the hologram is shown in Figure 2.25(c). If the reference waves are precisely as in Figure 2.25(a), the reconstructed waves at the right of the hologram are identical to those of the original object in Figure 2.25(a). As previously shown, the waves to the right of the hologram contain additional images which can be separated from that of the object.

In microwave holography [23], the field amplitude scattered from an object coherently illuminated from a transmitter is mapped over a prescribed recording aperture by a coherent detector which is scanned over the aperture. The detected bipolar signal, representing the complex envelope of the time-varying field, is added to a bias level sufficient to make the resultant always positive. The resulting signal is used to produce a film transparency with an amplitude transmittance function which is real and positive. The area probed by the detector represents the hologram aperture, the reference signal for the coherent detector represents the reference beam, and the signal scattered from the object is the object beam.

Microwave holography facilitates the possibility of independently scanning the transmitter and receiver. This process, termed *scanned holography,* exhibits some advantages in resolution [27]. When radar is used for imaging, the natural scanning process is one in which a collocated transmitter and receiver are made to scan a prescribed area. In radar terminology, this is the monostatic condition, in contrast with a bistatic condition in which the transmitter and receiver are moved relative to one another. The significant advantage of monostatic scanning is a two-fold improvement in resolution over the bistatic scanning. Figure 2.26 shows the bistatic and monostatic scanning cases. In Figure 2.26(a), the transmitter is stationary and the receiver is moved from position A to position B. Signals received from two extreme locations on the scan differ in two-way path length by Δx; in Figure 2.26(b), the transmitter and receiver are both moved from position A to position B and signals received from these two points vary in path length by $2\Delta x$. Because resolution is directly related to observed phase differences, the resolution performance is doubled for the monostatic scanned condition. As pointed out in [24], this type of scanning is equivalent to halving the wavelength used for bistatic scanning.

In any holographic imaging method, either conventional or scanned, the process is one of recording the diffraction pattern of the object falling onto a plane. The

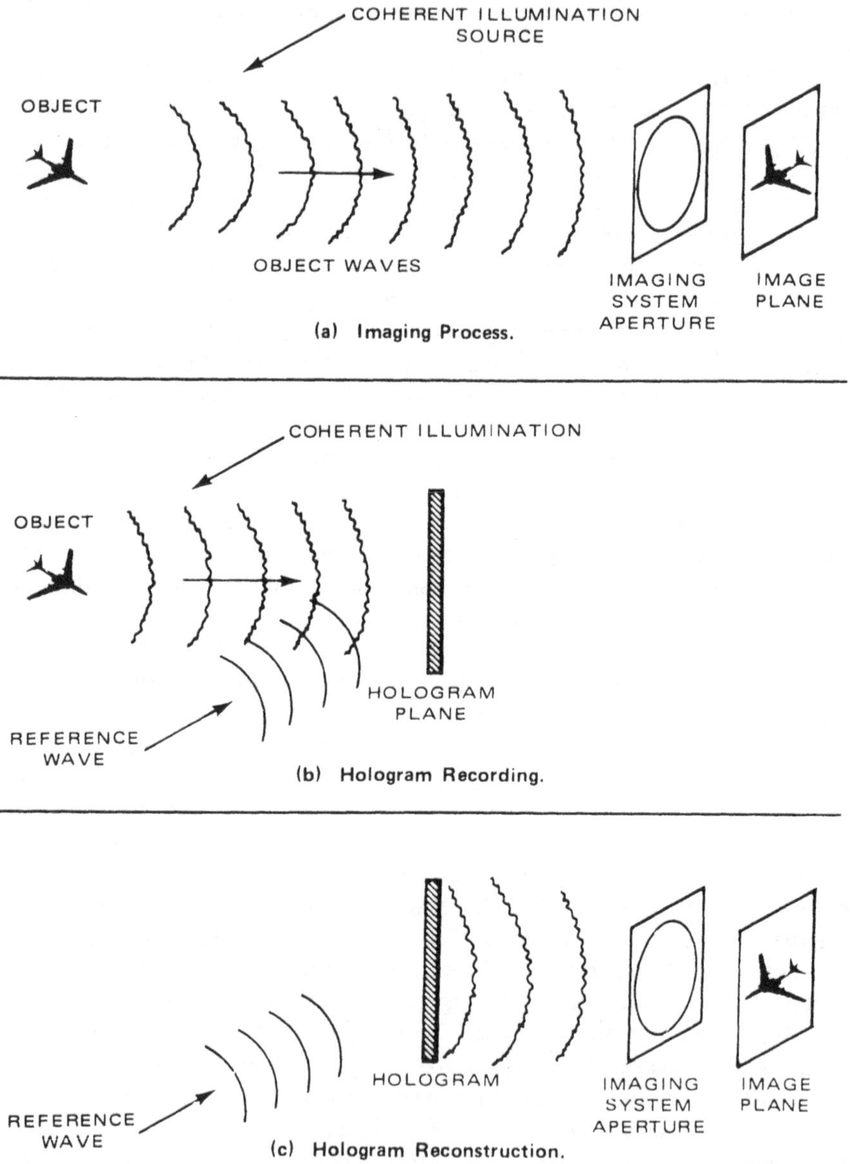

Figure 2.25 Coherent optical imaging and holography.

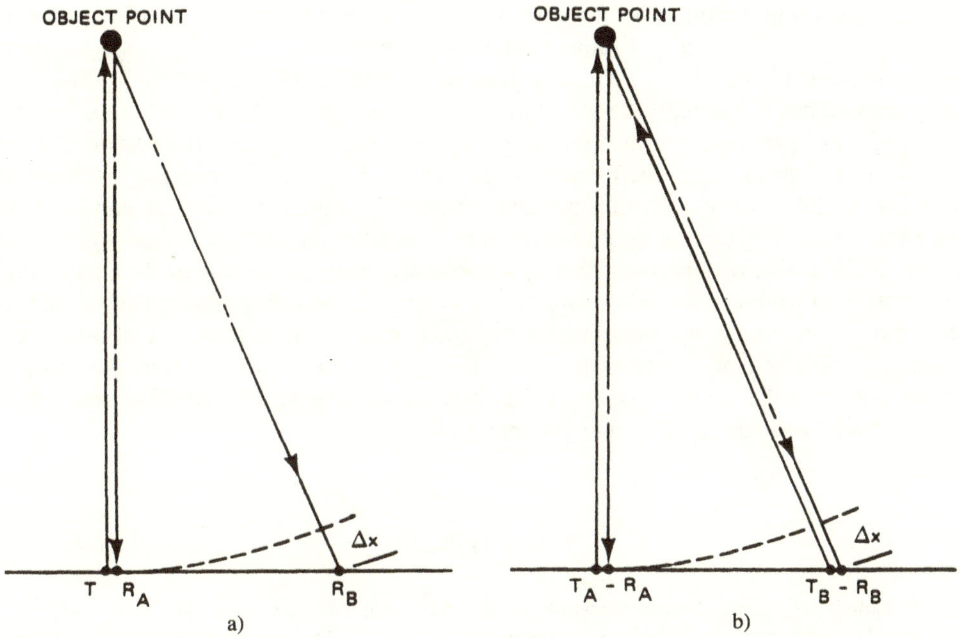

Figure 2.26(a) Bistatic field scanning geometry. **(b)** Monostatic field scanning geometry.

diffraction pattern is essentially the Fourier transform of the object illumination function; reconstruction of the image provides resolution in the cross-range direction similar to what would be achieved by an entrance pupil equal in size to the hologram. (This assumes a recording medium which is not limited in spatial frequency to less than the maximum spatial frequency of the diffraction pattern at the hologram.) Although the holographically reconstructed image is focused at all ranges, the image is basically two-dimensional; resolution in the axial dimensional is performed subjectively by human observers. This feature constitutes a significant difference between optical and microwave imaging; while two-dimensional imaging is satisfactory for visual observation of optical images, meaningful microwave images are generally required to be three-dimensional because radar sensors respond to relative spatial attitudes of object points in all three dimensions.

2.2.5 Relation of Imaging Methods

Imaging with microwave energy allows the achievement of direct resolution in the radial direction by using wide signal bandwidths and time-delay sorting. This feature is completely independent of resolution in the cross-range dimensions which makes use of diffraction phenomena.

Cross-range imaging by doppler, synthetic aperture, and holographic processing leads to fundamentally similar results because each of these processes depends on the physical properties of fields diffracted from irradiated objects [25]. The various approaches differ only in the initial viewpoint adopted to formulate the problem. Each of these processes utilizes the complex field incident on an observation surface which results from fields diffracted by the object. The angular interval, Θ, used to observe signals received from a rotating object for doppler processing can be considered to form a synthetic aperture which is a circular arc subtending an angle equal to Θ. If the fields observed on this aperture were encoded on a recording medium, the stored information would constitute a hologram. The hologram could be used to reconstruct an image of the object which would be identical to that obtained by doppler processing or by the response of the synthetic aperture focused and steered to all points of the object space. In the chapters to follow, we adopt the viewpoint which leads most directly to the required analysis.

REFERENCES

1. Skolnik, M.I., ed., *Radar Handbook,* 1st Ed., New York, McGraw-Hill, 1970, pp. 4–6.
2. Woodward, P.M. *Probability and Information Theory with Applications to Radar,* Oxford, Pergamon Press 1964, p. 105.
3. Thomas, J.B., *An Introduction to Statistical Communication Theory,* New York, John Wiley and Sons, 1969, pp. 96–99.
4. Rihaczek, A.W., *Principles of High Resolution Radar,* New York, McGraw-Hill, 1969, p. 127.
5. Jenkins, F.A., and H.E. White, *Fundamentals of Optics,* New York, McGraw-Hill, 1950, p. 293.
6. Kovaly, J., *Synthetic Aperture Radar,* Norwood, MA, Artech House, 1976.
7. Hovanessian, S.A., *Introduction to Synthetic Array and Imaging Radars,* Norwood, MA, Artech House, 1980.
8. Skolnik, M.I., ed., *Radar Handbook,* 1st Ed., New York, McGraw-Hill, 1970, Ch. 23.
9. Munson, D.C., J.D. O'Brien, and W.K. Jenkins, "A Tomographic Formulation of Spotlight-Mode Synthetic Aperture Radar," *Proc. IEEE,* Vol. 71, No. 8, August 1983, pp. 917–925.
10. Brookner, E., ed., *Radar Technology,* Norwood, MA, Artech House, 1977.
11. Walker, J.L., "Range-Doppler Imaging of Rotating Objects," *IEEE Trans. on Aerospace and Electronic Systems,* Vol. AES-16, No. 1, January 1980, pp. 23–52.
12. Wehner, D.D., *High Resolution Radar,* Norwood, MA, Artech House, 1987.
13. Ausherman, D.A., *et al.,* "Developments in Radar Imaging," *IEEE Trans. Aerospace and Electronics Systems,* Vol. AES-20, No. 4, July 1984, pp. 363–400.
14. Brown, W.M., and L.J. Porcello, "An Introduction to Synthetic Aperture Radar," *IEEE Spectrum.* September 1969, p. 52.
15. Brown, W.M., and R. Fredricks, "Range-Doppler Imaging with Motion Through Resolution Cells," *IEEE Trans. Aerospace and Electronic Systems,* AES-5, January 1969, pp. 98–102.
16. Hagfors, R., and D. Campbell, "Mapping of Planetary Surfaces by Radar," *Proc. IEEE,* vol. 61, No. 9, September 1973, pp. 1219–1225.
17. Sherwin, C.W., J.P. Ruina, and R.D. Rawcliffe, "Some Early Developments in Synthetic Aperture Radar Systems," *IRE Trans. Military Electronics,* April 1960, pp. 111–115.

18. Cutrona, L.J., E.N. Leith, C.J. Palermo, and L.J. Porcello, "Optical Data Processing and Filtering Systems," *IRE Trans. Information Theory,* Vol. IT-6, June 1960, pp. 386–400.
19. Cutrona, L.J., E.N. Leith, L.J. Porcello, and E.W. Vivian, "On the Application of Coherent Optical Processing Techniques to Synthetic Aperture Radar," *Proc. IEEE,* Vol. 54, July 1966, pp. 1026–1032.
20. Goodman, J.W., *Introduction to Fourier Optics,* New York, McGraw Hill, 1969.
21. Smith, H.M., *Principles of Holography,* New York, John Wiley and Sons, 1969.
22. Goodman, J.W., "An Introduction to the Principles and Applications of Holography," *Proc. IEEE,* Vol. 59, No. 9, September 1971, pp. 1291–1303.
23. Farhat, N.H., "High Resolution Microwave Holography and the Imaging of Remote Objects," *Optical Engineering,* Vol. 14, No. 5, September 1975, pp. 499–505.
24. Hildebrand, B.P., and K.A. Haines, "Holography by Scanning," *Journal of the Optical Society of America,* Vol. 59, No. 1, January 1969, pp. 1–6.
25. Leith, E.N., "Quasi-Holographic Techniques in the Microwave Region," *Proc. IEEE,* Vol. 59, No. 9, September 1971, pp. 1305–1318.

Chapter 3
Instrumentation and Signal Processing for RCS Imaging

The preceding chapter has established that the range and cross-range resolutions necessary for RCS imaging can be obtained synthetically by measuring, storing, and processing responses of objects irradiated by radars with angle and frequency diversity. While the fundamental physical limits are imposed by the measurement procedure and the instrumentation hardware, the operational utility and the optimum extraction of information from the measured data are governed by the signal-processing software.

This chapter presents the features essential to instrumentation systems for performing wideband measurements, and treats a number of software tools and concepts widely used in RCS imaging systems. The level of treatment is intended to present fundamental guidelines for designing or selecting these systems, while avoiding details of mechanization. Using these guidelines, the reader can design or specify the structure of an optimum system, or define objective criteria for selecting the optimum system from several options. In essence, we strive to establish those performance limits of RCS instrumentation systems which, being established by basic physical and mathematic principles, are inviolate to electronic or computer wizardry.

3.1 INSTRUMENTATION

3.1.1 Performance Criteria

The following sections discuss the criteria of sensitivity, dynamic range, resolution, and ambiguities. These notions, which determine the performance of RCS instrumentation systems, can be used to evaluate instrumentation options.

3.1.1.1 Sensitivity

The *sensitivity* of an instrumentation radar is defined as the minimum received signal level measurable with a given accuracy. Because the system sensitivity sets the measurement threshold, optimum receivers are highly desirable, particularly as the transmitted frequency increases and the scattered energy collected by the smaller receiving aperture decreases.

In terms of sensitivity, the optimum radar receiver is the matched filter. The input to the matched filter, a linear system with impulse response $h(t)$ and transfer function $H(f)$, is postulated to consist of the signal added to white Gaussian noise. As a consequence of linearity, the output of the matched filter also consists of a signal added to Gaussian noise, each influenced by the filter transfer function.

The structure of the filter is derived from the criterion of maximizing the signal-to-noise ratio (S/N) of the output. Properties of the matched filter [1–3] are summarized in the following statements, and shown in Figure 3.1.

a. The impulse response of the matched filter has the form of the time-reversed input signal.

b. The matched filter transfer function is the complex conjugate of the input signal spectrum; that is, the magnitudes of the filter transfer function and signal spectrum are identical, and the phases are of opposite sign. The matched filter exactly cancels the phase of the input signal spectrum and then performs a coherent integration over the signal duration.

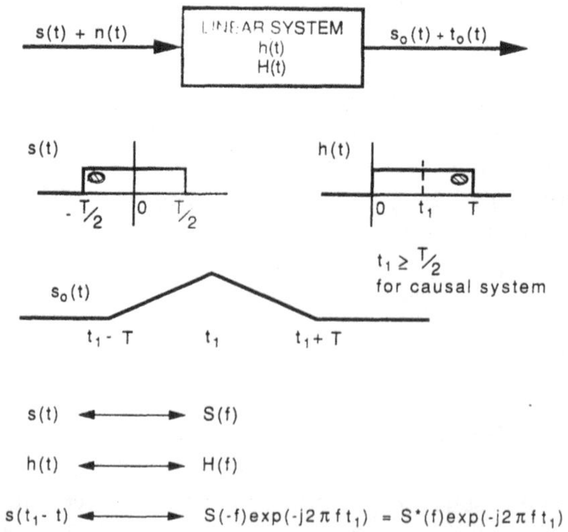

Figure 3.1 Matched filter with pertinent relations.

c. The output of the matched filter equals the autocorrelation function of the input signal.

d. The maximum power ratio of peak S/N occurs when the entire signal is contained in the matched filter memory and is given by:

$$S/N = 2E/N_0 \tag{3.1}$$

where E = signal energy and N_0 = single-sided power spectral density of noise.

The concise relation of (3.1) states that the sensitivity of the matched filter receiver is directly proportional to the signal energy. For a given transmitted power, the receiver sensitivity can be enhanced by increasing the signal duration, thus obtaining a proportional increase in the signal energy. The remarkable feature of (3.1) is its independence of signal waveform; evidently, the sensitivity of the matched filter receiver is determined strictly by the fundamental quantities of the noise power and signal energy. Although most practical receivers are not strict implementations of a matched filter, properly designed conventional receivers often approximate this performance to a close degree. For example, Skolnik shows [4] that a number of conventional bandpass filters used in pulsed-radar receivers provide S/N ratios within 1 dB of the matched filter.

Equation (3.1) establishes the upper limit of sensitivity, and can therefore be used as a standard of comparison for the optimality of practical receivers in terms of achievable S/N. The optimum S/N for any postulated system is determined by multiplying the received power (obtained from the radar range equation appropriate to the given application) by the signal duration to obtain the received energy. The single-sided power spectral density of the receiver noise, N_0, is given by kT_0F, where k is Boltzmann's constant, T_0 is the standard temperature of 290 K, and F is the receiver noise figure.

Although the matched filter achieves the optimum output S/N, the output signal waveform differs from the input; specifically, the output waveform equals the autocorrelation function of the input added to noise. In this type of receiver, signal fidelity is sacrificed to optimize S/N. Receivers designed for optimum pulse fidelity or other criteria will, therefore, be suboptimum in terms of S/N.

3.1.1.2 Dynamic Range

The dynamic range of an instrumentation system is the ratio of the largest to the smallest signal levels measurable with a given accuracy; the former is limited by the onset of saturation, the latter is limited by effects of noise.

We consider a model of a receiver which outputs the envelope of a sinusoidal signal added to Gaussian noise in response to a sinusoidal input. The statistics of the output of such a receiver are given by the classic Rician probability distribution

[5, 6], which describes the probability density of the envelope of signal-plus-noise as a function of the signal-to-noise ratio. This model applies to both coherent and incoherent receivers when the magnitude of the output envelope is of interest.

Figure 3.2 shows a receiver transfer characteristic in which the magnitude of the envelope is plotted on the ordinate in response to the amplitude of the sinusoidal input signal plotted on the abscissa. Because both input and output are normalized to the rms value of the zero-mean additive noise, the abscissa can be interpreted as S/N. The dashed line with unit slope represents the transfer characteristic of the ideal, noise-free receiver. For low S/N, the envelope is characterized by fluctuations induced by the noise; the solid line represents the mean of the output, and the two dashed lines straddling the mean represent the upper and lower standard deviations about the mean. For zero input, the mean and standard deviation of the envelope are 1.25 and 0.655 times the rms value of the noise, respectively. As the input increases, the mean of the output approaches the signal amplitude, while the standard deviation approaches the rms noise. As the input signal increases to large values, the effects of the noise become negligible and the receiver eventually approaches saturation, as indicated by the nonlinear response on the right of Figure 3.2. The input level which causes a deviation of the transfer function from linearity by a specified error, called the *compression point,* establishes the upper limit of the dynamic range. The lower limit of the dynamic range is established by a similar deviation from linearity in the mean of the envelope which is biased by presence of

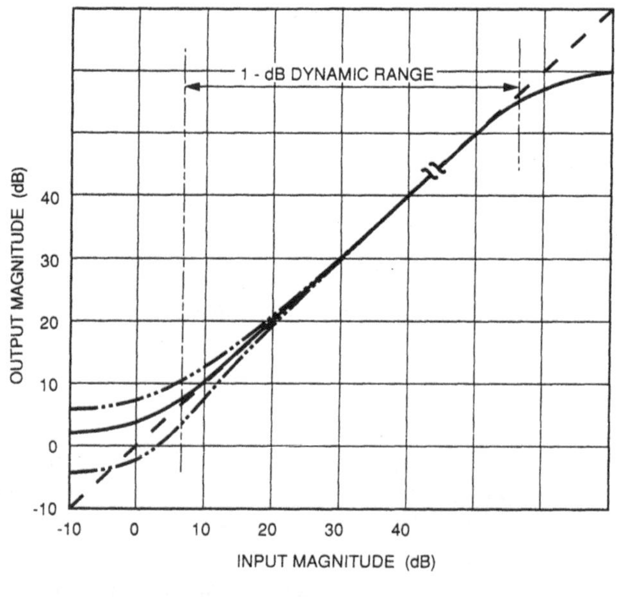

Figure 3.2 Receiver transfer characteristic.

the noise. Figure 3.2 indicates the dynamic range bounded by the 1-dB receiver compression at the upper end and by the 1-dB noise bias at the lower end.

3.1.1.3 Range Resolution

Resolution is the ability to distinguish closely-spaced features of a given quantity. In the context of radar imaging, *range resolution* is interpreted as the ability to distinguish contributions of isolated scatterers which are closely spaced in range. In the preceding chapter we demonstrated that responses of a complex object, observed as a function of frequency, can be processed to obtain a high degree of range resolution, which is inversely proportional to the frequency bandwidth.

Figure 3.3 shows a symbolic wideband radar capable of recording the response to a number of isolated scatterers as a function of frequency. For each discrete frequency f_k, the total received signal is a superposition of individual signals received from each scatterer, weighted by the scattering strength a_n and by the round-trip phase, $4\pi r_n f/c$. The superposition is supported by the linearity property of electromagnetic fields which add vectorially in free space and in the receiving sensor. The observed signal is therefore, represented by the summation:

$$V_k(f) = \sum_{n=1}^{N} a_n(r) \exp(-j4\pi fr/c) \tag{3.2}$$

K observations of $V(f)$ produce K complex pairs, representing the phase and magnitude of the received signal relative to the transmitted signal.

The objective of synthetic range imaging is to determine the range distribution of scatterers from the observed frequency response $V(f)$. The solution to this problem follows immediately after (3.2) is recognized as a Fourier transform involving the pair of variables f and $2r/c$ as shown in

$$V_k(f) = \sum_{n=1}^{N} a_n(r) \exp[-j2\pi(2r/c)f] \tag{3.3}$$

Figure 3.3 Symbolic wideband radar.

The required range distribution $a_n(r)$ is readily found by the inverse Fourier transform:

$$a_n(r) = \sum_{k=1}^{K} V_k(f) \exp[j2\pi(2r/c)f] \tag{3.4}$$

The preceding relations are equivalently expressed in terms of the round-trip delay rather than range by substituting $\tau = 2r/c$, in which case (3.4) becomes

$$a_n(\tau) = \sum_{k=1}^{K} V_k(f) \exp(j2\pi f\tau) \tag{3.5}$$

In the latter formulation, time and frequency constitute a Fourier transform pair, a result well known in electrical engineering. Determination of the range profile from the frequency response is subject to the resolution relations which govern Fourier transform pairs [7]; namely,

$$(2/c)B\Delta r = 1 \quad \text{and} \quad (2/c)L\Delta f = 1 \tag{3.6}$$

where Δr and Δf are the range and frequency resolutions, respectively, B is the frequency bandwidth, and L is the extent of the range profile being processed. The preceding relations are diagrammed in Figure 3.4, which shows the equivalent time and range responses with the corresponding Fourier transform relation indicated by double arrow. The uncertainty relations, which establish the range and frequency (time and frequency) resolutions, can be summarized as follows:

a. Fine range resolution is obtained by processing frequency responses spanning a wide bandwidth.

b. Fine frequency resolution is obtained by processing range responses spanning an extended distance.

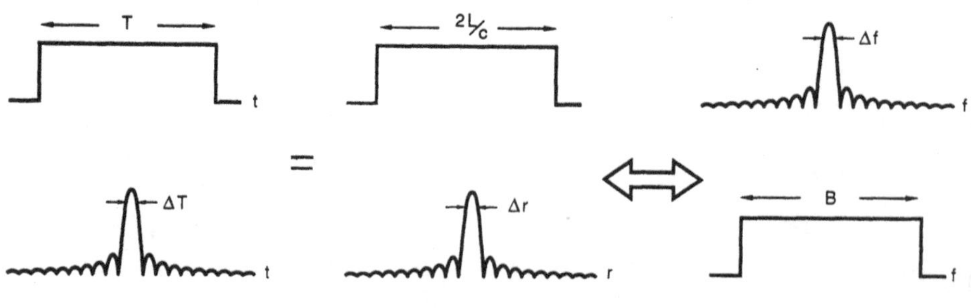

Figure 3.4 Uncertainty relations of Fourier transforms.

c. Either range or frequency, but not both, can be determined with arbitrary resolution.

The effects of increasing bandwidth on range resolution were shown in Figure 2.9 of Chapter 2. Figure 3.5 shows examples of frequency responses corresponding to arrays of point scatterers. If the frequency responses for infinite bandwidth could be measured, the processed range responses would exhibit delta-function resolution. When band-limited frequency responses are processed, however, the range responses corresponding to the point scatterers exhibit degraded range resolution. Figure 3.6 shows the dual relation in which frequency responses for limited range extents are processed. In each of the figures, all but the indicated number of range cells are set to zero. As the range extent is progressively reduced by range gating, the corresponding frequency resolution decreases. This indicates that arbitrary limiting of the range-gate extent blurs detailed features of the frequency response. Therefore, to recover closely-spaced resonant peaks is not possible for frequency responses obtained from range responses subjected to arbitrarily narrow range gates.

3.1.1.4 Ambiguities

When continuous signals are sampled, certain distortions can arise as a consequence of the sampling process. These distortions, known as *aliasing errors,* decrease as

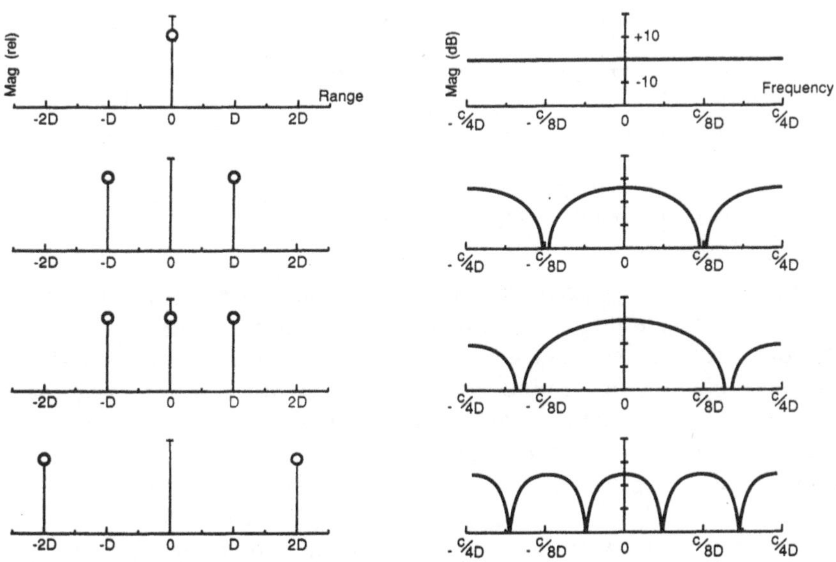

Figure 3.5 Examples of frequency responses for arrays of discrete scatterers.

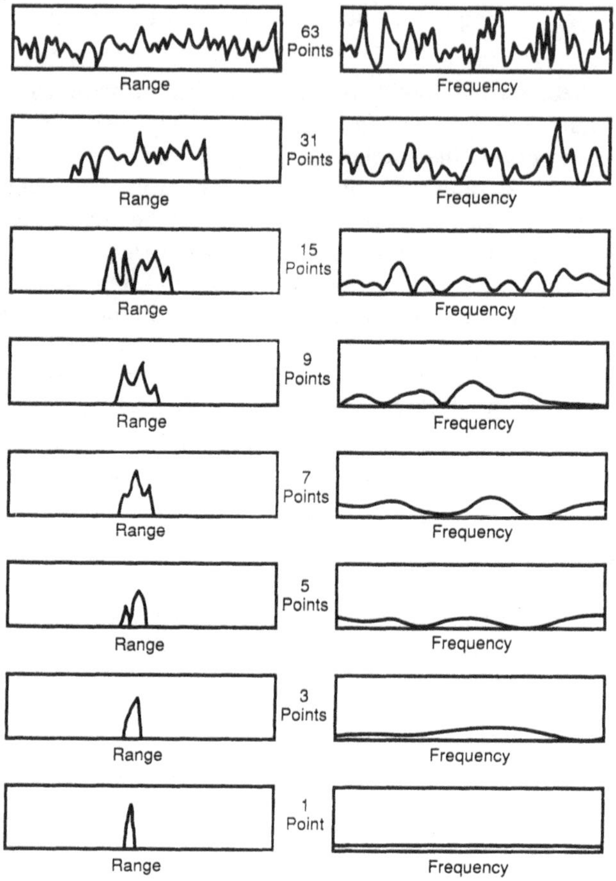

Figure 3.6 Frequency responses for gated range profiles.

the sampling increment or step size decreases. When sampled signals are Fourier transformed, aliasing errors cause ambiguous responses, which cannot be distinguished from true responses. Because the range and frequency responses of an object are found by Fourier transforming sampled data, ambiguous responses can result. All systems which use sampled signals are susceptible to aliasing errors, and must therefore be protected to avoid ambiguities.

Radar echoes from complex objects are composed of signals backscattered from individual scattering centers. While the magnitude of each of the signals is usually a relatively weak function of the irradiating frequency (determined by the physical structure of the scattering center), the phase of these signals is a strong function of frequency. The reason for this dependence is that the round-trip distance represents a different number of wavelengths for different frequencies. As a result of these two factors, signals scattered by complex objects are always a function of the irradiating frequency. When responses of complex objects are sampled at discrete frequency steps, the step size must be sufficiently small to limit variations in the phase of the received signal between steps. This constitutes the well known Nyquist sampling criterion, which arises in the sampling of continuous functions. If the criterion is satisfied, the frequency response throughout the entire band covered by the incremental frequency steps can be uniquely recovered from the sample values, and no information is lost in the sampling process. Under these conditions, there is a unique correspondence between the observed frequency response and the range distribution of the object. If the criterion is violated, the sampled frequency responses of objects at different ranges produce indistinguishable results. As a consequence of the violation, ambiguous range responses are created, a process called *aliasing*.

To illustrate these concepts, consider the variations of the signal received from a point target at range R as a function of the transmitted frequency f. Because the magnitude of the received signal is a weak function of frequency, it can be considered a constant, A. The phase relative to the transmitted signal is

$$\phi = 4\pi R/\lambda = 4\pi fR/c \tag{3.7}$$

where λ is the wavelength and c is the propagation velocity.

The signal is totally described by its phase and amplitude or by its real and imaginary components:

$$\mathrm{Re} = A\cos(4fR/c) \qquad \mathrm{Im} = A\sin(4fR/c) \tag{3.8}$$

The two alternative representations of the signal are shown in Figure 3.7. The complex-envelope signal describes the phase and amplitude modulation imposed by the object on the incident CW carrier as a function of the transmitted frequency. For a given range R, the signal exhibits cyclic variations with frequency, a complete cycle occurring for a frequency change $\Delta f = c/2R$. As the range increases, the complex envelope varies more rapidly as a function of frequency, with the cyclic rate being proportional to range. Figure 3.8 shows a cyclic signal sampled at a rate which reveals the true cyclic variation. The same signal, shown sampled at a lower rate in Figure 3.9, indicates the apparent cyclic variation to be lower than the true cyclic variation. This irreversible loss of information is known as aliasing. The sampling interval required to preserve all the information necessary for completely representing or reconstructing the signal is given by the sampling theorem or Nyquist criterion. Simply stated, the latter requires the phase of the highest frequency contained in a real signal to vary less than π radians between samples; the maximum allowable phase variation between samples of complex signals is 2π radians.

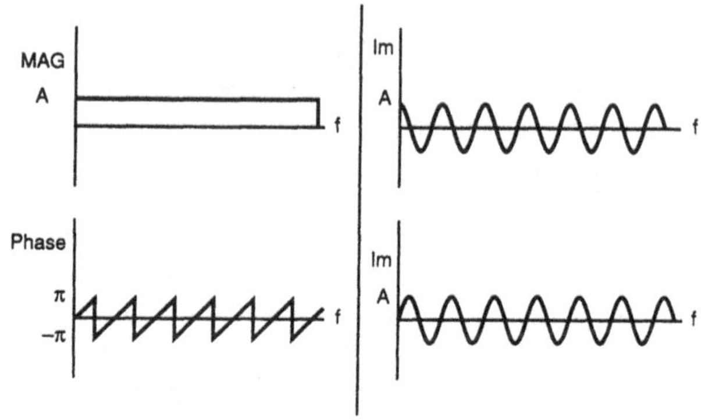

Figure 3.7 Alternate representations of complex envelope as a function of frequency.

The maximum alias-free range corresponding to a given frequency step size is obtained by solving (3.7) for a phase difference of 2π radians:

$$\Delta\phi = 2\pi = 4\pi R_U \Delta f/c \tag{3.9}$$

$$R_U = c/2\Delta f$$

R_U is the unambiguous range, and objects at ranges exceeding this value cause a folding of the true range into the unambiguous extent. Mathematically, the observed range can be expressed by the remainder of the true range divided by the unambiguous range.

3.1.2 Stepped-Frequency CW Systems

The stepped-frequency CW method of obtaining object responses over a wide range of frequencies is a relatively simple extension of CW measurements, in which the procedure is repeated for a number of discrete frequencies. The process is accomplished by using tunable, narrow band receivers operated in synchronism with programmable signal synthesizers controlled to step in uniform frequency increments.

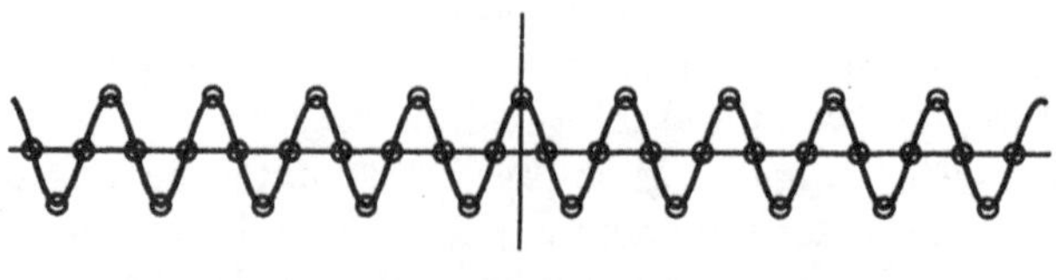

Figure 3.8 Cyclic signal sampled at a rate satisfying Nyquist criterion.

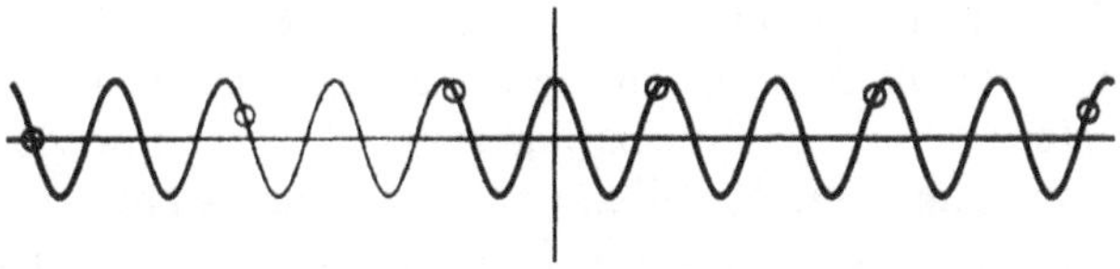

Figure 3.9 Cyclic signal sampled at a rate violating Nyquist criterion.

The principal advantages of stepped-frequency CW systems are (1) their relative simplicity, requiring only the addition of frequency tuning to a conventional CW system, and (2) their high degree of amplitude sensitivity, which is normally obtained from narrow band CW receivers. Figure 3.10 shows a block diagram of a stepped-frequency CW system.

If the dwell time for each frequency step approximately equals the inverse of the receiver bandwidth, the receiver operates essentially as a matched filter, and optimum S/N performance can be achieved. The dwell period for an optimum S/N in a receiver with bandwidth β is approximately $T = 1/\beta$. The total signal energy captured from a received signal with power P at the end of the dwell period is, therefore:

$$E = PT = P/\beta \tag{3.10}$$

The measured phase and magnitude can be subsequently processed by a Fourier transform to yield the range profiles of the object reflectivity with range resolution determined by the total bandwidth covered by the frequency steps. In a conceptual sense, this system provides optimum performance in terms of both range resolution and S/N.

Two practical considerations limit the utility of stepped-frequency CW systems: (1) their high level of transmitter signal leakage and (2) their susceptibility to aliased responses. These factors are described in the following paragraphs.

Because the transmitter and receiver are simultaneously active, the performance of CW systems is limited by signals directly coupled from the transmitter into the

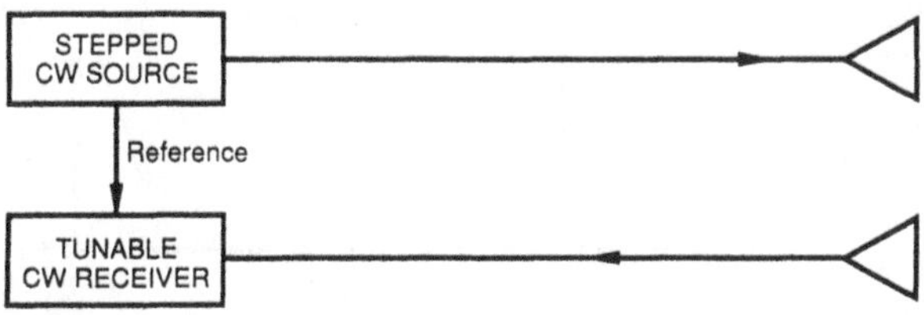

Figure 3.10 Stepped frequency CW system.

receiver channel, even in the absence of object echoes. The degree of coupling in systems using separate antennas is determined by the channel isolation; for systems using a single antenna, the coupling is determined by the impedance mismatch of the antenna. In either case, the amplitude of the coupled signal usually exceeds that of the object signal by several orders of magnitude and is frequency dependent. Figure 3.11 shows power levels that are typically encountered in a CW instrumentation system. The figure indicates that the receiver must provide sufficient dynamic range to accommodate the large transmitter coupling signal without distorting the relatively weak object signal. For the example shown, the required dynamic range of the receiver is 80 dB, of which the upper 20 dB is essentially wasted on the undesired coupling signal. The effects of the transmitter coupling signal can be separated from the object signal subsequent to the range processing. This is possible because the transmitter coupling signal appears at shorter ranges than the object signal, and can therefore be virtually eliminated by range gating. This feature, termed *software range gating,* allows the range-selective suppression of contributions of unwanted echoes from the total measured signal. Because the effects of the transmitter coupling signal can only be removed after processing, software range gating has no effect on the dynamic range requirements of the receiver, which must be capable of accepting the total received signal without saturation.

The unambiguous range of the stepped frequency CW system, given by (3.9), is established by the frequency step size, which causes a 2π-radian phase shift between steps. Because the frequency sampling process occurs at the transmitter, no antialiasing filtering is available prior to sampling. Consequently, spurious responses from ranges beyond the unambiguous range can be aliased into the range of interest

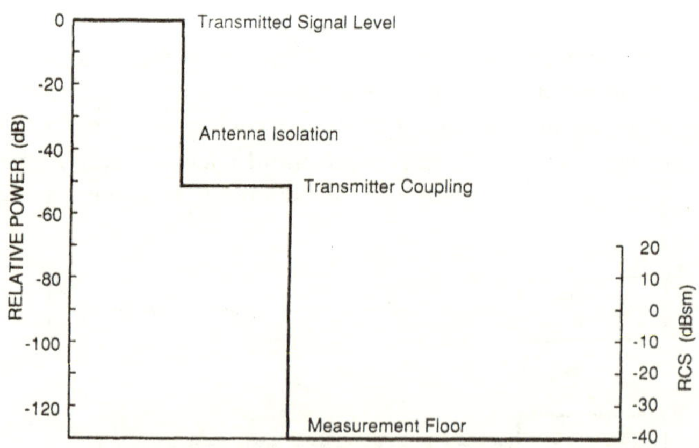

Figure 3.11 Typical power levels in a CW instrumentation system.

and be indistinguishable from desired object responses. Figure 3.12 shows the range response obtained from point scatterers A and B at ranges lesser and greater than the unambiguous range, respectively. Scatterer A is imaged at the proper range; scatterer B is aliased into the unambiguous range, and appears at range $R_B - R_U$. Spurious responses can arise from echoes of stationary clutter located at ranges beyond the unambiguous limit. Such spurious responses may also be object-related, such as delayed responses caused by reverberations within the object and interactions between the object and surroundings, or be induced by harmonics due to nonlinearities in the receiver hardware. Spurious responses not involving the object can be eliminated by a coherent background subtraction process in which signals received in the absence of the test object are determined as a function of frequency, and subsequently subtracted from signals received in the presence of the object. Spurious responses involving the object, however, are not eliminated in the background subtraction process.

In summary, stepped-frequency CW systems require a wide dynamic range of pre- and postprocessing signal instrumentation hardware to accommodate strong spurious echoes due to transmitter leakage. In addition, these systems are vulnerable to spurious responses that can be aliased into the range of interest and be indistinguishable from object responses. These systems should, therefore, be used only if we are able to establish *a priori* that no aliased responses can occur.

Stepped-frequency CW systems can be readily implemented using commercially available laboratory equipment consisting of computer-controlled, frequency-agile microwave synthesizers and narrow band receivers. High-quality narrow band receivers, tunable over wide frequency ranges, are embodied in modern vector network analyzers designed for component testing. These instruments, readily adapted to RCS measurements, provide an attractive alternative to a complete system design.

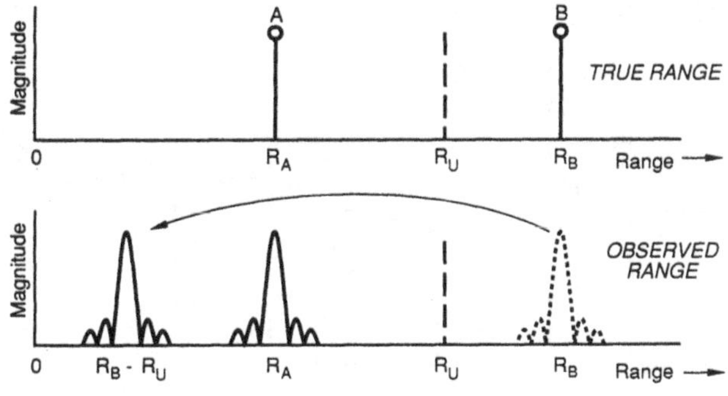

Figure 3.12 Ambiguous range responses.

3.1.3 Pulsed, Stepped-Frequency Systems

The shortcomings of stepped-frequency CW systems, namely their susceptibility to aliased responses and transmitter coupling, can be alleviated by pulsing the transmitter and time-gating the received signal to reject spurious echoes from outside the range of interest. Signals coupled from the transmitter can be rejected by the range gate because they arrive at times preceding object echoes; signals received from ranges exceeding the unambiguous limit, capable of inducing aliased responses, can likewise be rejected by the range gate because they arrive at times following object echoes.

Figure 3.13 shows a block diagram of a pulsed, stepped-frequency system identical to the stepped-frequency CW system, but augmented by the addition of pulse modulators. The coherent CW signal transmitted for a given frequency is modulated by a pulse train with an interpulse period sufficient to allow the reception of echoes from the maximum range prior to transmission of the next pulse. This precludes range ambiguities that would arise from second-time-around echoes.

Figure 3.14 shows conceptual waveforms of a pulsed system. The received waveform, consisting of the object echo and the transmitter coupling, is range gated to admit the former and reject the latter. Ideally, all signals except object echoes coinciding with the range rate are eliminated. Practically, however, residual signals are present due to the finite isolation of the pulse modulators, as indicated by the lower part of Figure 3.14. The design of optimum pulse modulators for RCS instrumentations involves limiting leakage responses to acceptable levels while maintaining the fidelity of the pulsed waveform [8]. The time-selective suppression of spurious echoes implemented by pulsing the transmitter and range-gating the receiver is known as *hardware range-gating*, distinguished from software range-gating which can be performed on the synthetic range responses subsequent to the Fourier transform process [9].

In addition to the stated advantages, the pulsed system can substantially increase the allowable size of the frequency steps, thus reducing the time required to sample object responses over a given frequency band. The frequency step required

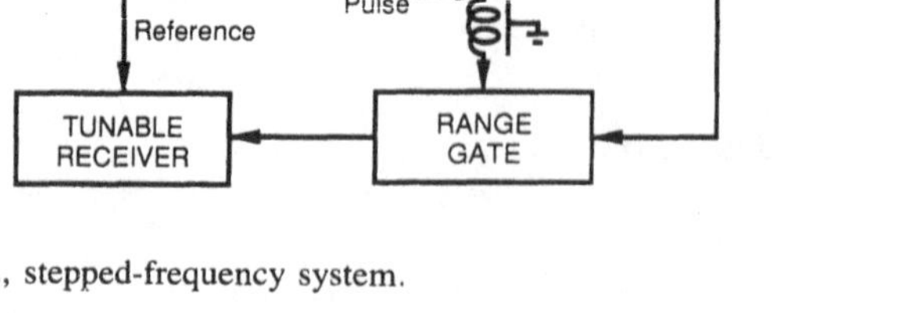

Figure 3.13 Pulsed, stepped-frequency system.

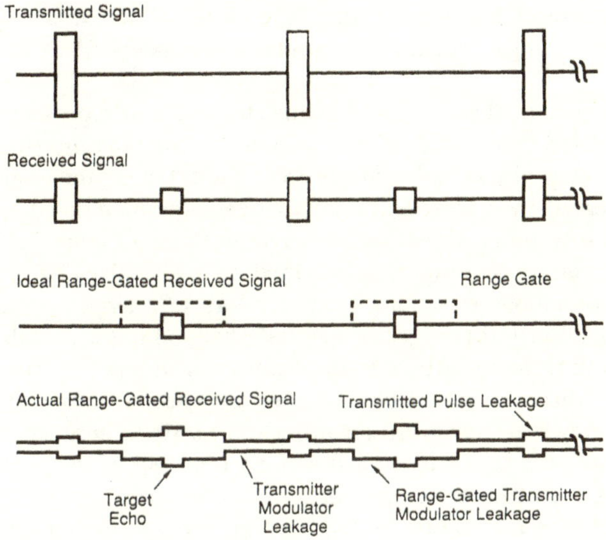

Figure 3.14 Conceptual waveforms of a pulsed system.

to preclude aliasing with a stepped-frequency CW system is dictated by the maximum object range. For example, as indicated by (3.9), measurements of an object at ranges of 10 and 100 m require step sizes of 15 and 1.5 MHz, respectively. Although the intrinsic RCS response of the object is not range dependent, the required frequency sampling becomes increasingly stringent as the observation range increases. Suppose, in contrast, that the frequency response is sampled by a pulsed, range-gated, stepped-frequency system using coarse frequency steps which place the object well beyond the unambiguous range. In the subsequent processing to obtain the range response, the range profile of the object is folded or aliased into the unambiguous range span, along with responses from all other ranges. However, because the hardware range-gating has *a priori* eliminated all echoes from outside the range gate, the object range response is immune to spurious, aliased responses. Because all echoes that potentially cause spurious responses are eliminated, the gated object response can be highly aliased without detrimental effects. Thus, the only requirement is that the unambiguous range exceed the range-gate extent, which ultimately can approach the object range extent. The preceding observation can result in substantial sampling economy in which the maximum frequency step size is given by

$$\Delta f = c/4L \tag{3.11}$$

where L is the range extent of the object.

The procedure of hardware gating, therefore, allows the frequency sampling required to obtain range profiles of objects to be independent of the observation range.

As a consequence of pulsing and gating, the waveform presented to the receiver is a pulse-modulated coherent carrier in which the complex envelope represents the response of the entire object to the carrier frequency. Unlike the unmodulated carrier, which is represented by a narrow frequency band, the modulated carrier occupies a bandwidth inversely proportional to the pulse duration. Figure 3.15 shows spectra imposed on the carrier by amplitude modulation with pulses of duration T. Figure 3.15(a) is the continuous spectrum corresponding to a single pulse, exhibiting the familiar sinc function $[(\sin \pi x)/\pi x]$ spectral shape. Figure 3.15(b) is the discrete spectrum resulting from an infinite train of pulses with repetition period T; the spectrum consists of discrete lines enveloped by the spectral shape of Figure 3.15(a). Figure 3.15(c) shows the spectrum resulting from a finite number of pulses, N. The spectrum consists of a series of sinc functions repeated at the pulse repetition frequency, each of extent inversely proportional to the total period of the finite pulse train, and bounded in magnitude by the spectral form of the single pulse.

To optimize the performance of pulsed systems, control of the duration and relative timing of the transmitted pulse and range gate is necessary. Figure 3.16 shows two point objects at ranges R_1 and R_2, representing up-range and down-range

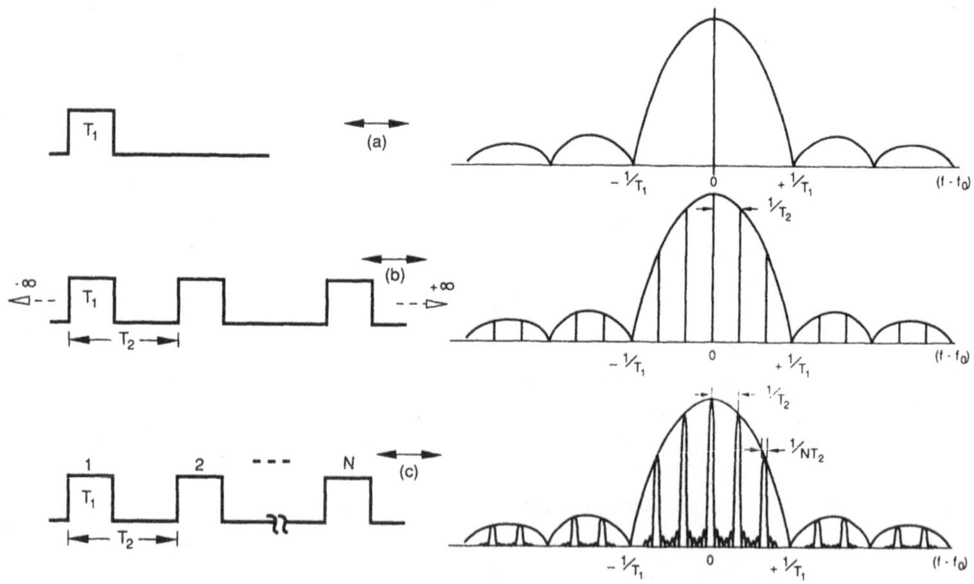

Figure 3.15 Pulsed signal spectra: (a) single pulse; (b) infinite pulse train; (c) finite train at N pulses.

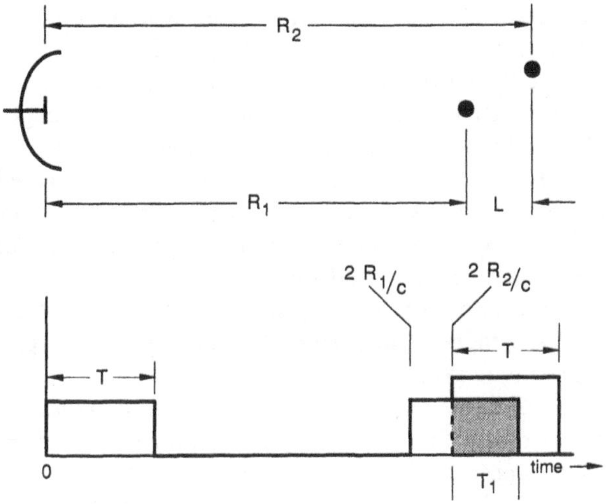

Figure 3.16 Pulsed response of dual-point object.

limits of an object with extent L. The accompanying timing diagram shows conceptual echoes of the transmitted pulse with duration T from each point. The time interval between the two responses is $2L/c$; the unshaded regions represent periods for which echoes from a single object point are received, and the shaded region represents the period for which echoes from both object points (and all points in between) are received simultaneously. If the duration of the transmitted pulse is less than $2L/c$, two distinct pulses are received; if the pulse duration exceeds $2L/c$, the pulses overlap, as indicated in the figure. During the overlap period, the received signal is a sinusoid with magnitude and phase given by the vector sum of the two individual echoes. Therefore, a measure of the complex amplitude of a CW response is identical to that obtained by a pulsed system during the overlap period. To allow a definite overlap period as illustrated in Figure 3.16, the following relation must be satisfied:

$$T \geq 2L/c + T_1 \tag{3.12}$$

where

T = the transmitted pulse duration,
L = the target length,
T_1 = the range gate duration.

The variation of signal energy captured by the range gate as a function of its time alignment relative to the received pulse is shown in Figure 3.17, where T and

T_1 represent durations of the pulse received from a point object and of the range gate, respectively. The range gate is considered fixed in its reference position, and the received pulse is delayed by τ. The lower plot, representing the area of overlap as a function of the relative time shift, illustrates that the range gate produces a uniform response for all received signals with relative delays between 0 and $T - T_1$. Received signals with delays greater than T produce no range-gated output, because there is no region of overlap. A system using transmitted pulse and range gate durations T and T_1 will, therefore, receive contributions from all object points in a range extent $c(T + T_1)/2$, and respond uniformly to object points in a range extent $c(T - T_1)/2$. If the transmitted pulse and range gate have equal durations, the lower plot of Figure 3.17 degenerates to a triangle, and the system provides the proper response for only a unique range.

When using (3.12) to establish the required transmitted pulse duration, practical effects of nonzero rise and fall times and time jitter of the pulses will offset the equality. In addition, a potentially significant difficulty arises in adequately selecting the object length, L, which has been postulated to correspond to up-range and down-range limits of extended objects. Because complex objects can exhibit delayed responses caused by multiple reflections, the proper measure of L should be the electrical object length; the latter is not generally known *a priori* and the pulse duration may, therefore, have to be determined empirically to avoid unintentional elimination of delayed object responses.

The modulated pulse train available at the output of the range gate must be processed by a receiver capable of accurately measuring the complex amplitude of the received signal. If the guideline of (3.12) is followed, the signal consists of a sinusoid modulated by a train of pulses with duration equal to the range gate, which can be described by a single complex number. The two categories of instrumentation used to extract the required information can be distinguished as wideband and narrowband receivers. The former can process a single pulse and have sufficient band-

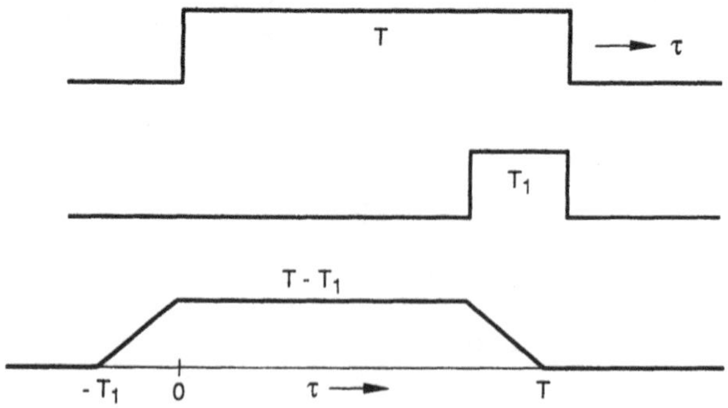

Figure 3.17 Signal energy captured by range gate.

width to retain features of the pulse modulation, while the latter can process a train of pulses and retain only a measure of the effective signal energy. Both types of receiver must be tunable over a wide range of frequencies; this tuning bandwidth is distinguished from the instantaneous bandwidth, which refers the frequency content of signals that can be processed simultaneously. Salient features of the two types of receivers are described in the following sections.

3.1.3.1 Wideband Receivers

This type of receiver is designed to optimally process a single pulse, characterized by the spectrum shown in Figure 3.15(c). The receiver bandwidth is sufficiently wide to accept the central spectral region which contains the majority of the received energy, and thus allows recovery of the pulse modulation waveform. Applications requiring maximum sensitivity in the presence of noise utilize receiver bandwidths equal to the inverse pulse duration. This provides optimum performance approaching that of the matched filter, developed in Section 3.1.1.1. For applications in which sharp leading and trailing edges of the pulse must be maintained to optimize the rejection of range clutter, the receiver bandwidth is increased over the matched filter bandwidth with a corresponding reduction in sensitivity. The basic function of the receivers, implemented as multiple-stage superheterodyne systems, is to translate the pulsed RF waveform to a faithful replica at an *intermediate frequency* (IF). Following the frequency translation, the signal is coherently demodulated by the process described in Section 3.1.3.3.

Because the maximum energy available to the receiver of a pulsed system is the product of the peak received energy times the range-gate duration, increasing the range-gate duration increases the sensitivity at the cost of increased clutter admitted to the receiver. The pulsed receiver conceptually captures the total signal energy presented to its input. Relative to a CW system with identical peak power capability, the pulsed system suffers a reduction in available energy, and therefore in sensitivity, by the duty factor. Practical applications of pulsed RCS systems often provide insufficient energy to adequately measure the received signal on the basis of a single pulse. By coherently summing responses to several pulses, the sensitivity of the receiver can be increased. The coherent integration process, performed digitally on stored data, is described in Section 3.2.1.

Optimum pulsed receivers for RCS instrumentation systems pose stringent design requirements in the areas of sensitivity, linearity, dynamic range, spurious modulations, and interference rejection. Receivers for these applications must have sufficient instantaneous bandwidth to accommodate the pulsed spectra, and be frequency agile to allow rapid tuning in discrete frequency steps. The performance of pulsed systems is taxed on large outdoor ranges operating at high frequencies (due to the low duty factors imposed by the maximum allowable PRF), by the limited powers of wideband coherent transmitters, and by the inherent small apertures of receiving antennas.

3.1.3.2 Narrow Band Receivers

The basic function of RCS instrumentation systems is to measure the object's equivalent response to a CW signal as a function of frequency. Although the output of a range-gated system is a pulsed waveform, the transmitted pulse and range gate durations are selected to provide a constant signal amplitude throughout the range-gate period. During this period, therefore, the gated signal represents a slice of the equivalent CW response.

Because these receivers coherently process the input signal over a period approximated by the inverse bandwidth, a narrow band receiver presented with a pulse train provides a measure of the signal amplitude, which is based on many pulses. If the signal amplitude does not vary appreciably during the integration period, the receiver output provides a reliable measure of the signal amplitude. This contrasts with wideband receivers, described in the preceding section, which can measure the signal amplitude based on a single pulse.

The spectrum of a carrier signal modulated by a pulse train, shown in Figure 3.15, consists of a distribution of spectral lines centered on the carrier frequency. Although the input waveform is pulsed, a narrow band receiver tuned to the center frequency, therefore, extracts the equivalent CW component of the pulsed waveform. Because the receiver only accepts the central region of the spectrum and rejects the remainder, however, only a fraction of the total available signal power is recovered.

Relative to a CW signal, the power contained in the central spectral line is reduced by the square of the duty factor. Processing the range-gated pulsed waveform through a narrow band receiver, therefore, constitutes an effective signal loss, which reduces the S/N. Stated in general terms, the loss is incurred because the narrow band receiver is not a matched filter for the pulsed waveform. Figure 3.18 shows the loss relative to a CW receiver as a function of the duty factor. Because the duty factor is inversely proportional to range, the loss incurred by using a narrow band receiver for pulsed applications increases with range. For applications involving short distances, typical of RCS measurement on indoor ranges, the loss is of the order of 15–25 dB; such losses can be compensated by RF amplification, thus retaining the intrinsic simplicity of tunable, narrow band receivers.

The loss of S/N in the gated narrow band receiver depends on whether the noise sources of the receiving system precede or follow the gate. For noise which precedes the gate, there is no loss in S/N because the combination of a gate matched to the pulsewidth and narrow band filter matched to the duration of the pulse train constitute a matched filter. In essence, the pulsing decreases the signal power by the same factor as the noise power.

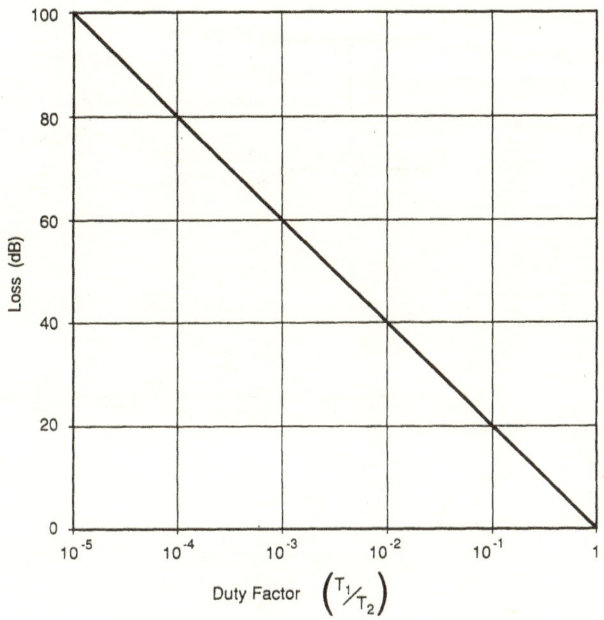

Figure 3.18 Loss of pulsed radar system using a narrow band receiver.

Optimum performance of narrow band receivers used in pulsed applications results when the receiver output is sampled at intervals approximating the inverse of the bandwidth and the signal does not change significantly during the integration period.

3.1.4 Swept Frequency, Linear-FM Systems

The stepped-frequency systems described in Sections 3.1.2 and 3.1.3 use receivers with instantaneous bandwidth sufficient to process signals for a particular frequency and are tunable over the frequency band necessary for synthetic range imaging. For these systems, the instantaneous bandwidth is relatively narrow when compared with the tuning bandwidth. RCS instrumentation systems can also be implemented using swept frequency, or linear FM systems and receivers with wide instantaneous bandwidth.

Reduced to its fundamental components, as shown in Figure 3.19, the linear FM system operates by transmitting a CW signal with center frequency f_0 linearly swept over a bandwidth B during a period T. The received signal is input to the RF port of a balanced mixer with the local oscillator (LO) port fed by a sample of the

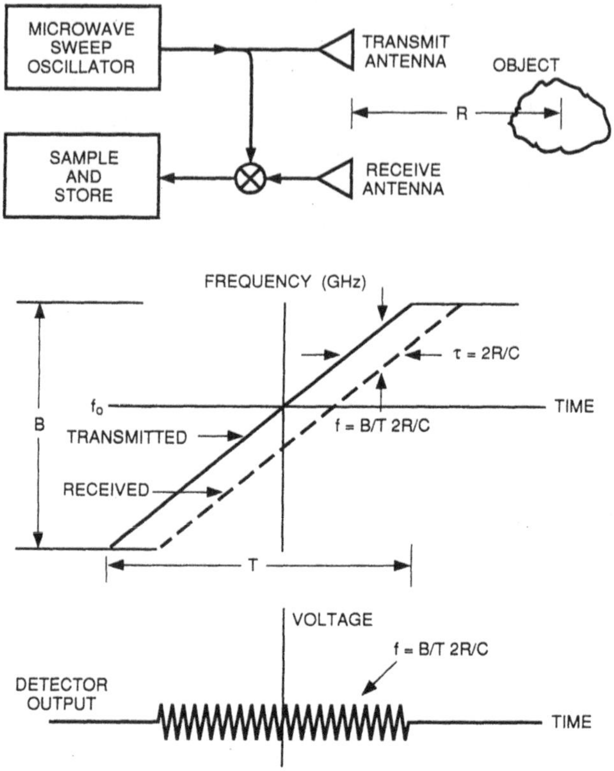

Figure 3.19 Conceptual linear-FM system.

transmitted signal. Because the transmitted signal serves as the LO, this arrangement constitutes a *homodyne receiver*. The output signal formed by the mixer is the product of the transmitted and received signals; for a single point object, the frequency of the output signal is proportional to range, consisting of the frequency deviation of the transmitter during the round-trip time. The output signal is sampled, digitized and stored for subsequent processing.

The following relations express the transmitted, received, and output signals for a point object:

$$e_T(t) = E_T \Pi\left(\frac{t}{T}\right) \sin(2\pi f_0 t + \pi B t^2/T) \tag{3.13}$$

$$e_R(t) = E_R \Pi\left(\frac{t-\tau}{T}\right) \sin[2\pi f_0(t-\tau) + \pi B T^{-1}(t-\tau)^2] \tag{3.14}$$

where τ is the round-trip time delay, and the function:

$$\Pi(x) = \begin{cases} 1, & \text{for } |x| < 0.5 \\ 0, & \text{otherwise} \end{cases}$$

The output of the mixer is described by the low frequency terms of the product of transmitted and received signals which, when using the relation $\tau = 2R/c$, reduces to

$$e_o(t) = E_o\Pi\left[\frac{t - R/c}{T - 2R/c}\right] \cos\left[\left(\frac{4\pi BR}{cT}\right) t + \left(2\pi f_0 - \frac{2\pi BR}{cT}\right)\frac{2R}{c}\right] \quad (3.15)$$

Because the signal duration greatly exceeds the signal round-trip travel time ($T \gg 2R/c$), the last phase term in (3.15) is negligible, and the terms in the first brackets can be well approximated by t/T; (3.15) can thus be properly approximated by

$$e_o(t) = E_o\Pi(t/T) \cos[(4\pi BRT^{-1}c^{-1})t + (4\pi f_0 c^{-1})R] \quad (3.16)$$

As expressed by (3.16), the output signal consists of a tone burst of duration T with frequency and phase both proportional to range. Arrays of object points at various ranges produce a linear superposition of signals with distinct temporal frequencies, which can be decomposed into individual components by a Fourier transform. Because the temporal frequency of the output signal is proportional to range, a spectral decomposition separates the contributions of object points on the basis of range.

The magnitude of the Fourier transform output provides a range profile of the object with a resolution which is determined by the signal duration T. The temporal frequency associated with a point object at range R is

$$f = \frac{2BR}{cT} \quad (3.17)$$

The spectral resolution, Δf, of a sinusoid observed over a period T is of the order of $\Delta f = 1/T$. Combining this relation with (3.17) and associating the spectral resolution Δf with the range resolution ΔR yields the fundamental relation:

$$\Delta R = \frac{c}{2B} \quad (3.18)$$

When these systems are operated using relatively long frequency sweep periods (10–100 ms) and short ranges, the output signal given by (3.18) is in the audio frequency range. This feature allows digitizing, recording, and processing the output signal by using conventional low-frequency, low-speed instruments and circuitry [10]. The homodyne receiver is a matched filter which performs bandwidth compression, while the bandwidth of the waveform input to the receiver can span several gigahertz, the bandwidth of the receiver output waveform is typically 10–100 Hz. The system essentially operates in a CW mode because signals are being received during transmission; however, the homodyne receiver differs from stepped-frequency systems because the frequency is continuously varied during transmission. This means that the receiver must have wide instantaneous bandwidth, in contrast with stepped-frequency systems in which the receiver has a relatively narrow bandwidth which can be tuned over a wide frequency range. For this reason, receivers used for these two classes of systems are totally incompatible. However, the two classes of systems are similar in that they perform a measurement of the object response as a function of frequency. While the linear FM system obtains data as a continuous function of frequency, the stepped-frequency system obtains the data for discrete frequency steps.

Potentially ambiguous responses can be reduced by using frequency-selective filters at the output of the homodyne receiver. Because the range-selective filtering is done prior to sampling the output signal, the possibility of aliased responses is virtually eliminated. Effects of transmitter coupling signals, which appear at short ranges relative to object echoes, can likewise be suppressed by frequency-selective filters, with the degree of rejection being established by the frequency roll-off of the range filters.

The difference between stepped- and swept-frequency systems is indicated by the simplified block diagram of Figure 3.20. In the swept system, the signal is processed as a continuous function of frequency up to the receiver output, where it is then sampled and digitized. In the stepped system, the signal is processed for discrete frequency steps and subsequently digitized. The basic difference is that in the swept system the sampling is done after detection, while in the stepped system the signal is sampled before detection. The FFT input for the swept system is a sequence of real numbers; input for the stepped system is a sequence of complex numbers. In spite of this difference, the same fundamental information is available from either system. The information content available from real and complex signals is discussed in Section 3.2.5.

3.1.5 I/Q Demodulation

Signals processed in radar and communication systems often consist of transmitted carriers which have been amplitude and phase modulated. In the simplest case, the transmitted signal with frequency ω_0 and arbitrary phase ϕ_0 is expressed by $v_T(t) = \cos(\omega_0 t + \phi_0)$. The received signal, with amplitude A, is shifted by a phase angle ϕ relative to the transmitted signal, and is expressed by $v_R(t) = A\cos(\omega_0 t + \phi_0 -$

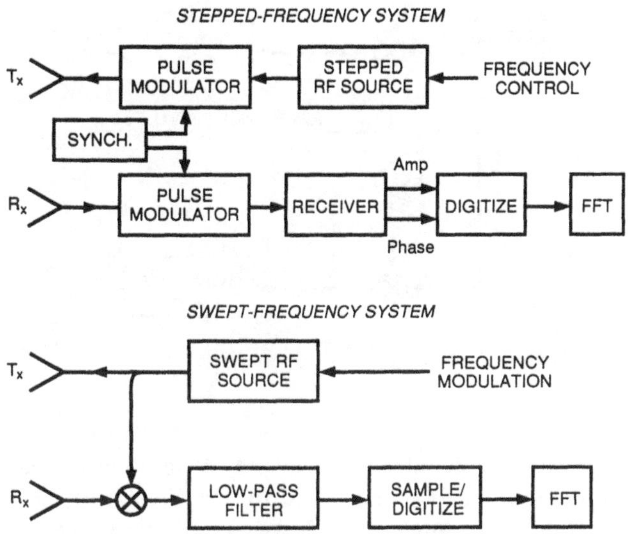

Figure 3.20 Stepped- and swept-frequency systems.

ϕ). In RCS instrumentation systems, A is proportional to the square root of the RCS and ϕ is the round-trip phase.

Because ω_0 and ϕ_0 are known or determinable, the information necessary to uniquely describe $v_R(t)$ consists of A and ϕ. In applications where the frequency ω_0 is sufficiently low, the signal can be recorded or processed directly in either analog or digital form, after suitable sampling and quantization. For high-frequency carriers, the direct recording of the signal is not practical, and signal demodulators are used to extract the quantities A and ϕ from the received signal. The circuit most frequently used to recover the required information is the coherent demodulator or *in-phase* and *quadrature* (I and Q) demodulator shown in Figure 3.21.

The received signal is input to one port of two multipliers; the other ports of the multipliers are fed by the reference signal and the same signal shifted in phase by 90 degrees. The product function can be implemented by actual analog multipliers in low-frequency applications, or by diode mixers in high-frequency applications. Assuming a direct multiplication, the output of the upper multiplier is

$$v(t) = A\cos(\omega_0 t + \phi_0 - \phi)\cos(\omega_0 t + \phi_0) \tag{3.19}$$

Expanding in a trigonometric identity yields:

$$v(t) = A/2\cos(2\omega_0 t + 2\phi_0 - \phi) + A/2\cos(\phi) \tag{3.20}$$

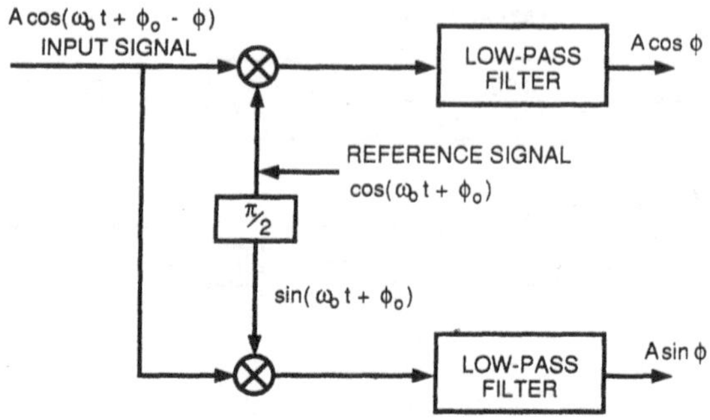

Figure 3.21 I/Q demodulator.

The first term represents a signal with frequency twice that of the carrier, and the second term is a dc component proportional to the product of the signal amplitude and the cosine of the phase angle. The low-pass filter rejects the high-frequency term, leaving an output which is proportional to $A\ \cos\phi$. By a similar development, the other output is shown to be proportional to $A\ \sin\phi$.

The upper part of the circuit is the in-phase channel, and the lower part is the quadrature channel. The two outputs, taken as a pair, define the vector quantity $A\ \exp(j\phi)$ by determining the two components $A\ \cos\phi$ and $A\ \sin\phi$. The terms "in-phase and quadrature" refer to the fact that two components of the received signal are recovered, one in phase with the reference and the other in phase-quadrature with the reference. The coherent demodulator, therefore, produces outputs proportional to two orthogonal components of the received signal phasor; these components uniquely define the phasor quantity in the same way that A and ϕ define the same quantity. Conceptually, no inherent advantage results from the I and Q method of demodulation as compared with measurements of A and ϕ or direct measurement of the received signal.

The received signal processed by the coherent demodulator can be expressed in more general terms by

$$v_R(t) = A\cos(\omega_0 t + \phi_0 - \phi) = \text{Re}[A\exp(-j\phi)\exp(j\omega_0 t + \phi_0)] \quad (3.21)$$

The quantity $A\ \exp(-j\phi)$, called the *complex envelope* of the signal, represents the modulation of the carrier signal which is expressed by the second exponential. In the general case, both A and ϕ can be time-varying functions, in which case $A(t)$ represents amplitude modulation and $\phi(t)$ represents phase modulation. Note that, although v_R is a real quantity, the modulation function is complex whenever phase modulation is involved. The real and imaginary components of the complex envelope are precisely the two quantities determined by the coherent demodulator. In the demodulation process, the modulated carrier is, in effect, shifted to zero frequency and the modulation, or baseband signal, is recovered.

The information-bearing part to the received signal is contained in the spectral bandwidth about the carrier, which is produced by the modulation. If the spectrum is downshifted ω_0 by using a conventional demodulator, the lower sidebands will be folded above zero frequency and added to the upper sidebands. The modulation envelope is highly distorted by this process, which cannot distinguish between modulation components above and below the carrier. This limitation is imposed because the conventional demodulator can only produce real signals, while the modulation envelope is complex. The coherent demodulator provides the translation to baseband while retaining both the real and imaginary components of the complex envelope. The process of translating to baseband reduces to a minimum the bandwidth of the signal containing the essential information. This minimization is of significant advantage in sampled signals because the I and Q channel outputs need only be sampled at the Nyquist rate determined by the modulation bandwidth. This requirement is in contrast with the direct sampling of the narrowband frequency-translated signal, which must be accomplished at a rate consistent with the maximum signal frequency.

3.2 SIGNAL PROCESSING

Synthetic methods to obtain a high degree of spatial resolution by necessity involve computations on stored signal samples. Signal processing is, therefore, an essential part of RCS imaging systems. The following sections describe tools and concepts essential to the application and understanding of signal processing for RCS imaging, consisting of coherent integration, Fourier transforms, windows, interpolation, and analytic signals.

3.2.1 Coherent Integration

RCS instrumentation systems are radars designed to extract signal information which faithfully describes the reflectivity characteristics of the object. As in all radar systems, the signal-to-noise ratio of the received signal directly affects the fidelity of the measured data. In RCS instrumentation systems, the S/N obtained by processing a single sample or pulse is often insufficient for adequate measurement of the object's cross section. The S/N can be enhanced by combining the response to several repeated samples. This process is termed *integration* because several detected samples are superimposed or added together. The notion that repeated observations of a deterministic signal made in the presence of random noise can lead to more accurate measurements of the signal appears intuitively acceptable.

When N coherent signal samples are coherently summed in the presence of random noise, the signal voltage increases by N while the rms noise increases by

$N^{1/2}$; as a result, an N-fold increase in the signal-to-noise power ratio is obtained. The S/N enhancement relies on the fact that the noise voltage competing with the signal is not correlated from sample-to-sample, and hence the sum does not increase linearly with N. If the signal is coherent, meaning that the N samples are identical in amplitude and phase, the superposition of replicated signals increases linearly with N. The S/N enhancement obtained by coherent integration depends on the sample-to-sample coherence of the signal and on the lack of sample-to-sample coherence of the noise.

By processing a large number of samples, it is theoretically possible to increase the S/N to arbitrary limits. Practical limits, however, are imposed by considerations of dynamic range, hardware linearity, and digital storage capacity. Because the integration gain depends on the signal coherence, variations in the phase of the received signal samples during the integration period reduce the maximum achievable integration gain. The loss relative to the ideal coherent integration gain is a function of the total phase shift occurring during the integration period. In RCS instrumentation systems using coherent integration, any variation in the integration gain constitutes a measurement error. Figure 3.22 shows a plot of integration loss as a function of the total linear phase shift incurred in the course of the integration period.

When systems employing coherent integration are used for RCS measurements, the object rotation rate poses a practical limit to the amount of coherent integration that can be utilized effectively. Because the RCS of extended objects is measured as a function of angular rotation, the phase of signals from scatterers displaced from the center of rotation varies in the course of the integration period. This phase vari-

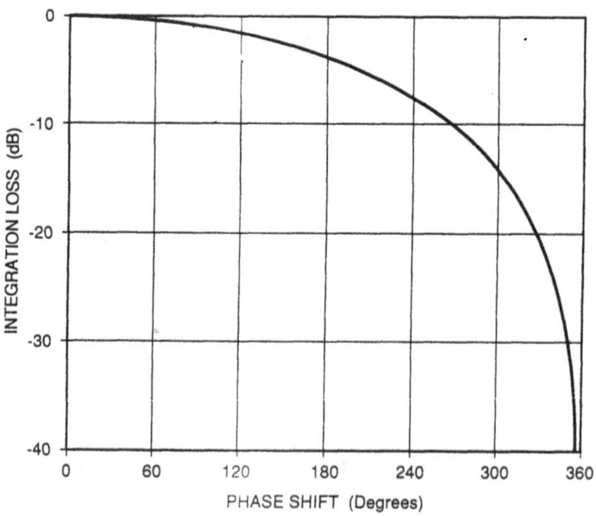

Figure 3.22 Coherent integration loss as a function of phase variation.

ation, caused by the varying round-trip distance, is directly proportional to the cross-range coordinate of the scatterer. As a consequence, echoes from a point object located at the center of rotation exhibit a constant phase during rotation, and therefore produce a maximum response at the output of the coherent integrator. Identical objects radially displaced from the center exhibit variable phase, resulting in a lower response at the output of the coherent integrator. The difference in integration gain obtained from the two object points constitutes a measurement error, which is a consequence of the coherent integration, coupled with the object motion during the integration period. While the single pulse echoes of the two object points are identical, the coherently integrated responses differ by an amount proportional to the integration time or number of samples integrated. Figure 3.23 shows a plot of the object rotation rate necessary to limit the integration error to 1 dB, in terms of normalized object extents and number of samples, N.

3.2.2 Fourier Transforms

The most ubiquitous signal-processing tool for high-resolution imaging by synthetic means is the Fourier transform; its utility owed to the decomposition of complex signals occurring in linear systems into elemental harmonic components.

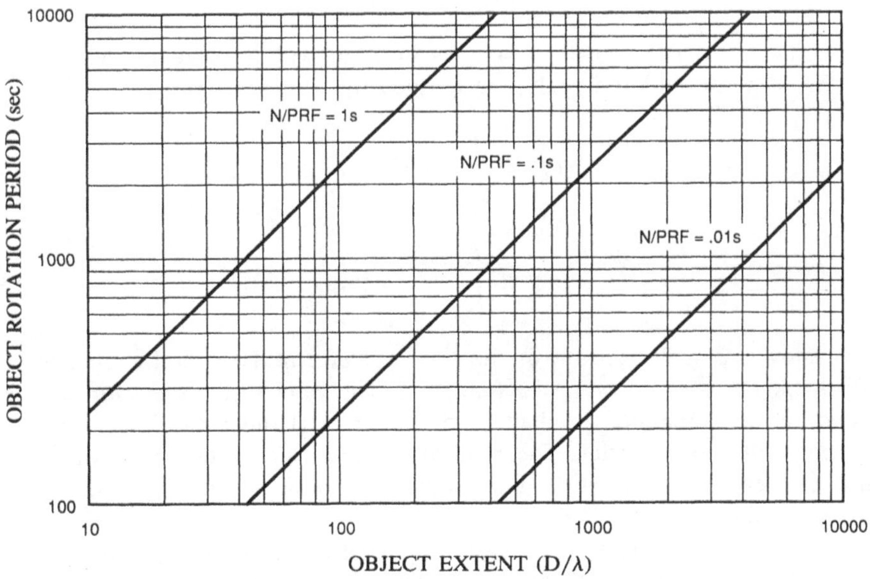

Figure 3.23 Object rotation period *versus* object extent as a function of integration time for 1.0 dB error.

Responses of a single point scatterer as a function of frequency and small rotation angles exhibit linear phase variations equivalent to sinusoidal variations in I/Q components. Responses of arrays of point scatterers which consist of complex superpositions of such elemental signals are, therefore, decomposed into linear-phase components uniquely associated with the location and magnitude of individual scatterers.

The Fourier transform, while not a fundamental law of nature, is a widely known mathematical tool applicable to a number of physical situation [11]. If an observed quantity can be represented by a Fourier transform in terms of another quantity, the latter can then be expressed in terms of the former by the corresponding inverse Fourier transform. The utility of this computational feature was indicated in (2.5) and (2.6), relating the transfer function and impulse response of linear systems. Because the transfer function is linked to the impulse response by a Fourier transform, the impulse response can be obtained from measured samples of the transfer function using an inverse Fourier transform.

The following sections briefly summarize features of: continuous Fourier transforms [11], discrete Fourier transforms [12], fast Fourier transforms [13], and chirp-z transforms [14]. The references are samples from a large number of works available on the subject.

3.2.2.1 Continuous Fourier Transform

The continuous Fourier transform and its inverse, widely used for the analysis of continuous time or analog signals, are defined by

$$X(f) = \int_{-\infty}^{+\infty} x(t) \exp(-j2\pi f t) dt \tag{3.22}$$

$$x(t) = \int_{-\infty}^{+\infty} X(f) \exp(j2\pi f t) df \tag{3.23}$$

$X(f)$ is known as the spectrum of the time waveform $x(t)$; f and t are continuous variables that can assume values between $\pm\infty$. The value of the spectrum for a given frequency is the integrated product of the signal $x(t)$ times the linear phase term expressed by the exponential. The process essentially tests the input signal to determine the degree of correlation with a cyclic signal as a function of the frequency of the cyclic signal.

Continuous transforms, forming the basis of analog linear systems, provide a vast basis from which physical insight can be derived. Because imaging by synthetic means necessarily involves sequences of sampled signals, however, the continuous Fourier transform is of limited computational significance.

3.2.2.2 *Discrete Fourier Transform*

The *discrete Fourier transform* (DFT) and its inverse, applicable to sampled signals, are defined by

$$X(\omega) = \sum_{n=0}^{N-1} x(n) \exp(-j\omega n) \tag{3.24}$$

$$x(n) = \frac{1}{2\pi} \int_{-\infty}^{+\infty} X(\omega) \exp(j\omega n) \tag{3.25}$$

$x(n)$ is a discrete sequence; ω is a continuous variable; $X(\omega)$, periodic in 2π, is defined by the range of from 0 to 2π. The units of ω are radians/sample. The occurrence of the summation in the forward transform and integral in the inverse transform result from formulating $x(n)$ as a discrete sequence while retaining $X(\omega)$ as a continuous variable. Equations (3.24) and (3.25) are approximations of the continuous Fourier transform where x is a sampled signal. Computing the spectrum of an N-point sequence for a particular value of ω involves N complex multiplications; computing K samples of the spectrum requires NK complex multiplications.

3.2.2.3 *Fast Fourier Transform*

Perhaps the most widely-exploited algorithm in modern signal processing is the *fast Fourier transform* (FFT) and its inverse, expressed by

$$X(k) = \sum_{n=0}^{N-1} x(n) \exp(-j2\pi kn/N), \tag{3.26}$$

$$\text{for } k = 0 \text{ to } N-1$$

$$x(n) = \frac{1}{N} \sum_{k=0}^{N-1} X(k) \exp(j^2\pi kn/N), \tag{3.27}$$

$$\text{for } n = 0 \text{ to } N-1$$

This algorithm, a variation of the DFT, operates on the sequence $x(n)$ to obtain samples of the spectrum for discrete values of $\omega = 2\pi kn/N$. Equation (3.26) operates on N samples of x to produce N uniformly-spaced samples of X, the discrete Fourier transform, for ω ranging from 0 to 2π. Because both x and X consist of discrete samples, both X and x are periodic in N. The term "fast" refers to the considerable computational efficiency resulting when the following restrictions are met:

a. Both k and n range between 1 and $N - 1$;

b. N is an integer power of two.

Under these conditions, N uniformly spaced samples of the Fourier transform of an N-point sequence can be calculated with $N \log_2(N)$ complex multiplications. This constitutes a significant computational saving over N^2 multiplications required for the DFT.

3.2.2.4 Chirp-z Transform

The *chirp-z transform* (CZT), a computational algorithm allowing greater flexibility than the DFT and the FFT, is defined by

$$X(k) = \sum_{n=0}^{N-1} x(n)A^{-n}W^{nk}, \quad \text{for } k = 0 \text{ to } K - 1 \tag{3.28}$$

where

$$A = A_0 \exp(j2\pi\theta_0) \text{ and } W = W_0 \exp(j2\phi_0).$$

Under the special case of $A = 1$, $K = N$, and $W = \exp(-j2\pi/N)$, the CZT is identical to the FFT. The implementation of the CZT removes the restriction of k ranging from 0 to $N - 1$ and allows the computation of the Fourier transform for uniformly spaced increments of X spanning any region. In addition, the restriction of N and K to integer powers of 2 is eliminated, and N and K need not be equal. The principal advantage of the CZT in RCS imaging applications is that it allows a zooming feature for improving the definition of spectral regions of interest. The enhanced flexibility of the CZT decreases the computational efficiency available with the FFT. Computation of K samples of the spectrum of an N-point sequence requires $L \log_2(L)$ complex multiplications, where L is the smallest integer power of two exceeding $(N + K)$.

3.2.3 Windows

Spectral leakage is a result of processing finite-duration records when signals being analyzed are not periodic with the observation period. Figure 3.24 shows a periodic extension of a sinusoid that is not periodic in the observation interval. When this signal is processed by a Fourier transform, the discontinuities at the extremities of the observation interval cause spurious spectral responses known as *leakage*. Because

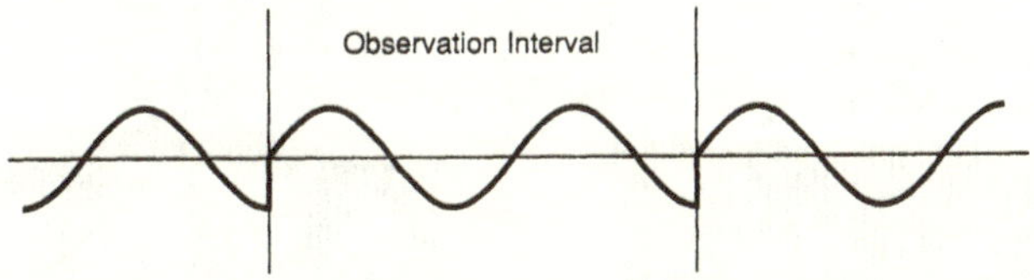

Figure 3.24 Periodic extension of signal not periodic in observation interval.

the FFT essentially considers sequences to be periodic, the condition depicted always prevails except in pathological cases when the signal is periodic in the observation interval. To reduce the effects of spectral leakage, weighting functions are applied to data. The multiplicative weighting functions, known as *windows,* are tapered smoothly to zero at the extremities so that the continuity of the periodic extension is enhanced.

The design of tapered windows [15] involves a trade-off between the suppression of leakage sidelobes and reduced spectral resolution. Figure 3.25(a–i) from [15] show examples of tapered windows and corresponding spectral responses. The rectangular window, shown in Figure 3.25(a), exhibits the highest spectral resolution and the highest sidelobe level of −13 dB. As the windows taper more smoothly to zero, the sidelobe levels decrease at the cost of decreased resolution. The Blackman-Harris windows, shown in Figure 3.25(h) and 3.25(i), exhibit sidelobe performance in excess of −70 dB.

3.2.4 Interpolation

A consequence of processing signals with discrete Fourier transforms is that the spectra are only described at discrete intervals. Figure 3.26 shows examples of spectra of a complex sinusoid. The signal for the upper plot is exactly periodic in the observation interval; the signal for the lower plot, having a slightly different frequency, is not periodic in the observation interval. The solid curve represents the spectrum as a continuous function of frequency, and the dotted values represent discrete samples of the spectrum obtained from a 16-point FFT. Although the amplitude of the complex sinusoid is identical in both cases, the granularity of the FFT output causes variations in the observed peak spectral response.

Variations in the observed magnitude of a spectral component depend on the position of the actual signal frequency relative to the discrete frequency values for which the spectrum is computed. The range of potential variations is termed *scalloping loss,* referring to the estimate of spectral peaks always being lower than the true value, except in the unique case when the signal is periodic in the observation interval. The application of tapered windows broadens the central lobe of the spectrum relative to the granularity, thus decreasing the scalloping loss. Figure 3.27 shows

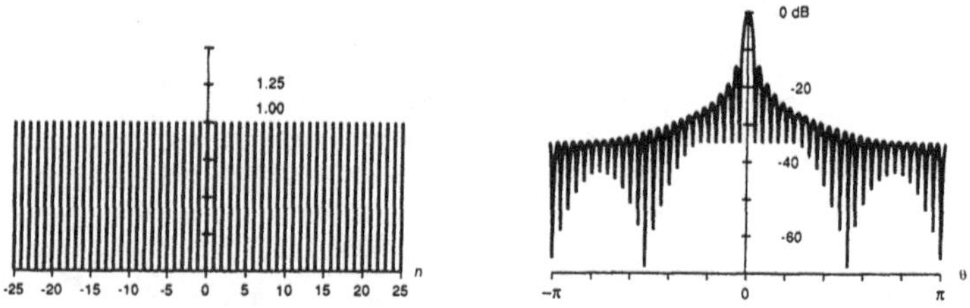

(a) Rectangular window: log-magnitude of transform.

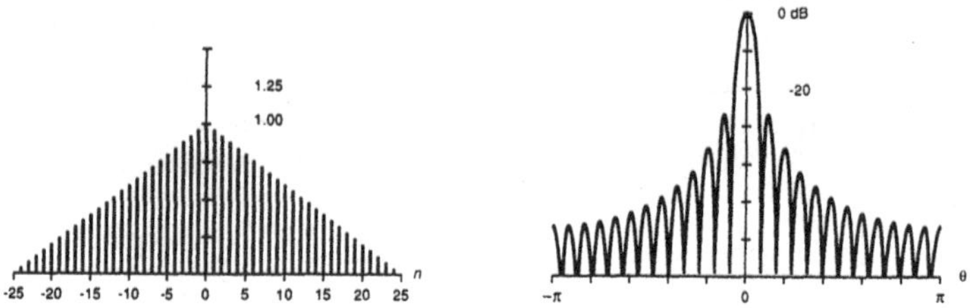

(b) Triangular window: log-magnitude of transform.

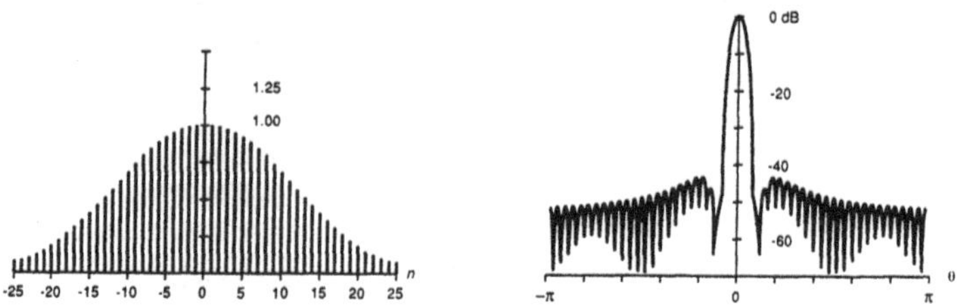

(c) Hamming window: log-magnitude of Fourier transform.

Figure 3.25 Tapered windows and corresponding spectral responses.

spectra of the same signals considered previously (without window), but with a Hann window applied. The 3.92 dB worst-case scalloping loss of the rectangular window is reduced to 1.42 dB by the Hann window. Scalloping losses corresponding to other tapered windows are tabulated in [15]. Figure 3.28 shows the reduction in scalloping loss resulting from interpolation for the rectangular and Hann windows.

The preceding discussion indicates that the accuracy of spectral estimates can be increased by finer granularity. This improvement can be accomplished by computing additional spectral samples or by interpolating to estimate the spectrum between the available values. The basis of interpolation [16–17] is summarized in Fig-

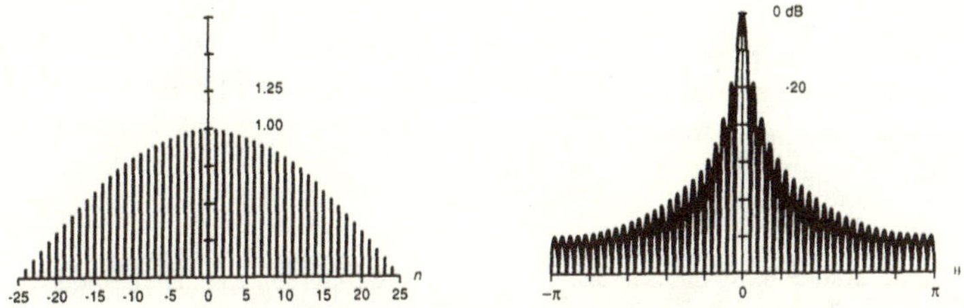

(d) $\cos(n\pi/N)$ window: log-magnitude of transform.

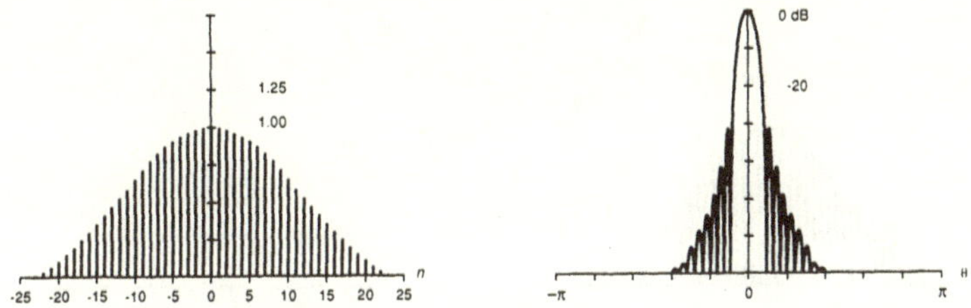

(e) $\cos^2(n\pi/N)$ window: log-magnitude of transform.

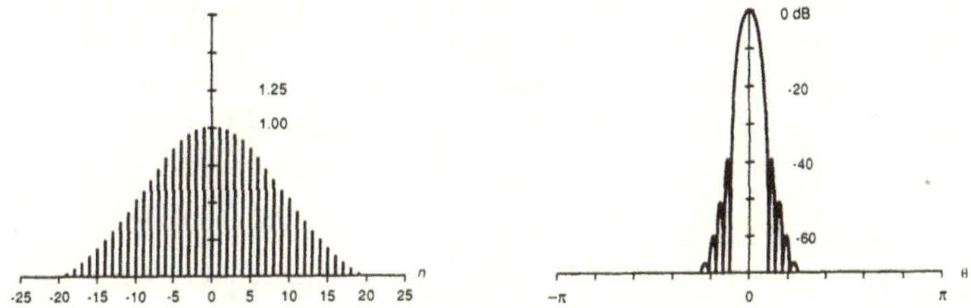

(f) $\cos^3(n\pi/N)$ window: log-magnitude of transform.

Figure 3.25 (continued).

ure 3.29. A continuous, band-limited signal and its spectrum are shown at the top of the figure, in which horizontal arrows denote a Fourier transform operation. The consequence of sampling the signal, indicated on the second and third row of figures, is to cause replicas of the original signal spaced at the sampling frequency. The band-limited restriction ensures that the spectrum is contained in a finite bandwidth, thus preventing the replicas from overlapping. The spectrum of the original signal can be recovered from that of the sampled signal by low-pass filtering the sample to reject the spectral replicas. The operation equivalent to multiplication by an ideal rectangular function in the spectral domain is a convolution of the sampled signal with a sinc(·) function. The original signal for any particular time is, therefore, recovered from the summed product of the sampled values times the sinc(·) weighting function.

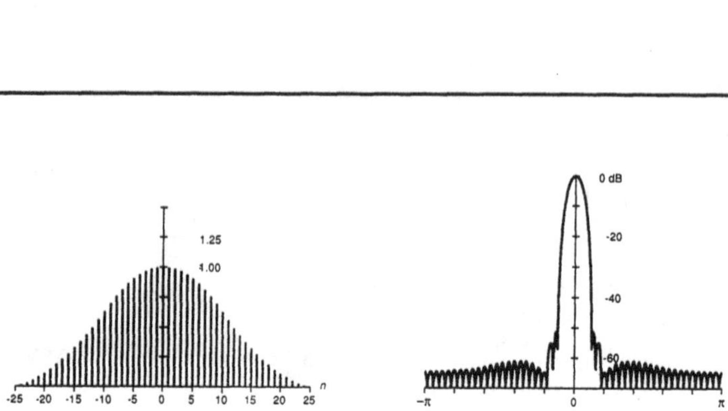

(g) Exact Blackman window: log-magnitude of transform.

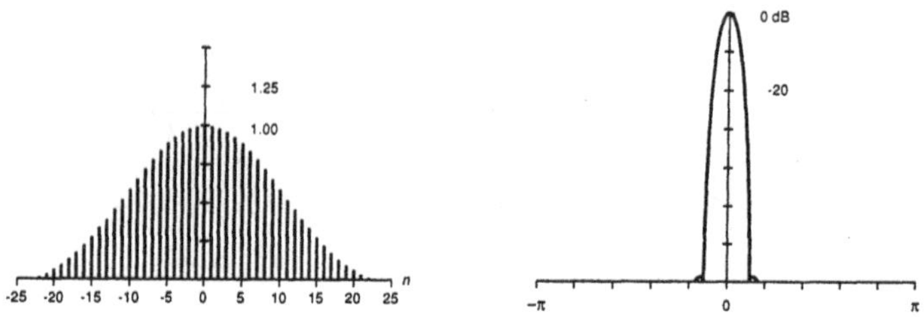

(h) Minimum 3-term Blackman-Harris window: log-magnitude of transform.

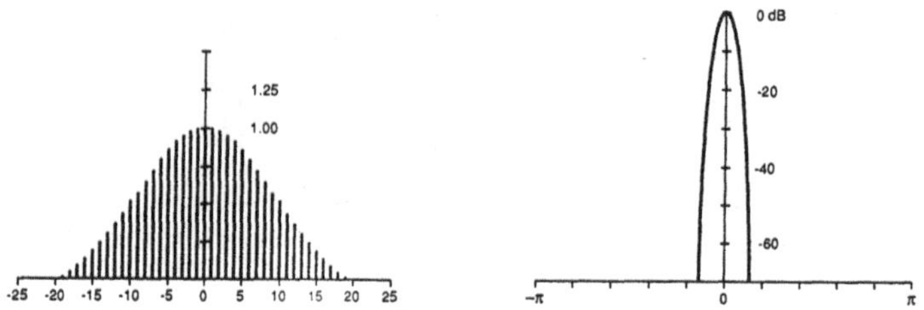

(i) 4-term Blackman-Harris window: log-magnitude of transform.

Figure 3.25 (continued).

The preceding development establishes the feasibility of precisely interpolating sampled signals, subject only to the restriction that the signals are band-limited and can, therefore, be adequately sampled. Note that the ideal interpolation filter is global in nature, that is, values between available samples are determined by all neighboring samples. Simple schemes such as linear interpolation between adjacent samples are suboptimum.

In view of the preceding discussion, we can make a frequently overlooked distinction between resolution and granularity. *Resolution,* describing the spectral width of the signal, is determined by the intrinsic signal properties, the observation length, and the window. *Granularity,* describing the discretization of spectral samples, is determined by the number and spacing of the computed samples. While

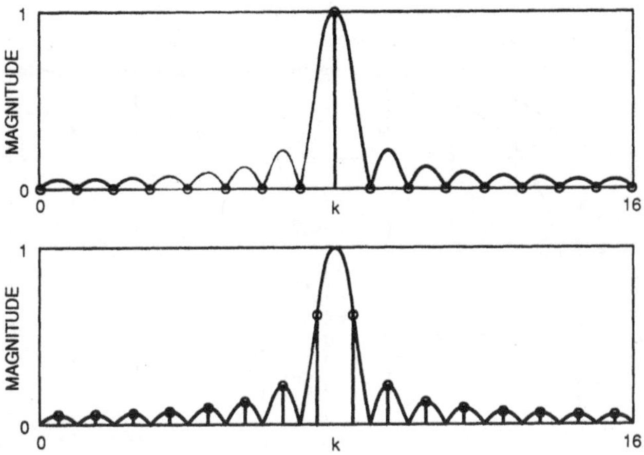

Figure 3.26 Spectra of complex sinusoids for rectangular window.

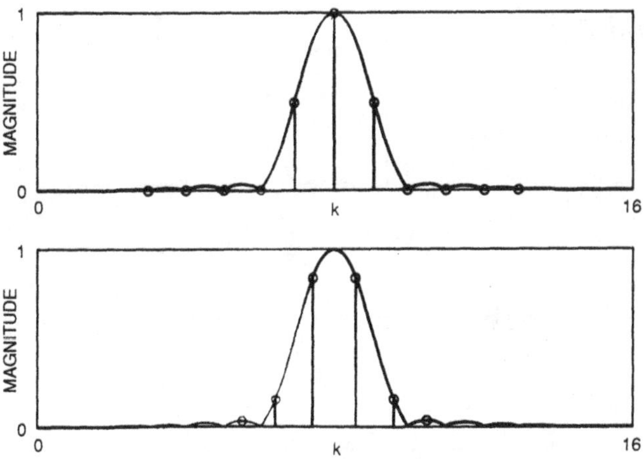

Figure 3.27 Spectra of complex sinusoids for Hann window.

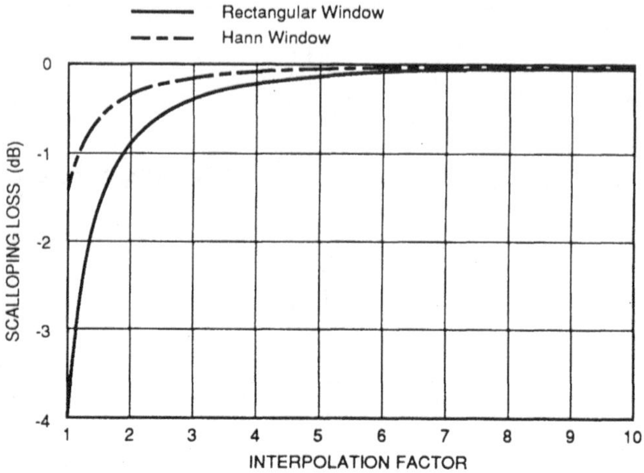

Figure 3.28 Reduction of scalloping loss by interpolation.

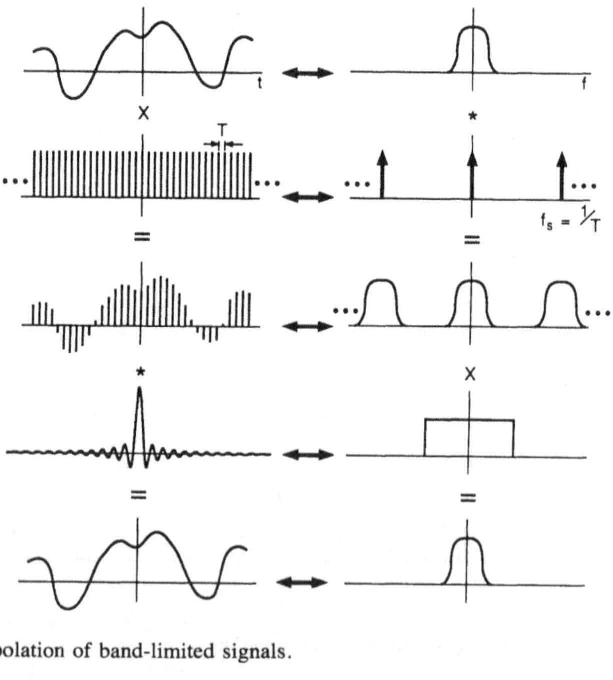

Figure 3.29 Interpolation of band-limited signals.

granularity can always be refined by interpolation methods, the intrinsic resolution available for the Fourier transform process cannot be enhanced by additional computations.

3.2.5 Analytic Signals

Radar signal analyses routinely deal with complex signals to describe modulated carrier signals of the form:

$$g(t) = \text{Re}\{[A(t) \exp(j\phi(t))] \exp(j2\pi f_0 t)\} \tag{3.29}$$

The modulation described by the bracketed terms is a complex quantity, but the actual signal is real. Although a complex description of the modulation is compact, no loss of information is incurred by describing the real signal $g(t)$. In this sense, complex signals are not intrinsically better descriptors than real signals.

The vast majority of practical signals satisfies the narrowband assumption, meaning that the frequency content of the signal is confined to a relatively narrow region around a carrier or center frequency. Under this condition, we can see that real or complex descriptions of the signal are equivalent.

Consider, for example, processing N samples of a band-limited signal with an FFT. If the signal is real, the spectrum will exhibit conjugate symmetry and provide $N/2$ unambiguous spectral samples. If N samples of a similar complex signal are processed, N unambiguous spectral samples result. This apparent increase of information results at the cost of requiring $2N$ values to describe the N complex signal samples. On this basis, therefore, the complex representation is not more efficient.

The preceding assertions can be mathematically supported by the notion of analytic signals and the Hilbert transform [18], described in the following paragraphs.

A narrowband signal can be described by:

$$g(t) = \text{Re}[\tilde{g}(t) \exp(j2\pi f_0 t)] = \text{Re}[g_p(t)] \tag{3.30}$$

The term $\tilde{g}(t)$, representing the modulation, is known as the complex envelope; $g_p(t)$ is a complex signal known as the pre-envelope. The spectrum of the real signal, denoted by $G(f)$, is two-sided, as shown in Figure 3.30.

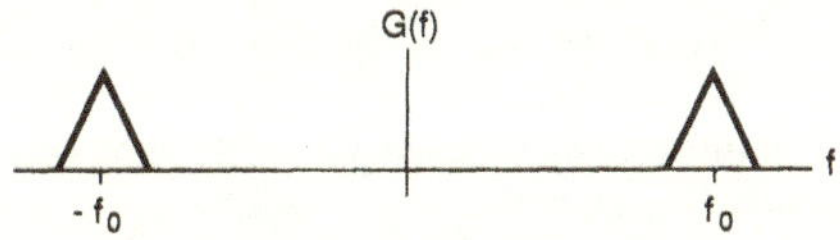

Figure 3.30 Spectrum of narrow band signal.

The spectrum of the complex signal $g_p(t)$ can be obtained by suppressing the negative frequencies of $G(f)$. The spectrum of the complex envelope is obtained by suppressing the negative frequencies and shifting $G(f)$ to the left by f_0. The pre-envelope, or analytic signal, can be obtained directly from the real signal by:

$$g_p(t) = g(t) + j\hat{g}(t) \tag{3.31}$$

where $\hat{g}(t)$ is the Hilbert transform of $g(t)$, defined by:

$$\hat{g}(t) = g(t)*(1/\pi t) \tag{3.32}$$

where $*$ denotes convolution.

The spectral operation corresponding to the convolution is a phase shifting of the positive frequencies by 90 degrees and of the negative frequencies by −90 degrees. The addition of $g(f) + j\hat{g}\,(t)$ results in the suppression of the negative frequencies and doubling the positive frequency part of $G(f)$. The spectrum of the complex envelope is obtained by shifting the spectrum of the analytic signal to the left by f_0, expressed by:

$$\tilde{G}(f) = \mathcal{F}[\tilde{g}(t)] = \begin{cases} 2G(f+f_0), & f > -f_0 \\ 0, & f < f_0 \end{cases} \tag{3.33}$$

The preceding analyses indicate that the complex signal equivalent to a real, narrow band signal can be formally recovered without approximation and without loss of information.

REFERENCES

1. Skolnik, M.I., ed., *Radar Handbook,* 1st Ed., New York, McGraw-Hill, 1970, p. 3–8.
2. Thomas, J.B., *An Introduction to Statistical Communication Theory,* New York, John Wiley and Sons, 1969, pp. 186–194.
3. Rihaczek, A.W., *Principles of High-Resolution Radar,* New York, McGraw-Hill, 1969, pp. 28–34.
4. Skolnik, M.I., ed., *Radar Handbook,* 1st Ed., New York, McGraw-Hill, 1970, p. 5–25.
5. Whalen, A.D., *Detection of Signals in Noise,* New York, Academic Press, 1971, pp. 101–105.
6. Thomas, J.B., *An Introduction to Statistical Communication Theory,* New York, John Wiley and Sons, 1969, pp. 160–167.
7. Bracewell, R., *The Fourier Transform and its Applications,* New York, McGraw-Hill, 1965, pp. 160–161.
8. Kao, P.S., "Automated Step-Frequency Radar Cross-Section Measurements," *IEEE Antennas and Propagation Symposium Digest,* 1982, pp. 237–239.
9. McCarter, G., "Software or Hardware Gating—What's Best for RCS Measurements," *Microwave Systems News,* Vol. 19, No. 11, November 1989, pp. 60–65.

10. Mensa, D.L., "Wideband Radar Cross Section Diagnostic Measurements," *IEEE Trans. Instrumentation and Measurement,* Vol. IM-33, No. 3, September 1984, pp. 206–214.
11. Bracewell, R., *The Fourier Transform and its Applications,* New York, McGraw-Hill, 1965.
12. Oppenheim, A.V., and R.W. Schafer, *Digital Signal Processing,* Englewood Cliffs, NJ, Prentice-Hall, 1975.
13. Brigham, O.E., *The Fast Fourier Transform,* Englewood Cliffs, NJ, Prentice-Hall, 1974.
14. Rabiner, L.R., R.W. Schafer, and C.M. Radar, "The Chirp-z Transform Algorithm," *IEEE Trans. Audio and Electroacoustics,* Vol. AU-17, No. 2, June 1969, pp. 86–99.
15. Harris, F.J., "On the Use of Windows for Harmonic Analyses with the DFT," *Proc. IEEE,* Vol. 66, No. 1, January 1978, pp. 51–83.
16. Schafer, R.W., and L.R. Rabiner, "A Digital Signal Processing Approach to Interpolation," *Proc. IEEE,* Vol. 61, No. 6, June 1973, pp. 692–702.
17. Crochiere, E.C., and L.R. Rabiner, "Interpolation and Decimation of Digital Signals," *Proc. IEEE,* Vol. 69, No. 3, March 1981, pp. 300–331.
18. Whalen, A.D., *Detection of Signals in Noise,* New York, Academic Press, 1971, pp. 55–70.

Chapter 4
Two-Dimensional Range and Cross-Range Imaging of Rotating Objects

This chapter combines methods for obtaining high degrees of resolution in range and cross range to produce two-dimensional microwave images. Wide signal bandwidths and synthetic aperture, or doppler processing, are combined to resolve objects rotating about an axis normal to the line-of-sight into two-dimensional resolution cells, as shown in Figure 4.1. Resolution in range, obtained by use of wideband signals and time-delay sorting, does not require object motion. Backscattered signals associated with isorange cells, typified by the shaded area in Figure 4.1, can be isolated by this method.

As the object is rotated, scattering centers crossing through the range-resolution cells cause signals associated with each range cell to vary; these signals can be processed to obtain resolution in cross range. The result of these combined processes is a two-dimensional image of the object formed in the radial (range) and normal to the radial (cross-range) directions lying in a plane normal to the axis of rotation.

Any of the instrumentation systems described in Chapter 3 can be used to acquire the data necessary for two-dimensional imaging by the following general procedure.

1. A transducer with sufficient angular beamwidth to uniformly irradiate the object is held fixed and used to transmit and receive microwave signals.
2. The complex object response is measured for a fixed aspect angle as a function of frequency by an instrumentation system connected to the transducer.
3. The object is rotated and the measurement is repeated for a number of discrete angular increments.
4. The frequency responses for each aspect angle are processed by Fourier transforms to obtain range profiles.
5. The signal contained in each range cell as a function of rotation angle is processed by Fourier transforms to obtain the cross-range profile corresponding to each range cell.

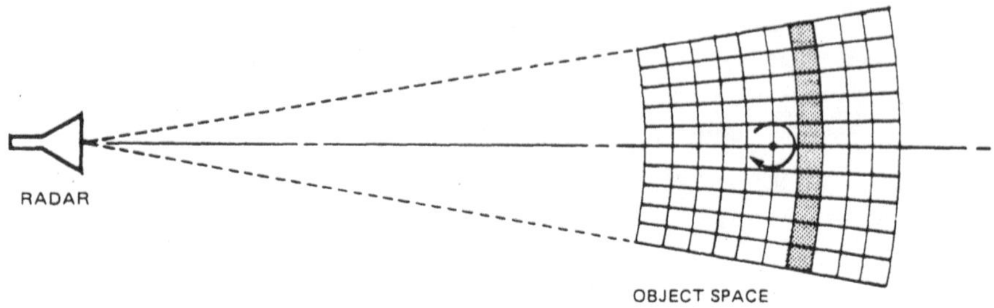

Figure 4.1 Two-dimensional resolution of rotating object.

6. The results are plotted to display visual images formed by the magnitude of the resulting signal as a function of range and cross range.

Because both range and cross-range resolutions are obtained synthetically, the signal processing required to produce an image from the measured data can pose a significant computational challenge; therefore, efficient algorithms such as the FFT are used wherever possible. The following sections develop specific methods for range and cross-range processing of experimental data.

4.1 RANGE PROCESSING

As developed in Chapter 2, synthetic range processing involves a Fourier transform operation applied to the measured frequency response of an object. If the object is held stationary while the frequency response is measured, the Fourier transform is the correct operation and involves no approximations. In a practical measurement, the frequency response is sampled for N discrete frequency steps Δf, spanning a bandwidth B such that

$$B = (N - 1)\,\Delta f \tag{4.1}$$

N complex samples of the frequency response can be processed by an N-point FFT to yield N samples of the range profile. The N outputs of the FFT represent $N - 1$ range increments, establishing the granularity of the range profile which equals $c/2B$. The range corresponding to the first index of the FFT output is zero; that for the Nth (last) index is $c/2\Delta f$.

To reduce errors resulting from spectral leakage and scalloping losses, range profiles are processed with the application of tapered windows [1]. Figures 4.2 and 4.3 show theoretical point-object responses for rectangular and Hann windows. The plots indicate the envelope of resulting range samples, which can fall anywhere in the envelope at uniform spaces of $c/2B$.

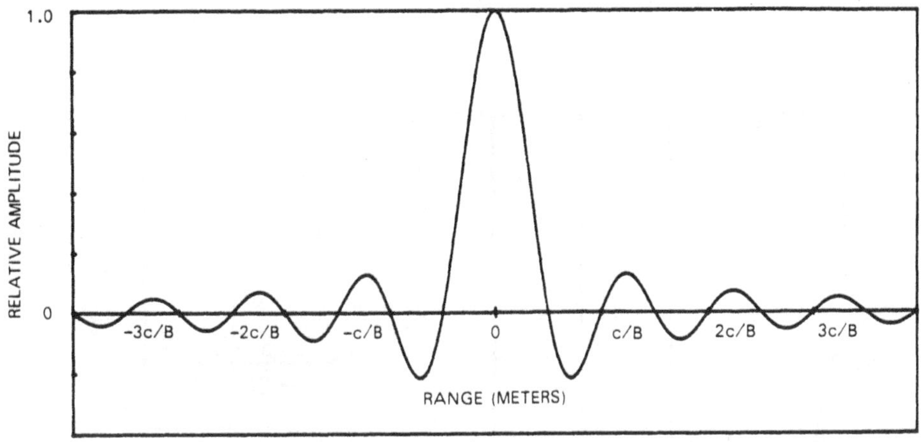

Figure 4.2 Range response of point reflector (rectangular window).

Figure 4.4 shows a range profile for a missile body viewed along its longitudinal axis, obtained using a bandwidth of 1.5 GHz and a Hann window. The granularity is 5 cm, the resolution is 20 cm, and the plot displays 128 range cells in histogram form to emphasize the granularity. The range profile constitutes a one-dimensional image of the object; the magnitude of each cell of the range profile results from the coherent summation of all signals from object elements contained in the particular range cell. Several scattering centers can be associated with the physical structure of the object. Figure 4.5 shows an ensemble of range profiles of

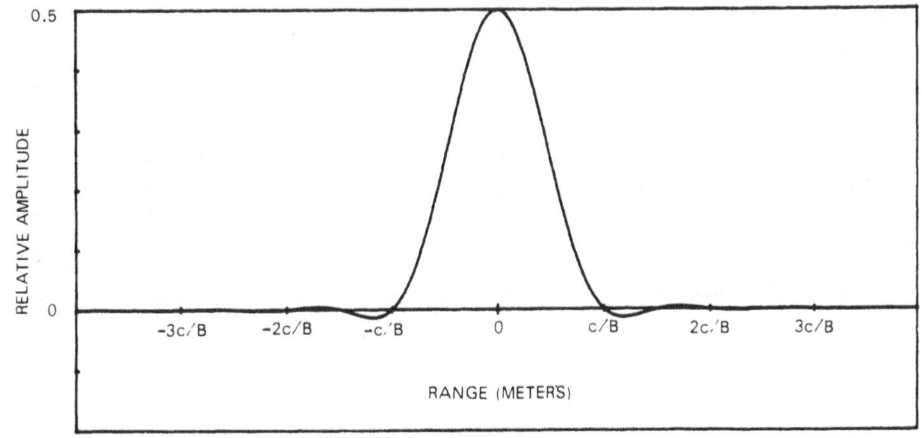

Figure 4.3 Range response of point reflector (Hann window).

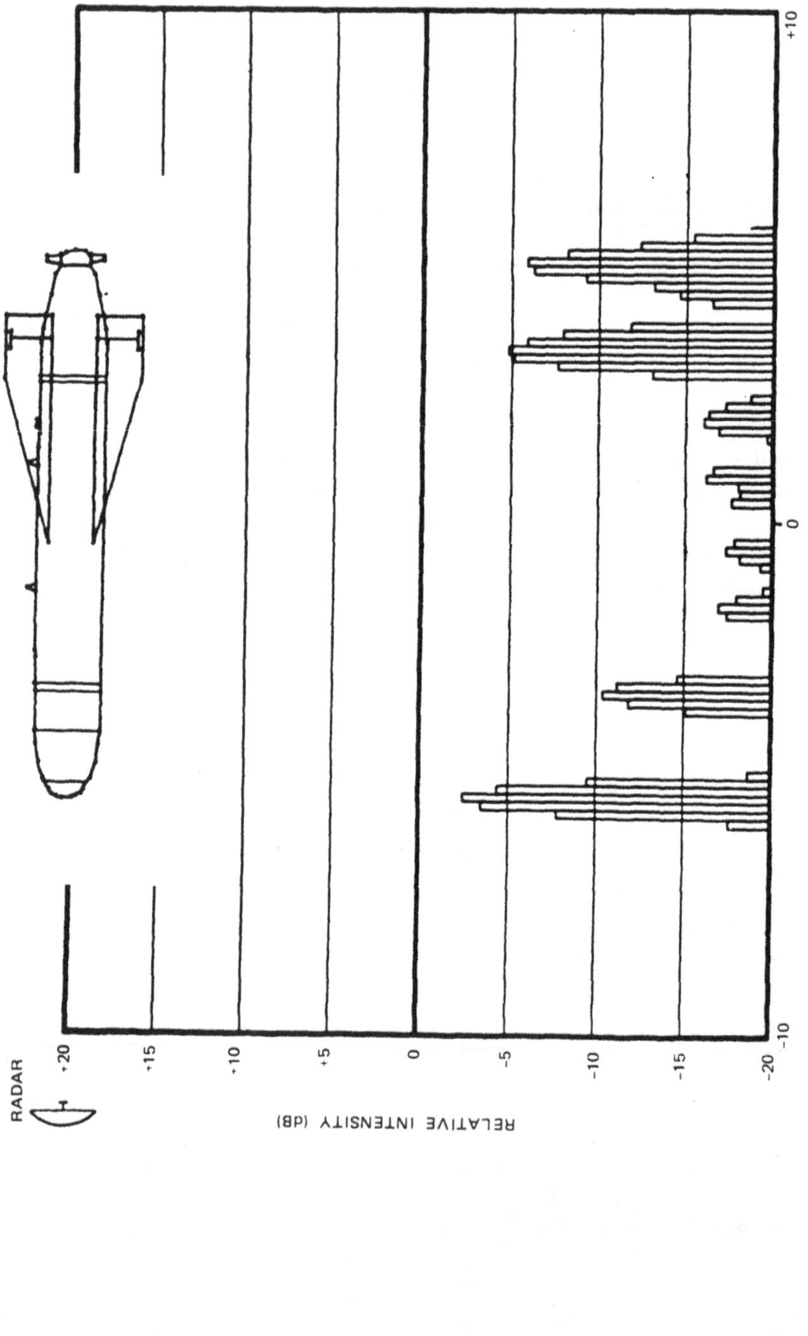

Figure 4.4 Range profile of missile body.

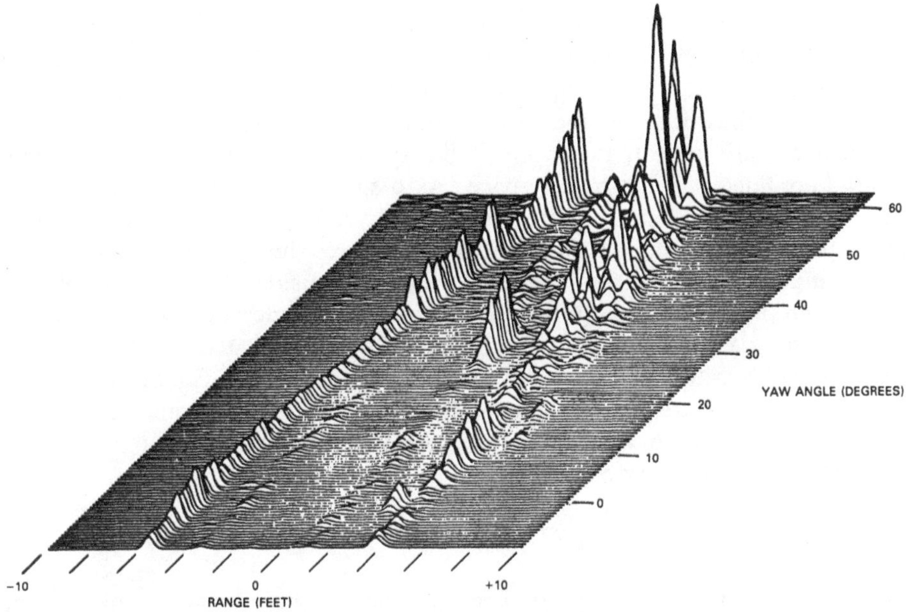

Figure 4.5 Ensemble of range profiles for missile body.

the same object obtained for a number of discrete viewing angles. The zero-degree yaw angle profile corresponds to the plot of Figure 4.4. The ensemble plot, drawn by connecting discrete range samples using straight-line segments, reveals the angular dependence of individual scatterers. The scattering from the front part of the object is shown on the left of the plot to persist over a large range of aspect angles; this contrasts with other features of the object which are highly aspect-dependent.

4.2 CROSS-RANGE PROCESSING

A high degree of resolution in cross range can be achieved by forming a large aperture synthesized by scanning a sensor over a surface, storing the received signals, and subsequently coherently processing these signals. Resolution in range is assumed *a priori* to have been accomplished by time-delay sorting methods; the signals of interest are, therefore, assumed to be backscattered from sections or slices of the object space contained within a specific range-resolution cell.

When large apertures are required, a significant practical advantage results from forming the synthetic aperture by holding the sensor fixed and rotating the object through a range of angles that would be subtended by the aperture. This is the ISAR configuration described in Chapter 2. The synthetic aperture is formed by recording the output of a single, fixed sensor for a number of object angular positions, and subsequently coherently summing the resulting signals. The process of rotating the object generates an equivalent circular-arc aperture centered on the object center of rotation, which thus becomes the focal point of the aperture.

To image the object, the beam of the synthetic aperture must be steered to arbitrary coordinates in the object space. As diagrammed in Figure 2.18, a phase correction must be applied to each sampled field value prior to the summation. The required phase corrections are those necessary to force the summed signals to an in-phase condition. To derive the required correction, consider the geometry of Figure 4.6, in which the complex amplitude of the field backscattered from an object point is sampled along a circular aperture with radius R_0, where $R_0 \gg x$. Because the entire aperture is equidistant from the object center, the response to a central point is maximum, and the aperture is said to be focused on the object center. The distance between an arbitrary point on the aperture, identified by the angle θ, and an arbitrary point in the object space (x, y), is

$$r(\theta,x,y) = R_0[1 + (2/R_0)(y\cos\theta - x\sin\theta) + (x^2 + y^2)/R_0^2]^{1/2} \qquad (4.2)$$

Field values reflected by a point at (x, y), sampled at the aperture, exhibit a phase angle corresponding to the two-way path length. The normalized amplitude of the received field is

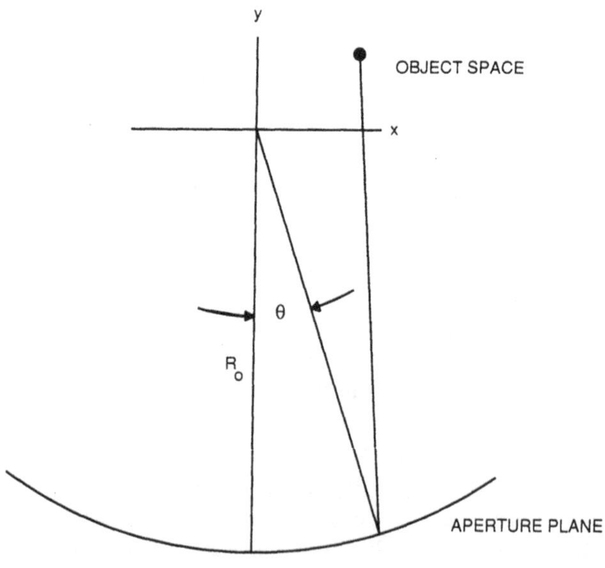

Figure 4.6 Circular aperture focused at object center.

$$V(\theta,x,y) = \exp\{-j4\pi(R_0/\lambda)[1 + 2(y\cos\theta - x\sin\theta)/R_0 + (x^2 + y^2)/R_0^2]^{1/2}\} \tag{4.3}$$

Equation 4.3 expresses the phase variations observed across the aperture. When signals sampled across the aperture are coherently summed to synthesize the response, the resultant magnitude is reduced from the resultant of a similar sum for an object point at $x = y = 0$. The coherent summation of signals expressed by (4.3) for a point-object at (x, y) represents the response of the aperture focused to $x = y = 0$ to a point-object located at (x, y). If a phase term equal to the argument of the exponential of (4.3) is subtracted from each signal sample prior to summation, the phase variations will be corrected, and the summed signals will be in-phase. The result of the summation will then be a maximum; this operation provides diffraction-limited focusing of the aperture. The response of the focused synthetic aperture steered to a point object at (x, y), is thus:

$$G(x,y) = \sum_{\theta} V(\theta,x,y)\exp\{j4\pi(R_0/\lambda)[1 + 2(y\cos\theta - x\sin\theta)/R_0 + (x^2 + y^2)/R_0^2]^{1/2}\} \tag{4.4}$$

The focusing operation constitutes a significant computational burden; (4.4) indicates that a distinct phase correction must be applied for each combination of θ and (x, y). If the argument of the exponential (4.4) is expanded in a series with high-order terms neglected, and the following approximations are made:

$$R_0 \gg x,y, \quad \sin\theta \cong \theta, \quad \cos\theta \cong 1$$

then (4.4) reduces to

$$G(x,y) \cong \exp[j4\pi(R_0 + y)/\lambda]\sum_{\theta} V(\theta,x,y)\exp(-j4\pi x\theta/\lambda) \tag{4.5}$$

The phase term outside the summation has no effect on the resultant magnitude, which can be expressed as

$$|G(x,y)| \approx \left|\sum_{\theta} V(\theta,x,y)\exp(-j2\pi\theta\, 2x/\lambda)\right| \tag{4.6}$$

Equation (4.6) represents a DFT sum, indicating that an approximate image can be obtained by Fourier transforming the field values sampled along the aperture. The assumptions required for the DFT approximation are large observation distances relative to the object dimensions, and small angles subtended by the aperture. The DFT approximation allows a significant reduction in computations required to determine the response of the aperture steered over the entire range of x. The steering process is inherent to the DFT, which applies to the field values sampled along the aperture a phase term that is linear in θ and proportional to the coordinate to which the aperture is steered.

The aperture beam can be steered to any arbitrary value of x, each requiring a distinct summation. If an FFT is utilized for computation speed, the number of samples in the output transform equals the number of data samples. Thus, if N data samples are transformed, the aperture is steered to N different values of x. The propriety of the transform operation, of course, requires sampling consistent with the Nyquist criterion. The use of (4.4) to process sampled field values steers the synthetic aperture beam to any point on the object plane and produces diffraction-limited focusing; the use of (4.6) (the DFT approximation) steers the aperture beam, which remains focused to the diffraction limit at the object center, but becomes defocused when directed to other points in the object plane.

The nature of the focusing error is illustrated in Figure 4.7, which shows equivalent geometries of the synthetic apertures formed by the various processes. The unsteered aperture, inherently focused to the object center, is represented by sampled locations along the circular arc centered on the object center. The steered focused aperture, formed by applying a phase correction to the sampled field values to bring them to an in-phase condition, is a circular arc centered on the coordinates (x_0, y_0) to which the aperture is focused. The differential path length representing the required phase correction is denoted by Δ. The DFT-steered aperture is formed by applying to the sampled field values discrete phase corrections proportional to x and θ. This phase correction precludes tangency between the DFT-steered and steered focused apertures, and results in phase errors corresponding to the differential path length denoted by ε. Processing the data with an N-point DFT is equivalent to generating N steered beams, each formed by a phase correction proportional to x. Each of the N beams is only approximately focused in the vicinity of the x coordinate due to the phase errors inherent to the DFT approximation.

To determine the form of the steered aperture response, consider the sampled aperture geometry of Figure 4.8. The phase of the signal reflected from a point object located at x_0, y_0, and received at an aperture coordinate θ is:

$$\phi(x_0,y_0;\theta_n) = \frac{-4\pi}{\lambda}\{[R_0^2 + 2R_0(y_0 \cos\theta_n - x_0 \sin\theta_n) + x_0^2 + y_0^2]^{1/2}\} \quad (4.7)$$

The DFT of the sequence of signals observed across the aperture is represented by*

*For this development, the amplitude variations of the observed signal are ignored, and a point reflector is assumed to provide unit-magnitude received signals at all aperture element positions. This approximation is consistent with the far-field assumption.

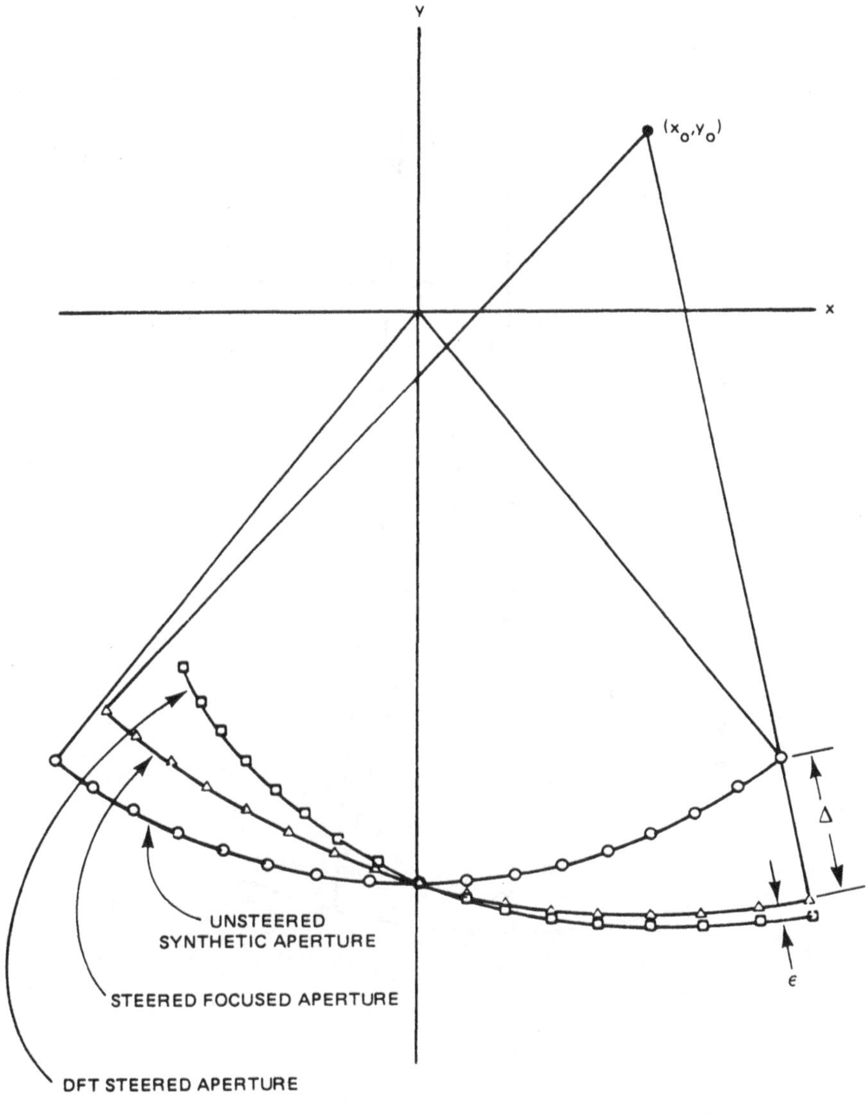

Figure 4.7 Geometry of synthetic apertures.

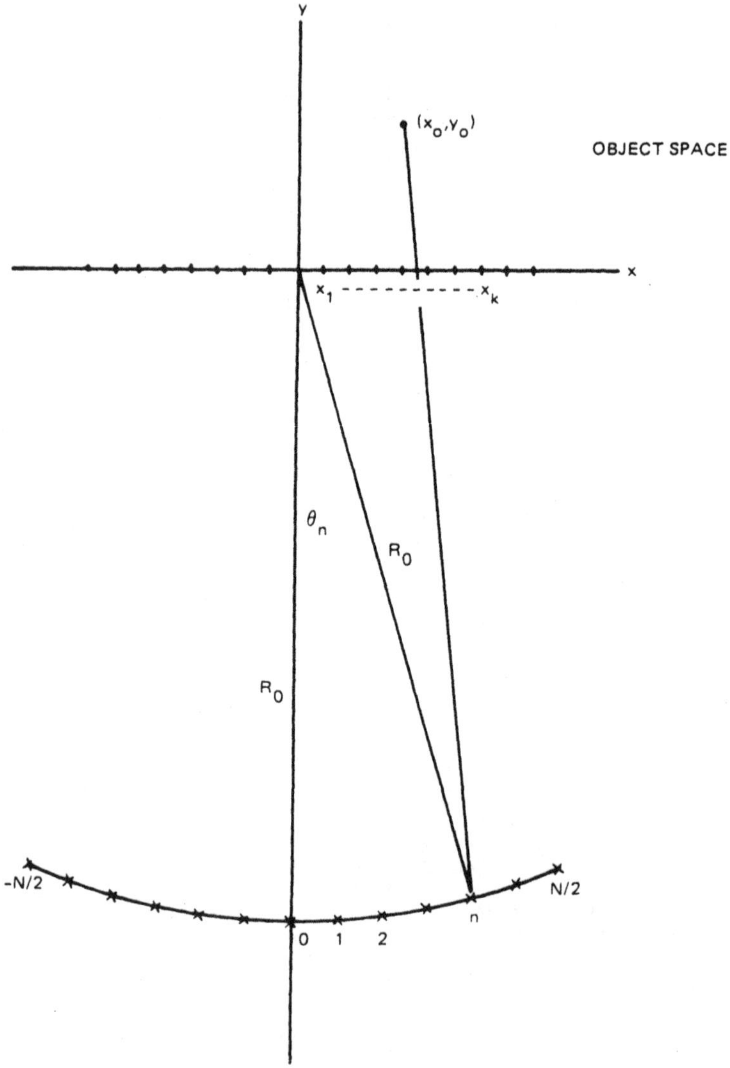

Figure 4.8 Sampled aperture focused at object center.

$$G(x_0,y_0;\, x_k) = \sum_n \exp[\,j\phi\,(x_0,y_0;\theta_n)]\,\exp(-j2\pi\, 2x_k\theta_n/\lambda) \tag{4.8}$$

where $\phi(x_0,y_0;\theta)$ is defined in (4.7).

Equation (4.8) is the response of the aperture steered to x_k to a point object located at x_0,y_0. Calculating the response of the steered beams can be simplified by allowing the observation distance, R_0 to approach infinity. Rewriting the bracketed terms of (4.7) in a binomial expansion and assuming $R_0 \gg x_0 + y_0$ provides an expression for the phase variation across the aperture in the limit of infinite observation distance. Factoring terms of (4.8) not involving θ outside the summation yields:

$$\begin{aligned} G_\infty\,(x_0,y_0;x_k) = {} & \exp\{-j4\pi[R_0/\lambda + (x_0^2 + y_0^2)/R_0^2]\} \\ & \cdot \sum_n \exp\{-j4\pi/\lambda[(y_0 \cos\,\theta_n - x_0\,\sin\theta_n) + x_k\theta_n]\} \end{aligned} \tag{4.9}$$

The subscript of G indicates that the response applies to an infinite observation distance, indicating the far-field response.

The general form of the response for the central beam ($x_k = 0$) is shown in Figure 4.9, which plots the intensity $|G^2|$ of (4.9) for $x_k = 0$ and a total aperture angle of 30°. The contours are lines of constant intensity for uniform increments of 0.1 times the peak value. The peak response of the unsteered beam occurs for $x_0 = y_0 = 0$ because signals reflected from this point and received across the aperture are in phase. As the location of the object point deviates from center, phase errors developed across the aperture cause a reduction in the magnitude of the response. The precise shape of the response for each beam is determined by plotting (4.9) for each beam.

We can develop a tractable expression for estimating the size of the main response of the beams by establishing a phase-error criterion. The response of an aperture without phase error has a maximum amplitude and minimum beamwidth; when the aperture is subject to phase errors, the magnitude is reduced and the beamwidth broadened. The limit of allowable phase errors is somewhat arbitrary, but a generally accepted criterion [2] allows maximum phase errors of $\pi/2$ radians at the edges of the aperture. If the error is quadratic across the aperture, the peak magnitude of the response is decreased by less than 15 percent, and the spatial response widened by less than 10 percent. By adopting the $\pi/2$ radians maximum phase error, we can determine the boundaries of point-object locations in the $x - y$ space, which result in aperture phase errors equal to the prescribed limit. Points inside these boundaries will be imaged with minor degradation; points outside these boundaries will be imaged with a magnitude reduction determined by the deviation from the boundary. The phase of signals reflected from a point object located at x_0, y_0 and received at the end and center of the steered aperture are given respectively by the argument of the inner exponent of (4.9) for $\theta_n = \Theta/2$ and $\theta_n = 0$. The phase difference is then

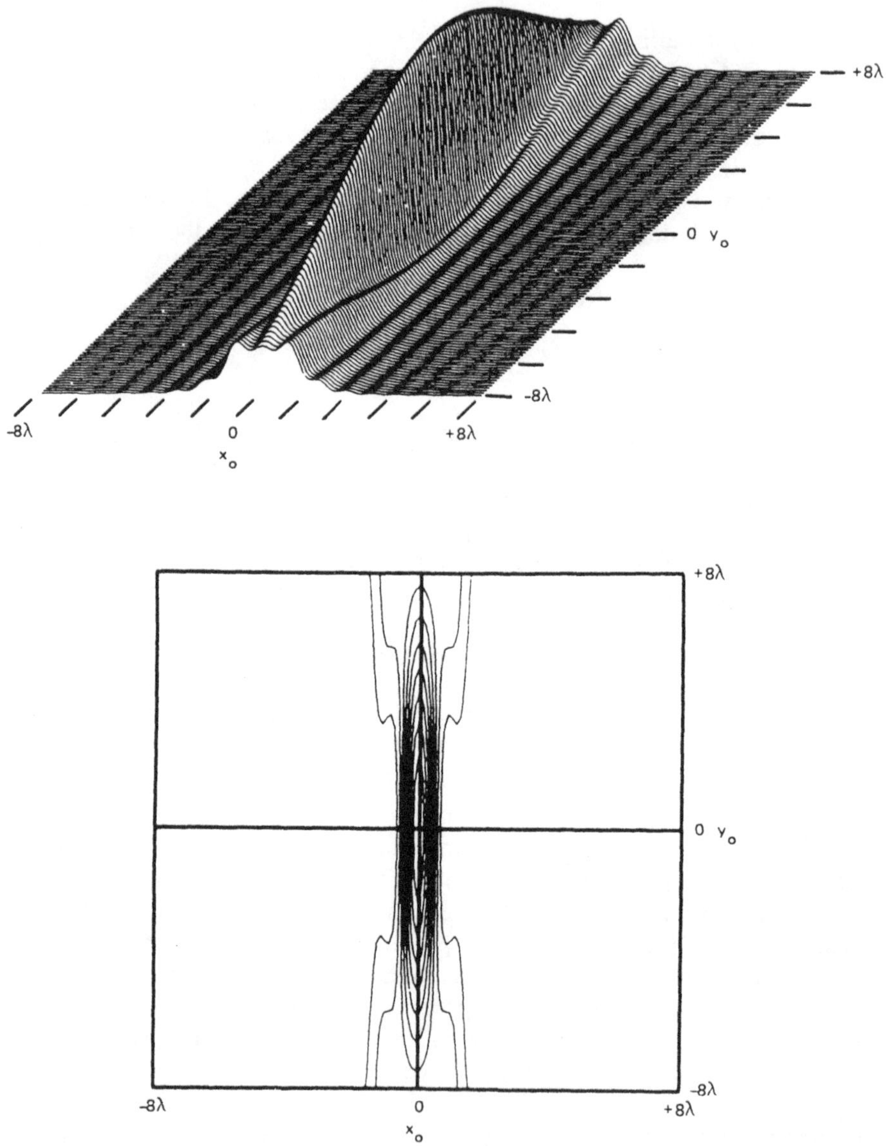

Figure 4.9 Intensity of far-field response for unsteered beam of 30° aperture.

$$\frac{4\pi}{\lambda}\{-y_0[1 - \cos(\Theta/2)] - x_0 \sin(\Theta/2) + x_k(\Theta/2)\} = \frac{\pi}{2} \qquad (4.10)$$

where Θ is the total angle subtended by the aperture for which a maximum phase error of $\pi/2$ radians can occur at either edge of the aperture.

Figure 4.10 shows the focal zone pattern defined by (4.10) when the N data samples are processed by an N-point DFT or FFT to produce N steered beams. The focal zones are contiguous without overlap; they can be made to overlap by computing additional beams, accomplished by padding the sampled data with zeros and computing an M-point DFT where $M > N$. The following observations can be made from Figure 4.10: (1) the depth (y extent) of the focal zones remains unchanged with x_k, and (2) the peak response for a beam steered to x_k occurs for a point object location x_0 such that $|x_0| > |x_k|$. The maximum response of a particular beam occurs for a location slightly offset from the location to which the beam is steered; the offset, which increases with x_k, results from performing the DFT using equal increments in θ rather than in $u = \sin\theta$.

The lateral extent of the object space in which points can be imaged can be defined somewhat arbitrarily by the condition of having the x_k coordinate to which the beam is steered fall on the boundary of the focal zone of the kth beam. This condition is shown in Figure 4.11 in which the overall extent of the object space is indicated by the large rhombus. Responses for beams of a 30° aperture steered to values of x_k (denoted A through E in Figure 4.11) are shown in Figure 4.12 through 4.16, and were obtained by using (4.9). The offset in the response of the steered

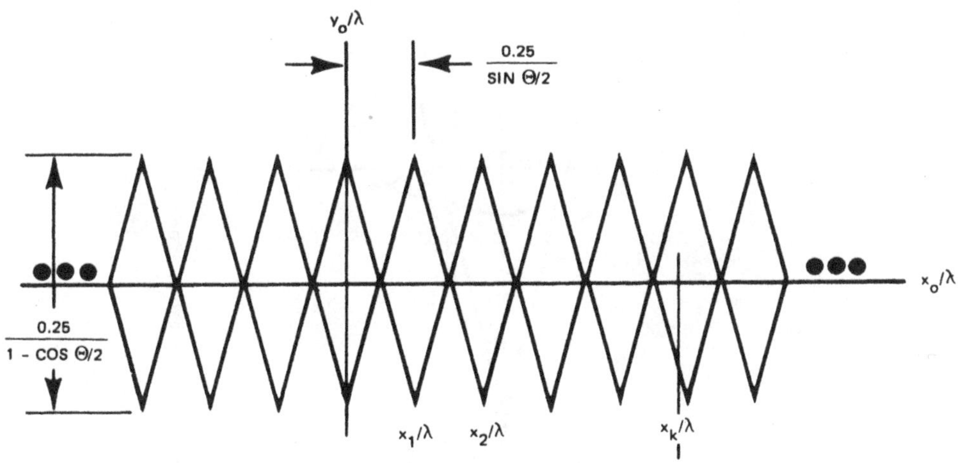

Figure 4.10 Far-field focal zones of DFT-steered beams.

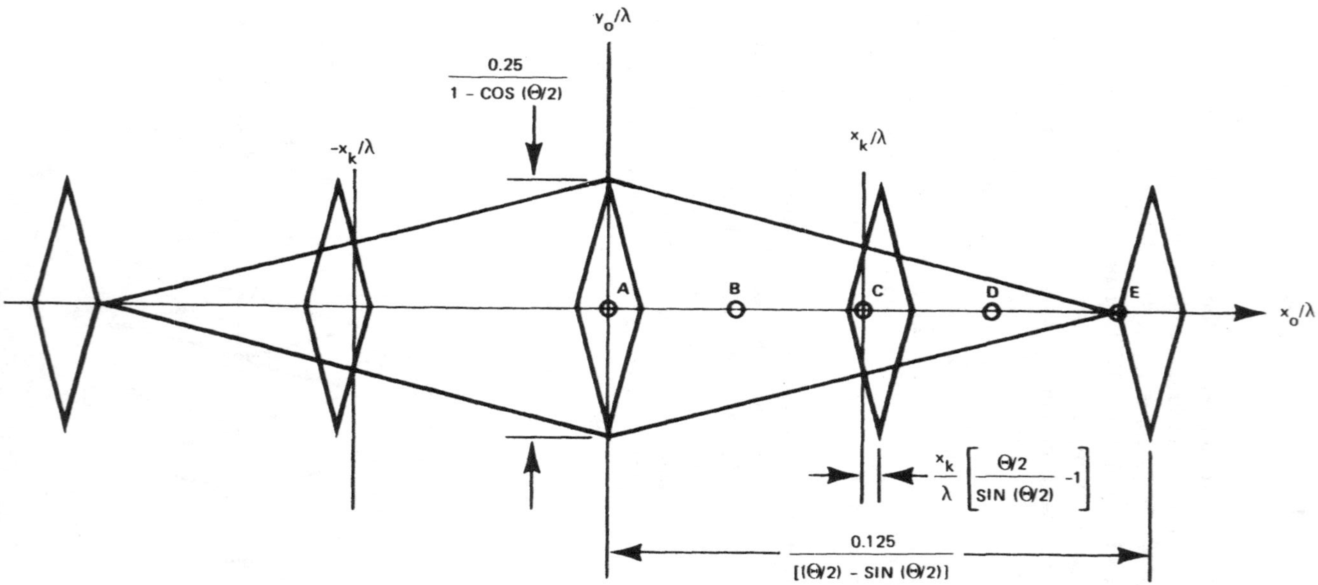

Figure 4.11 Spatial extent of object field for DFT-steered aperture.

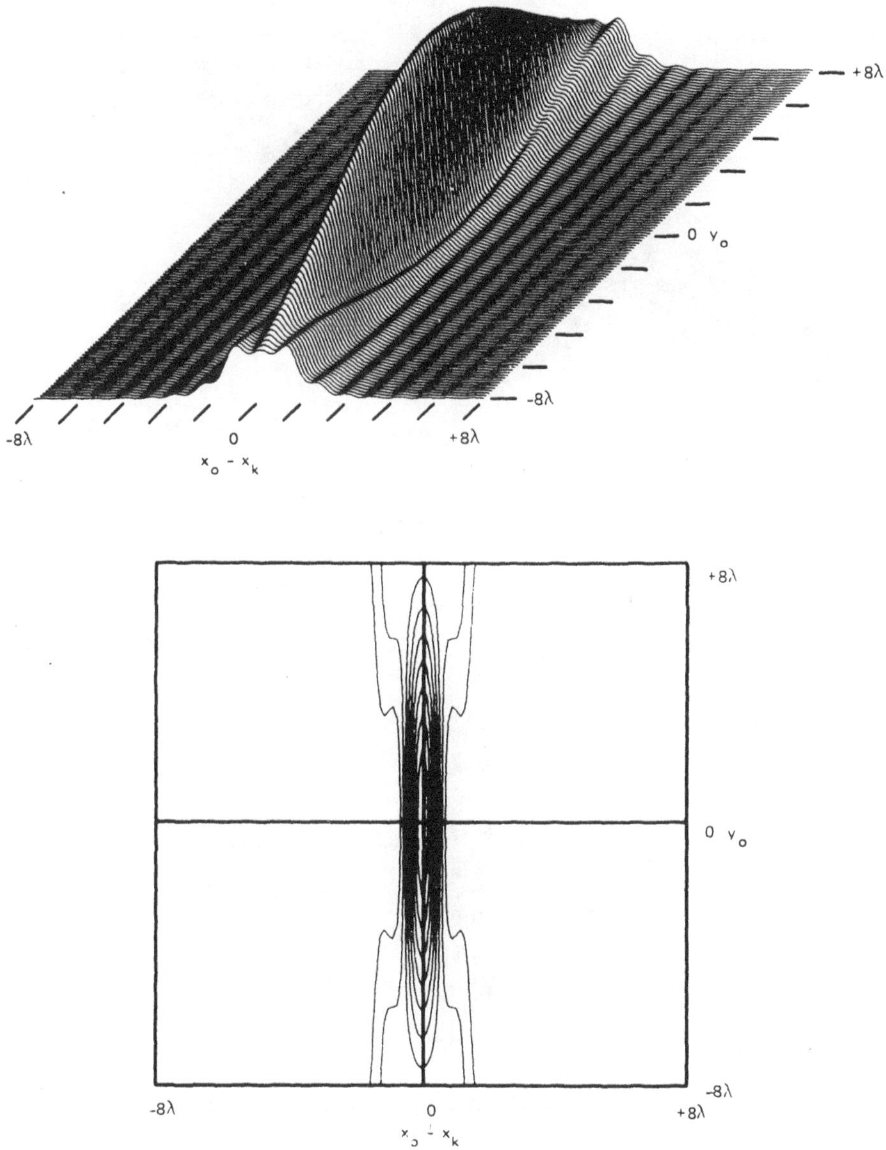

Figure 4.12 Response of 30° aperture beam steered to point A.

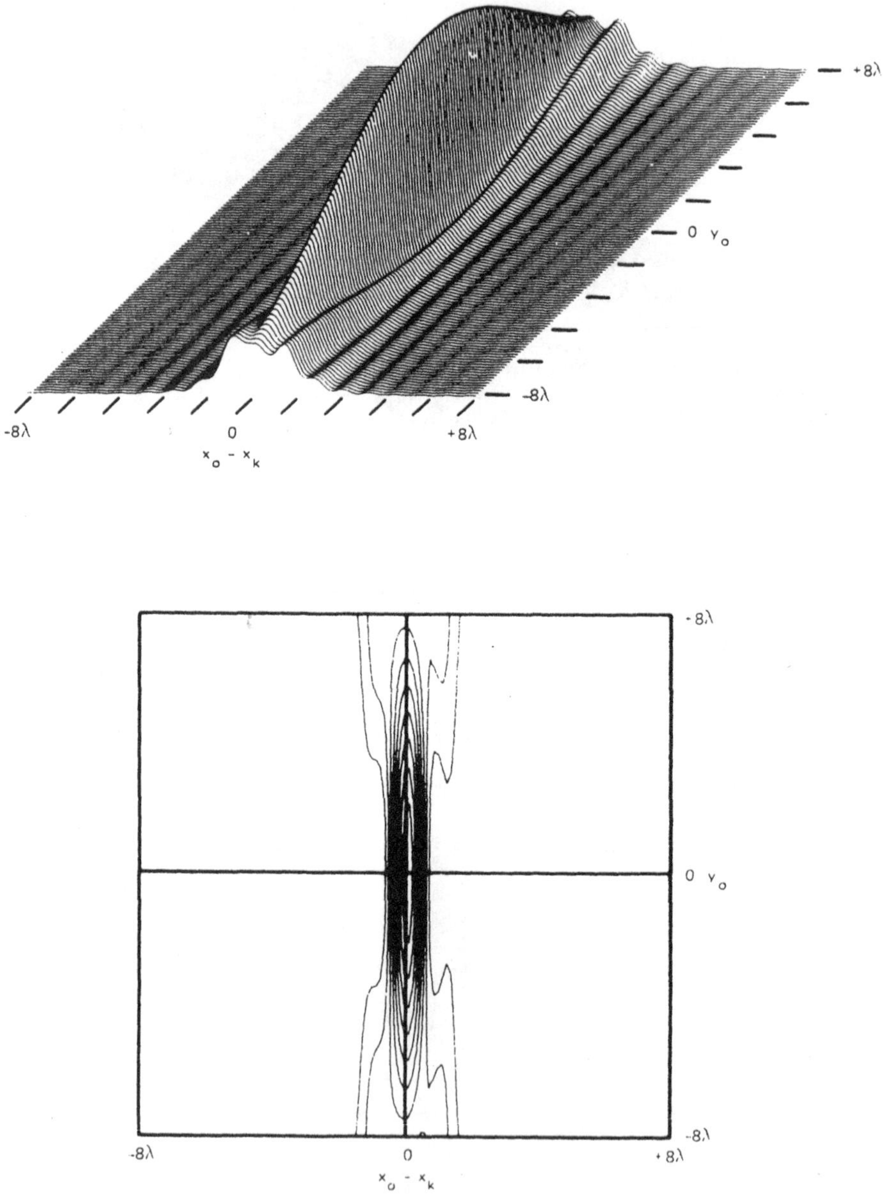

Figure 4.13 Response of 30° aperture beam steered to point B.

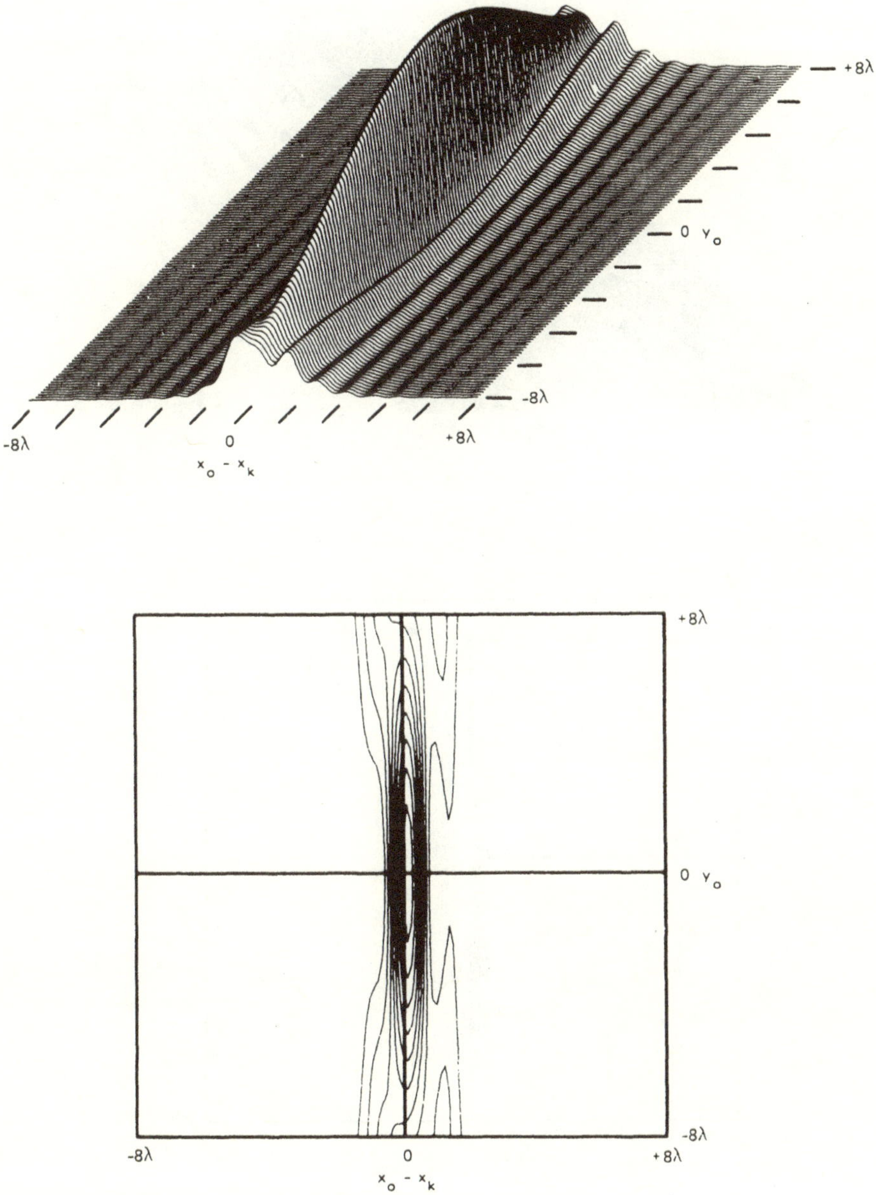

Figure 4.14 Response of 30° aperture beam steered to point C.

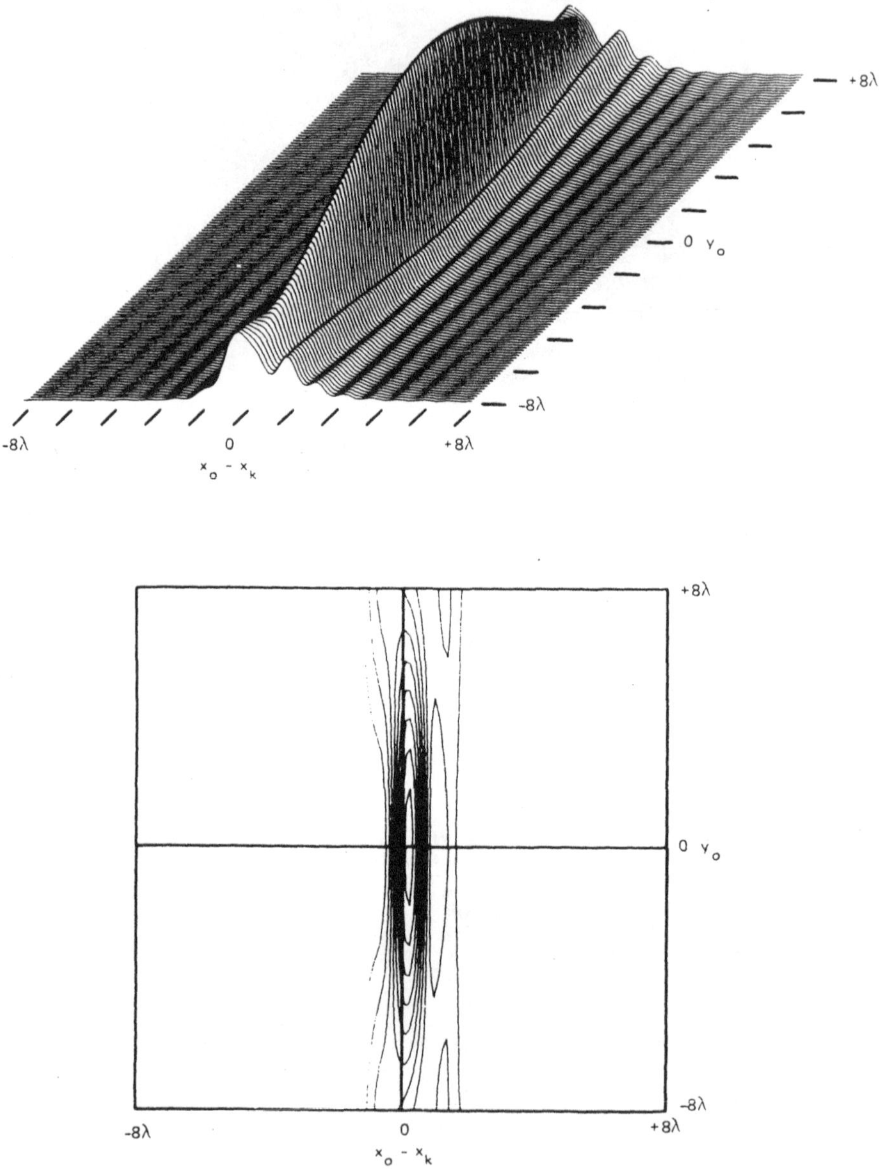

Figure 4.15 Response of 30° aperture beam steered to point D.

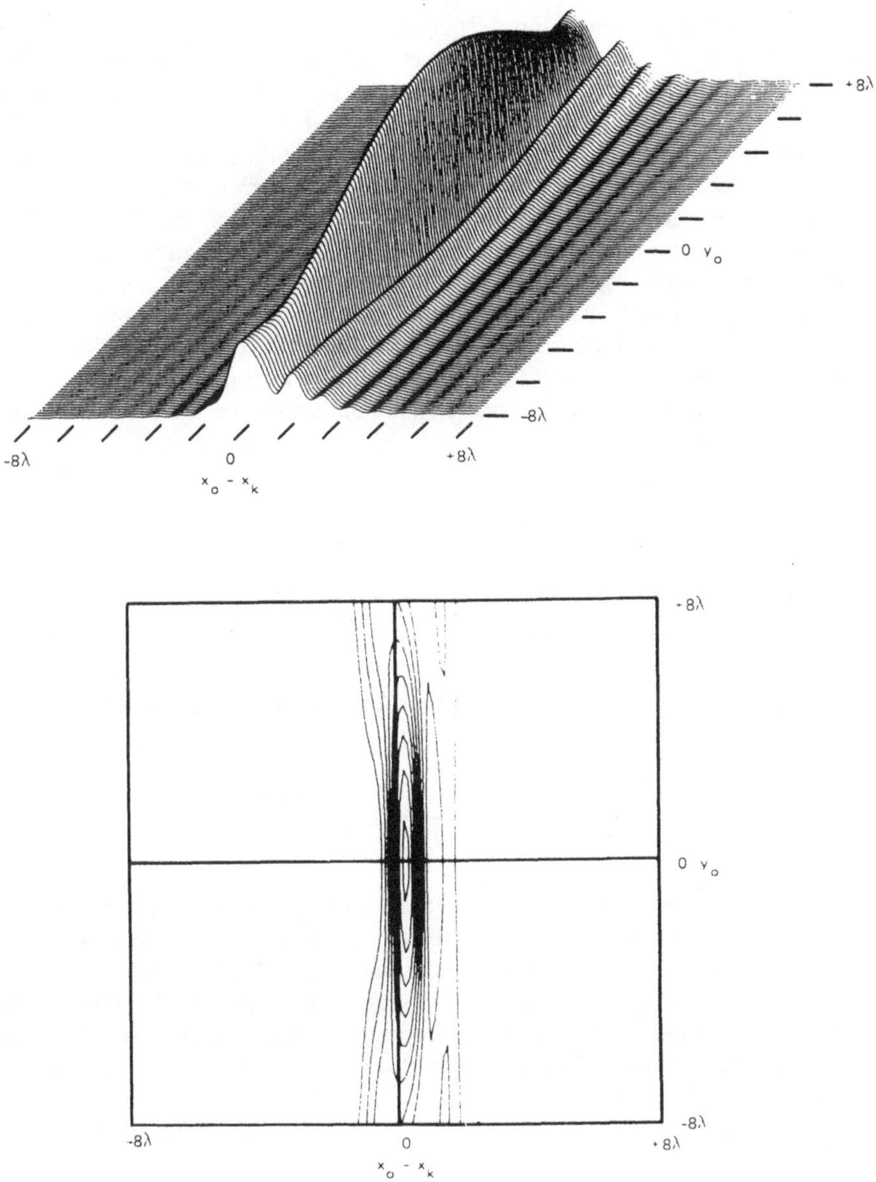

Figure 4.16 Response of 30° aperture beam steered to point E.

beams can be noted in these figures. Figures 4.17 and 4.18 show responses for the aperture steered to an x coordinate 2 times and 4 times that for point E, respectively. These figures illustrate the severe degradations for beams steered to large lateral offsets.

In the preceding analysis, the allowable degradation has been established by the arbitrary selection of $\pi/2$ radians maximum phase error. This is relatively conservative because it limits the degradation to a 15% reduction in intensity. The general dependence of the focal zones on the aperture angle is valid for aperture angles less than 1 radian. For larger apertures, the phase errors become excessive, and the $\pi/2$ radian error criterion is no longer valid. For the majority of applications, this restriction is satisfied, and the dimensions of the object field can be quickly estimated for a given aperture angle, using the relations shown in Figure 4.11. In the limit of small aperture angles, the lateral and longitudinal extent of the focal zone for the central beam become, respectively:

$$\Delta x = \frac{0.25\lambda}{\sin(\Theta/2)} \approx \frac{\lambda}{2\Theta} \tag{4.11}$$

$$\Delta y = \frac{0.25\lambda}{1 - \cos(\Theta/2)} \approx \frac{0.25\lambda}{1 - [1 - (\Theta^2/4)]} = \frac{\lambda}{\Theta^2} \tag{4.12}$$

By noting the following inequality between the lateral and longitudinal field extents cited in Figure 4.11,

$$\frac{0.125}{(\Theta/2) - \sin(\Theta/2)} \geq \frac{0.125}{1 - \cos(\Theta/2)} \tag{4.13}$$

we establish that the maximum radial distance of a point-object which can be imaged with tolerable degradation is limited by the depth of field established by the longitudinal extent of the focal zone. Because the lateral (cross-range) resolution and the depth of field both depend on the aperture angle, a fundamental relation between lateral resolution and maximum object dimension can be established. The ratio of field depth to the width of the lateral resolution cell is given by the aspect ratio of the focal zones in Figure 4.11.

The number of lateral resolution cells contained in the field-depth extent is

$$N = \frac{\sin(\Theta/2)}{1 - \cos(\Theta/2)} \tag{4.14}$$

The total extent represented by N cells of dimension Δx is

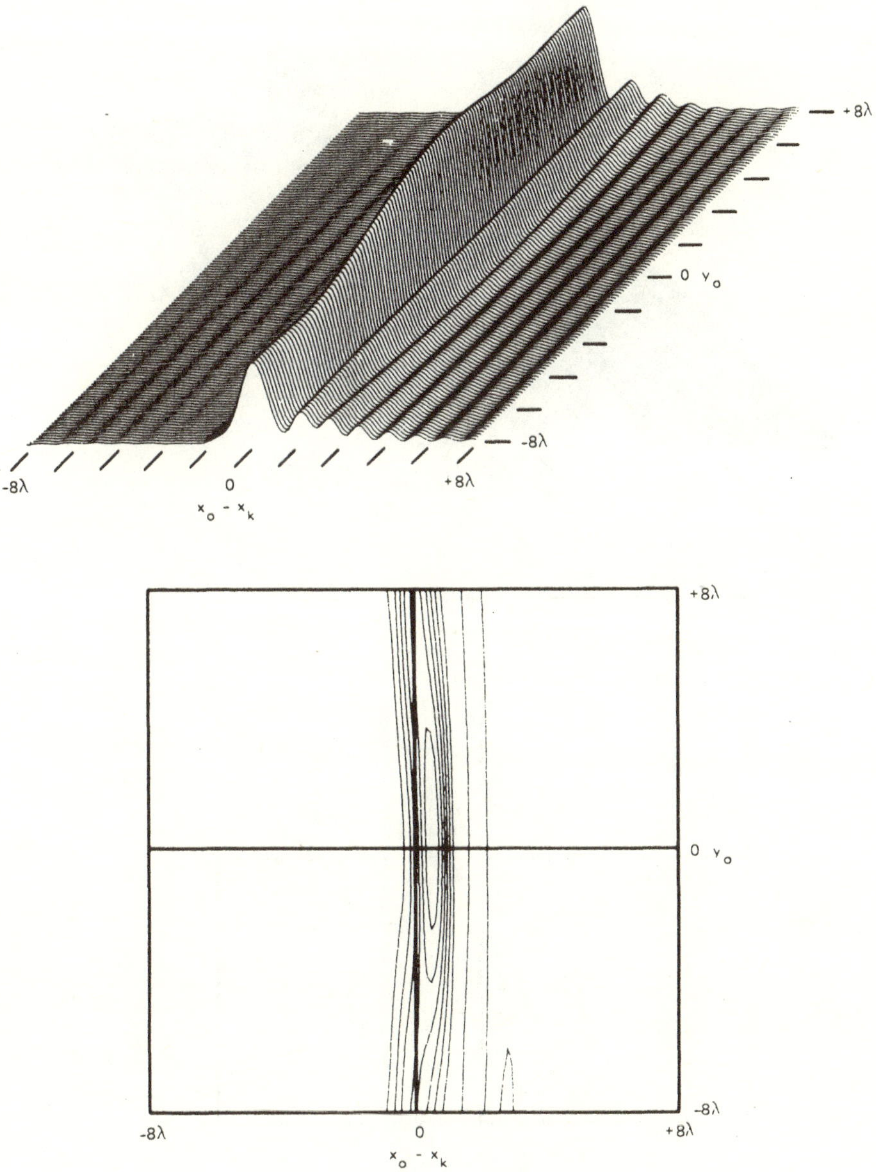

Figure 4.17 Response of 30° aperture beam steered to point 2E.

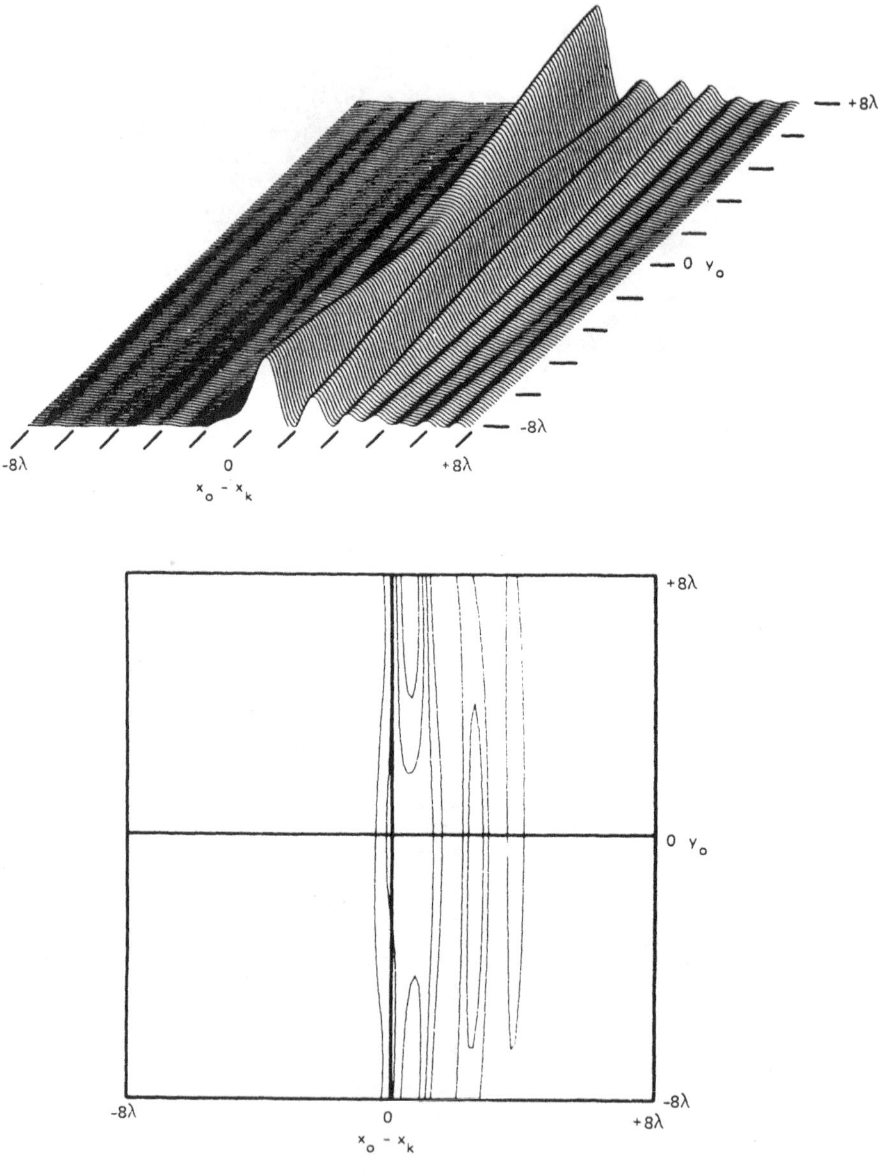

Figure 4.18 Response of 30° aperture beam steered to point 4E.

$$D = N\Delta x = \frac{0.25\lambda}{1 - \cos(\Theta/2)} \tag{4.15}$$

Combining (4.15), (4.14), and (4.11) to eliminate Θ yields:

$$\frac{D}{\lambda} = \frac{0.25}{1 - \cos\left[\sin^{-1}\left(\frac{0.25\lambda}{\Delta x}\right)\right]} \tag{4.16}$$

and

$$N = \frac{0.25}{\frac{\Delta x}{\lambda}\left\{1 - \cos\left[\sin^{-1}\left(\frac{0.25\lambda}{\Delta x}\right)\right]\right\}} \tag{4.17}$$

Equations (4.16) and (4.17), plotted in Figures 4.19 and 4.20, respectively, establish the limits of object size and the number of resolution cells. The above limitations are fundamental effects resulting from sensitivity of the Fourier transform to phase errors inherent to the circular motion of rotating object points.

The preceding analysis was based on the assumption of infinite observation distance. This implies that phase errors are due only to the DFT approximation to steering a circular aperture. A finite observation distance induces an additional source of phase error that can be analyzed by using the phase from the exponent of (4.8) and omitting the subsequent simplifying assumptions. The phase error for a finite observation distance, R_0, is again equated to $\pi/2$ radians to define the focal zones:

$$\frac{4\pi}{\lambda}\{[R_0^2 + 2R_0(y_0\cos(\Theta/2) \mp x_0\sin(\Theta/2)) + x_0^2 + y_0^2]^{1/2}$$
$$- [R_0^2 + 2R_o y_0 + x_0^2 + y_0^2]^{1/2} \pm x_k(\Theta/2)\} = \frac{\pi}{2} \tag{4.18}$$

A computer program was used to perform a numerical search to define the boundaries established by (4.18). The results can be summarized as follows. Each beam exhibits a focal zone very similar to the far-field case. However, the focal zones are aligned with the axes of the beams, which are now radially disposed about the center of the synthetic aperture (which is also the location of the physical aperture). This constitutes the principal difference from the far-field case, in which the beams are parallel to the y-axis. The consequence of this effect is a radial distortion of the object space from rectangular coordinates to a set of polar coordinates with

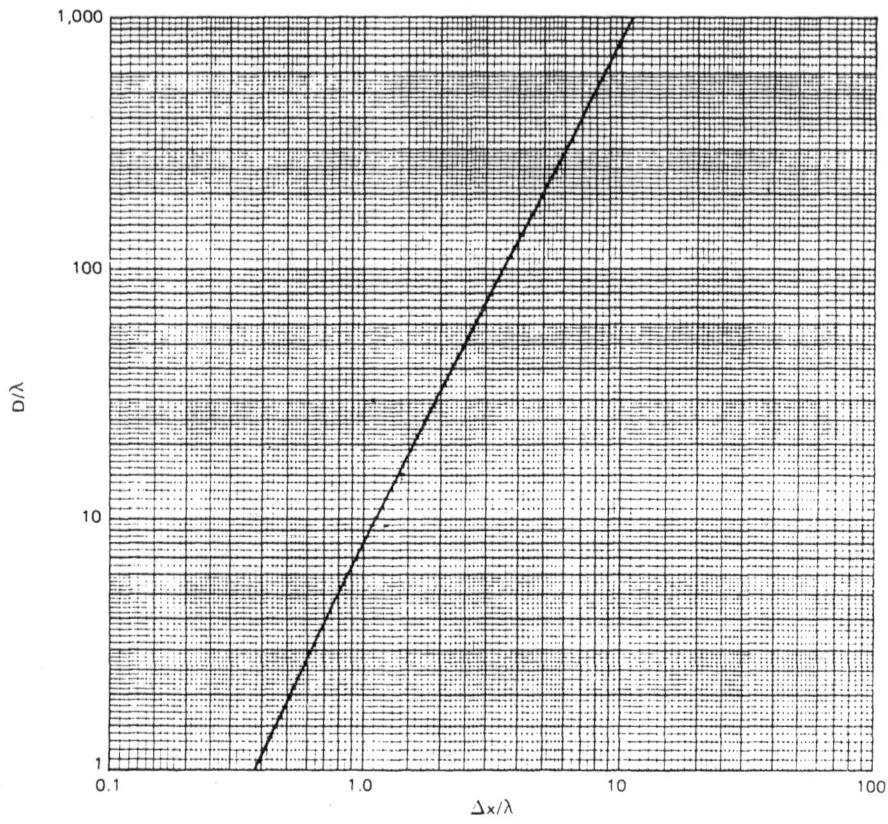

Figure 4.19 Maximum object field, D, *versus* lateral resolution, Δx.

origin at the point of observation. In most applications, this distortion is a minor effect which can be corrected after range and cross-range imaging is performed. Figures 4.21 through 4.25 show plots of near-field focal zones. The conditions are identical to those used for the far-field responses shown in Figures 4.12 through 4.16, but the observation distance R_0, is 25 m.

4.3 EXPERIMENTAL METHODS FOR TWO-DIMENSIONAL IMAGING

Having established conceptual aspects of the imaging process and instrumentation options, this section considers the practical task of collecting and processing experimental data suitable for producing two-dimensional images. The latter are ob-

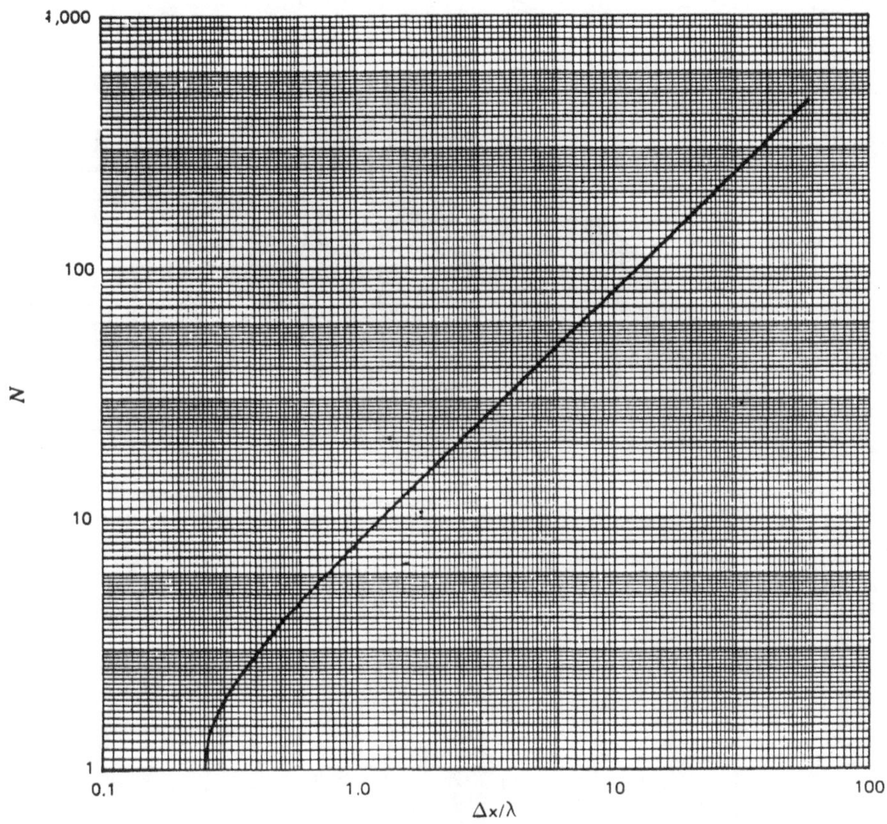

Figure 4.20 Maximum number of resolution cells, N, *versus* lateral resolution, Δx.

tained by combining methods discussed previously to resolve the object in range and cross range. The following sections describe: the data calibration method, the selection of hardware and software parameters, the experimental procedures, and examples of images obtained from experimental data.

4.3.1 Calibration Method

The data used for imaging are obtained with an RCS instrumentation radar having sufficient linearity and dynamic range to acquire calibrated RCS data. As in all RCS instrumentation systems, the absolute RCS of the test object is obtained by a ratiometric measurement of the object response relative to the response of a calibration

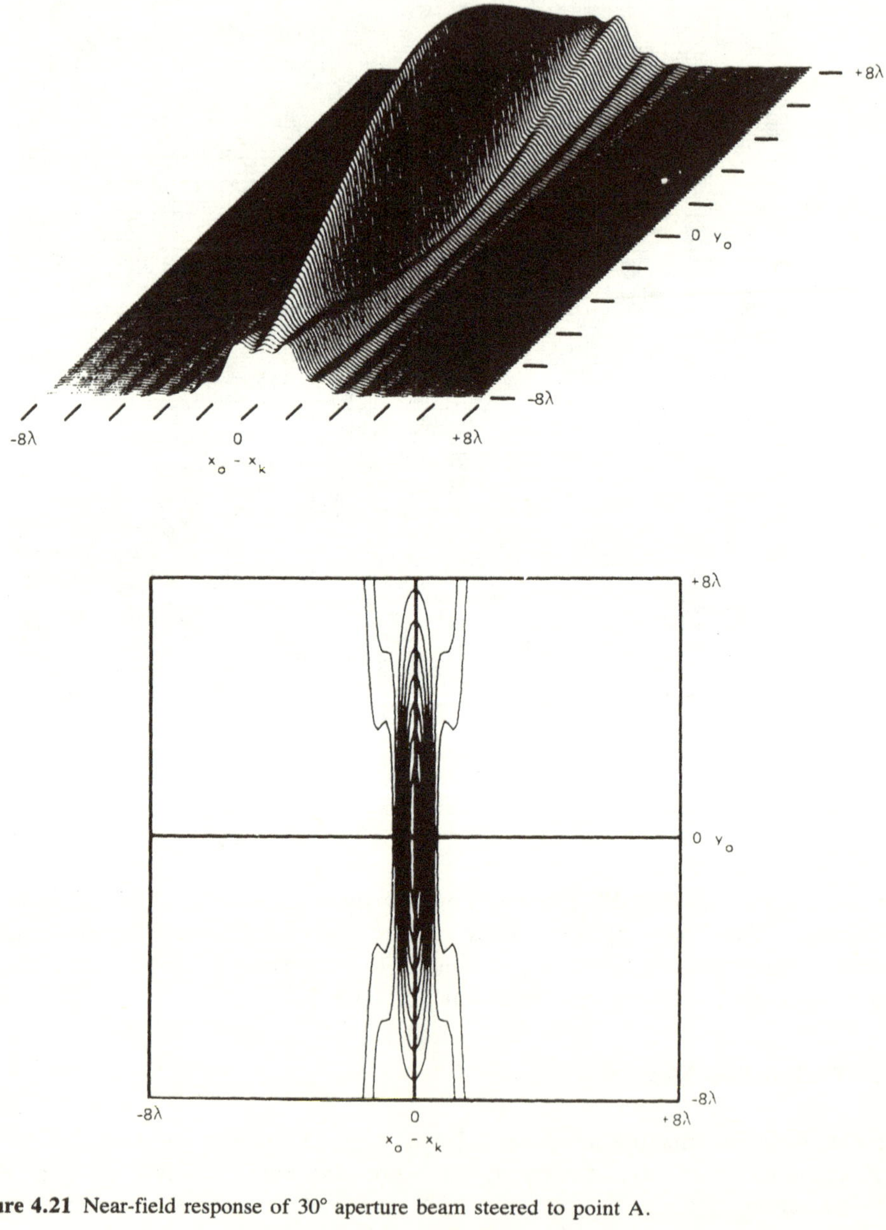

Figure 4.21 Near-field response of 30° aperture beam steered to point A.

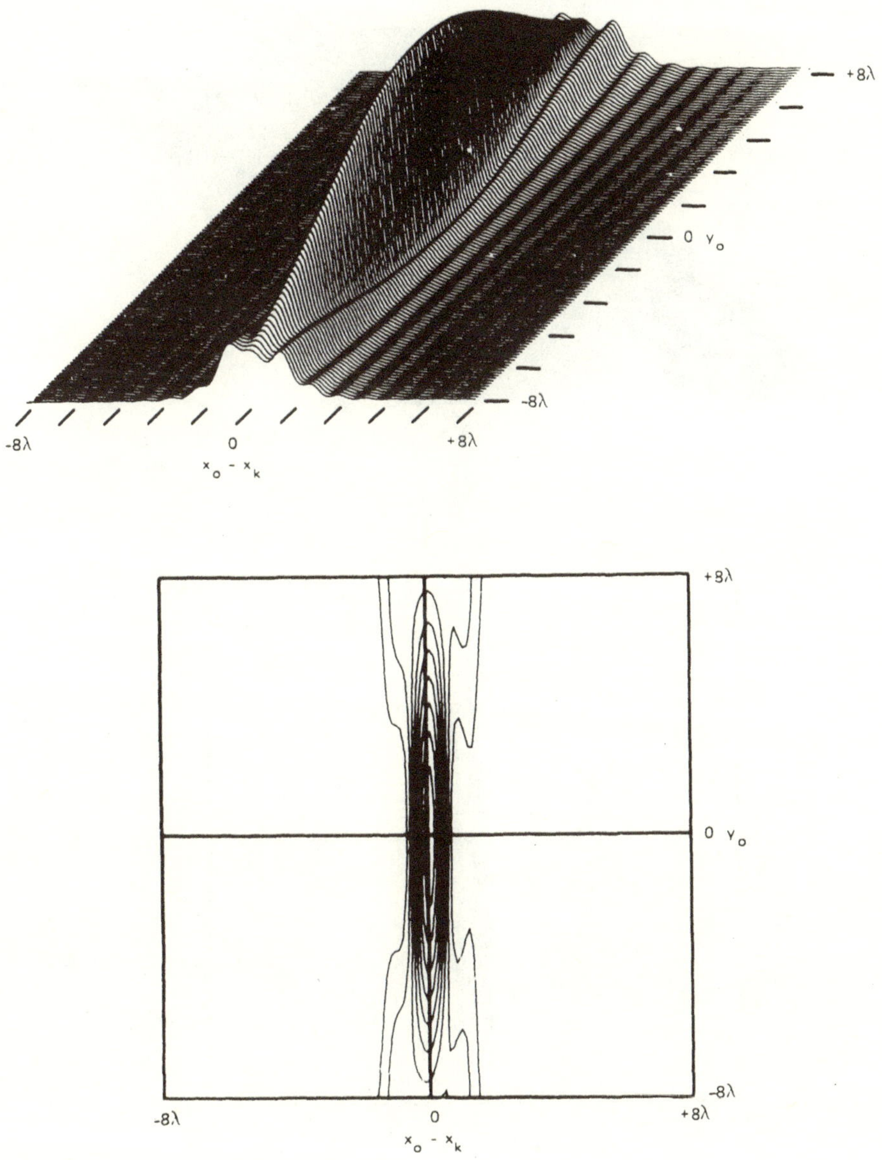

Figure 4.22 Near-field response of 30° aperture beam steered to point B.

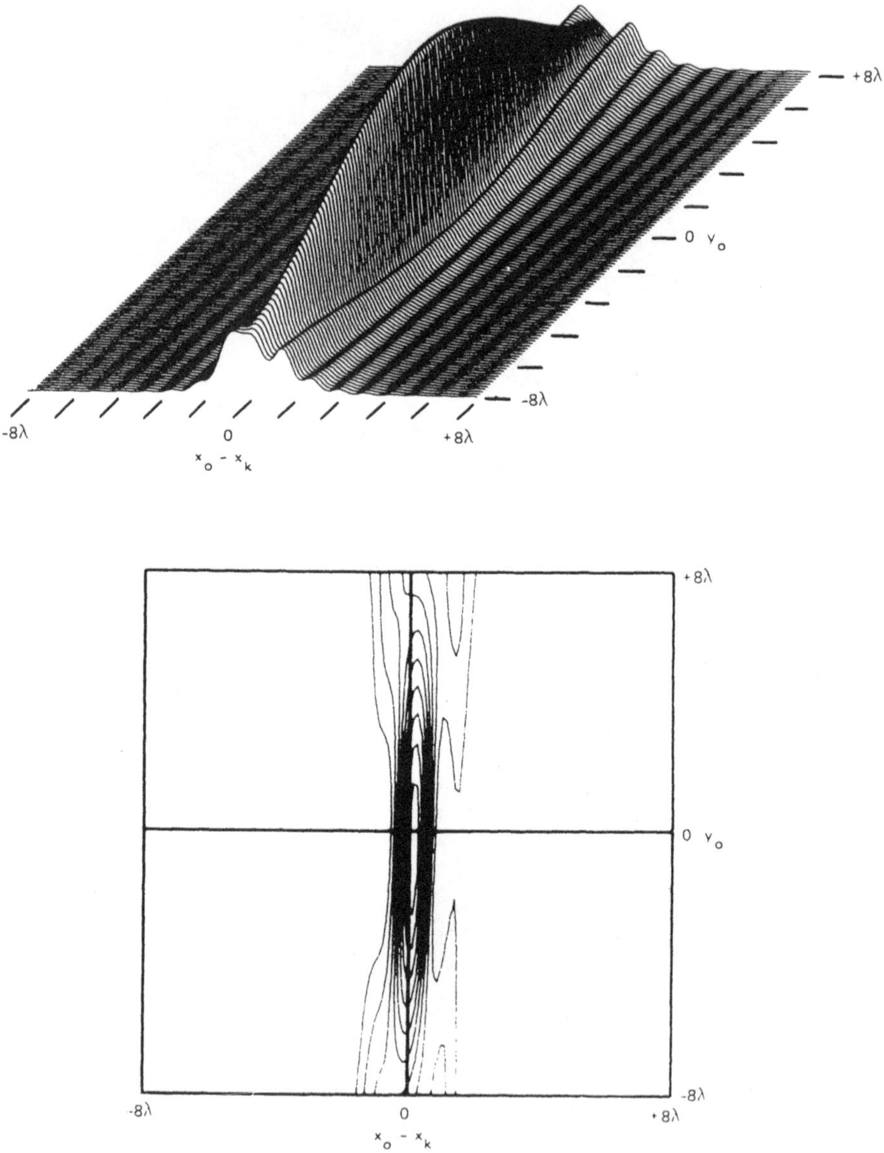

Figure 4.23 Near-field response of 30° aperture beam steered to point C.

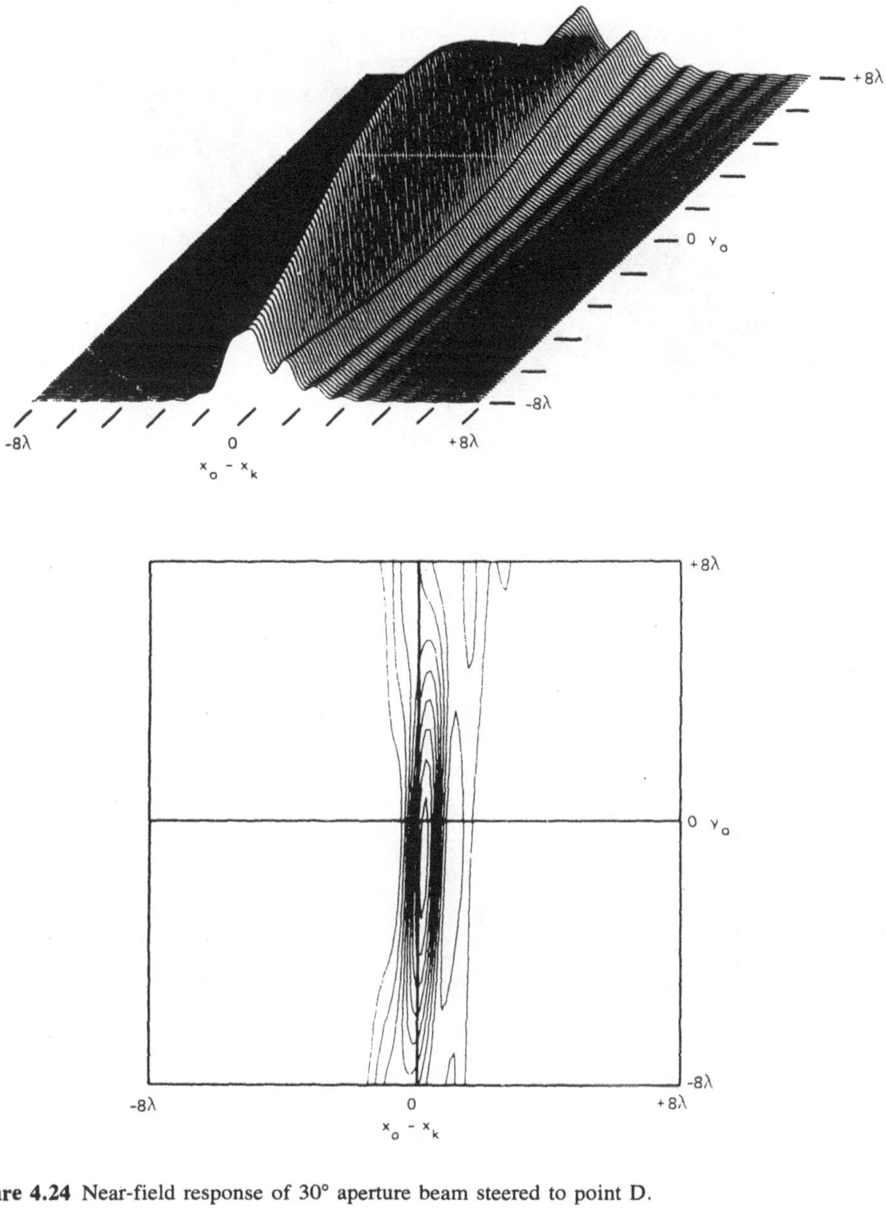

Figure 4.24 Near-field response of 30° aperture beam steered to point D.

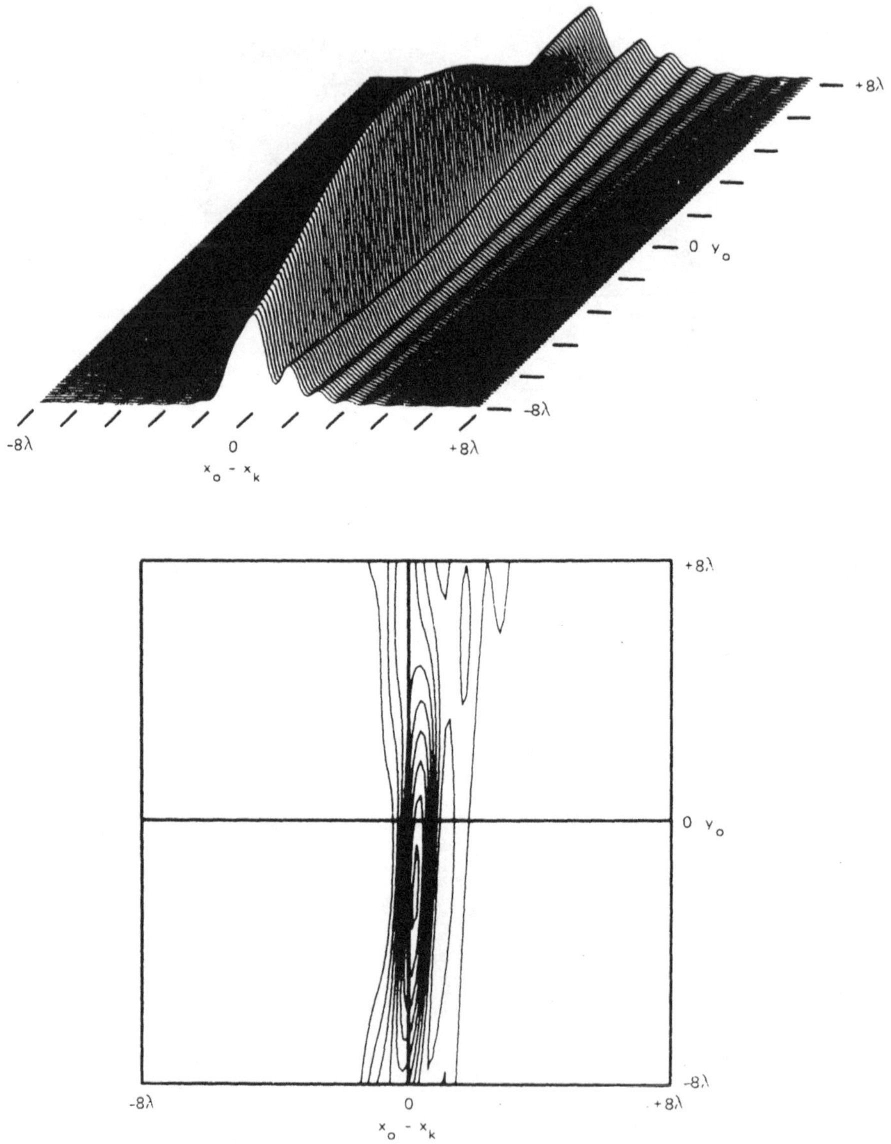

Figure 4.25 Near-field response of 30° aperture beam steered to point E.

shape of known RCS. In the current context, the term "radar cross section" implies the complex data represented by the magnitude and phase (I/Q) components of the scattered field.

The accuracy in measurements contaminated by fixed additive clutter background can be improved by subtracting the contaminating signal. The effectiveness of this technique presumes that the contamination is additive and common to both measurements of the calibrator and test object. Because this assumption is not strictly verifiable *a priori*, the technique of background subtraction is somewhat controversial among RCS practitioners.

The general calibration procedure involves separate measurements of complex frequency responses of the background, calibration, and test object, denoted respectively by $V_B(f)$, $V_C(f)$, and $V_T(f)$. If the complex amplitude of the calibrator is known and denoted $E_C(f)$, the complex response of the object is computed by:

$$E_0(f) = E_C(f)\frac{V_T(f) - V_B(f)}{V_C(f) - V_B(f)} \tag{4.19}$$

All terms in (4.19) are vector quantities, and $V_B(t)$ is eliminated when the background subtraction process is omitted. In either case, (4.19) results in a frequency response of the object calibrated in terms of absolute RCS.

4.3.2 Selection of Parameters

Instrumentation and signal-processing parameters are selected to establish the scale, resolution, and granularity of the resulting image in range and cross range. N-point FFTs used to process data sampled in frequency and angle increments Δf and $\Delta\theta$ result in range and cross-range granularities of

$$\delta R = \frac{c}{2\,(N-1)\,\Delta f} \tag{4.20}$$

and

$$\delta x = \frac{c}{2f\,(N-1)\sin(\Delta\theta)} \tag{4.21}$$

The range and cross-range image extents are determined by $(N-1)$ increments of granularity, yielding

$$R_E = \frac{c}{2\,\Delta f} \tag{4.22}$$

and

$$x_E = \frac{c}{2f\sin(\Delta\theta)} \tag{4.23}$$

The maximum achievable resolutions, defined by the separation between the peak and first null of the point-spread functions for unwindowed data, are

$$\Delta R = \frac{c}{2B} \tag{4.24}$$

for range and

$$\Delta x \simeq \frac{\lambda}{2\Theta} = \frac{c}{2f\Theta} \tag{4.25}$$

for cross range.

In (4.24) and (4.25), B and Θ are the effective bandwidth and total radian angular aperture, respectively. Equations (4.20) to (4.23) are used to establish the scaling of the resulting image size, while (4.24) and (4.25) provide estimates of the physically-achievable resolution. As indicated in these equations, image extent is controlled by the frequency and angular step size; granularity, controlled by the FFT size, can be increased by appending zeros to the available data; resolution is controlled by the total bandwidth and angular aperture.

4.3.3 Experimental Procedures

Experimental data used to generate the images shown in the next section were collected using a linear-FM system, described in Chapter 3 and shown in the block diagram of Figure 4.26 [3]. The system was operated with a set of horn antennas irradiating objects from a distance of 22 m in an anechoic chamber using a radiated power of approximately 3 mW. An angular encoder connected to the object rotator triggers the linear modulation of the transmitted signal frequency at prescribed angular increments. For the examples shown, the transmitted frequency is swept over a bandwidth of 3.15 GHz in a period of 102.4 ms. The output of the receiver, nominally a 4–5 kHz tone burst, is sampled at a 20 kHz rate and the 2048 samples obtained during the sweep period are digitized and stored. A sequence of 2048 samples is recorded for each angular increment until the object rotation is completed [3].

The angular increment between frequency sweeps is selected to satisfy the sampling criterion for scatterers with a maximum cross range of 3 m; this provides an unambiguous cross-range image extent of 6 m. The angular increments, computed

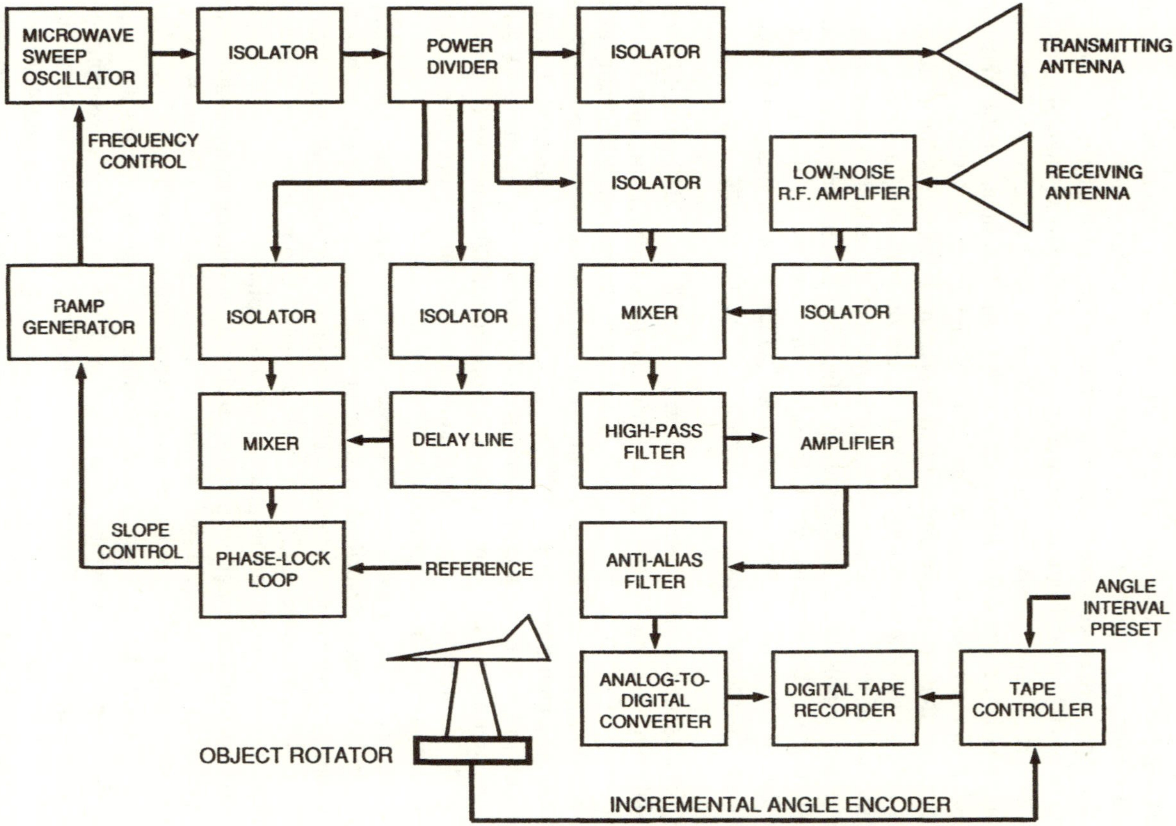

Figure 4.26 Block diagram of experimental systems.

with (4.23), are 0.14 degrees for images obtained from 10 GHz data; angular increments for images obtained at other frequencies vary inversely with frequency.

Range processing for each angle increment is accomplished by: setting the first and last 512 of the 2048-point sequence to zero, weighting the central 1024 samples with a Hann window to reduce range sidelobes, and performing a 2048-point FFT. Because the input sequence is real, the magnitude of the FFT output has even symmetry with only 1024 unambiguous samples, corresponding to an unambiguous range of 50 m. The resolution determined by the effective 1.57-GHz bandwidth and the Hann window is 19 cm, and the granularity is 4.7 cm. The index of the FFT output corresponding to the center of the image space is identified, and 128 complex samples centered on that index are retained to represent a range extent of 6 m. The sequences involved in the preceding operations are diagrammed in Figure 4.27.

Setting half of the 2048 available samples reduces the effective bandwidth, and therefore the achievable range resolution, by a factor of 2. This arbitrary degradation is imposed to meet the focusing-error criterion analyzed in the discussion leading to (4.16) and (4.17). Image focusing methods which remove this restriction, and thus allow the increased resolution without blurring the image, are presented in Chapter 5.

The computational procedure just described performs a 2048-point FFT of the recorded data and discards all but 128 spectral samples because the object occupies a limited portion of the total range space. To discard the unused 1920 samples represents a computational inefficiency, which can be overcome by zoom FFT techniques [4, 5] in which the required 128 samples are computed directly. This is accomplished by implementing a digital bandpass filter operating on the 2048-sample sequence, decimating to 128 samples, and computing a 128-point FFT. This operation, properly implemented, provides results identical to the previous procedure with significant savings in computations.

Data obtained by using the linear-FM system could also be collected with stepped-frequency instrumentation described in Chapter 3. By using a pulsed, range-gated, stepped-frequency system, the required 128 complex frequency samples could be obtained directly, and subsequently processed by an FFT to produce 128 range samples identical to those described in preceding paragraphs.

An ensemble of 128 range responses, each corresponding to an angular increment, are placed in columns of a two-dimensional array shown in Figure 4.28. Each row represents the angular extent of the synthetic aperture, which is 18° for a center frequency of 10 GHz. The first and last 32 samples in each row are set to zero, and the central 64 samples are weighted by a Hann window to reduce cross-range sidelobes in the subsequent FFT. This procedure, similar to that followed for the range processing, provides identical resolution and granularity of 19 cm and 4.7 cm, respectively. The final operation is a 128-point FFT applied to the complex data in each of the 128 rows.

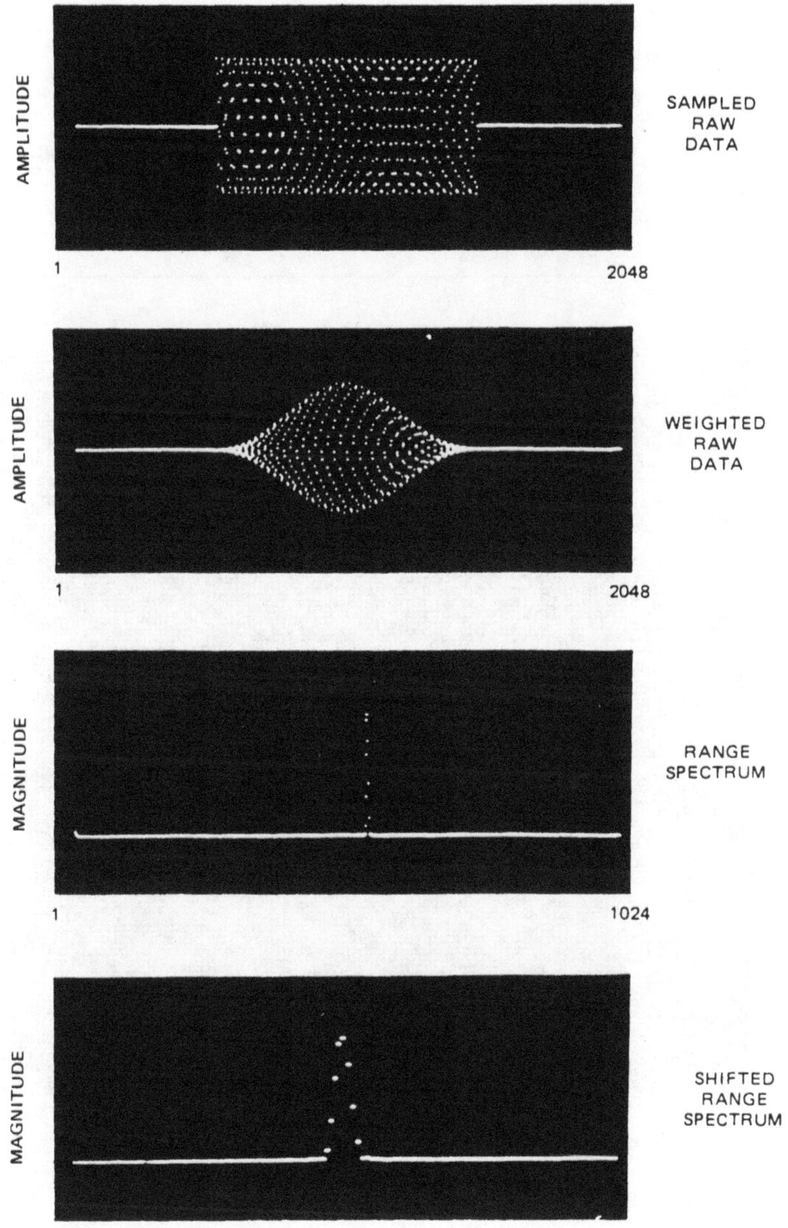

Figure 4.27 Structure of data for range processing.

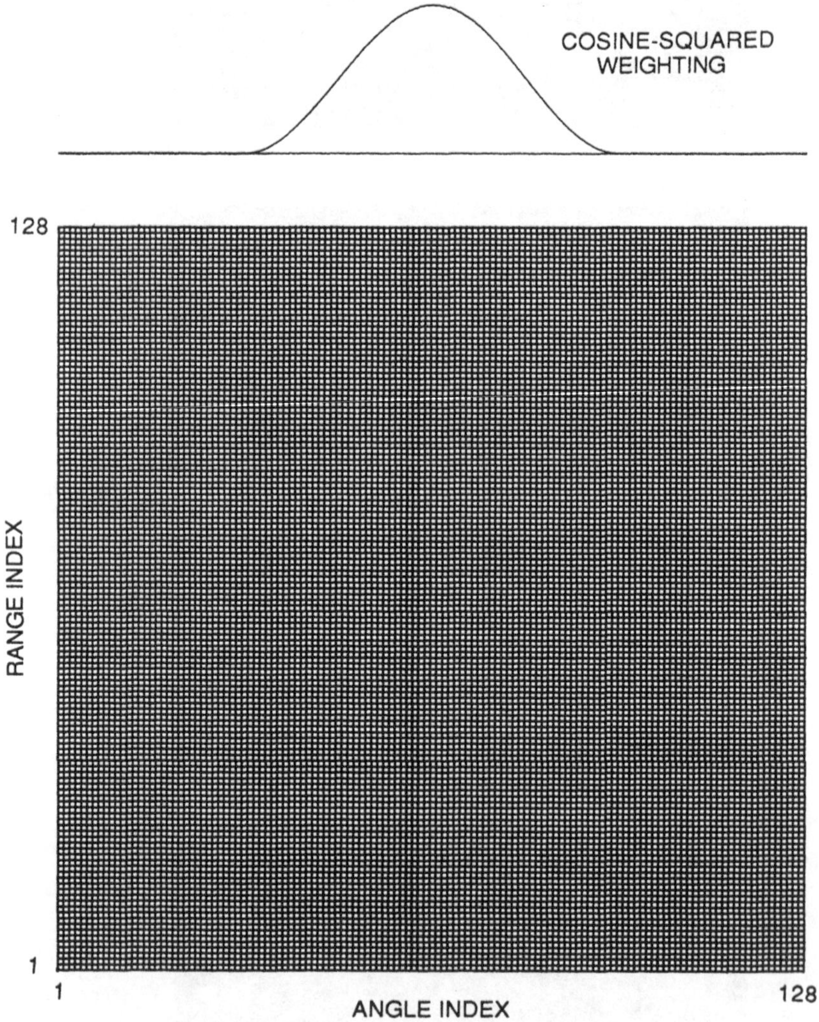

Figure 4.28 Configuration of data array for cross-range processing.

The structure of the array in Figure 4.28 is summarized as follows. The samples forming each column are obtained by Fourier transforming the real input data, and each sample represents the magnitude and phase of the sum of signals reflected from all points contained in a particular range cell. Each row, in turn, represents the variation in the content of a particular range cell as a function of rotation angle. Fourier transforming the content of each row steers the synthetic aperture to 128 different locations, and thus describes a cross-range profile for each range cell. The result is a two-dimensional array which represents the range (vertically) and cross range (horizontally) of the object's reflectivity density. The weighting applied prior to both transforms suppresses the sidelobes in both range and cross range, and appended zeros enhance the spectral definition by decreasing the granularity. This process maps the reflectivity distribution of the object in a 6 m by 6 m square field with equal resolution in range and cross range of the order to 20 cm.

To test the computational process, simulated data were generated to represent the idealized signal reflected from a linear array of four independent point objects, shown in Figure 4.29. The composite received signal is given by the sum of four signals, each expressed by (3.16), with the range corresponding to each point. The simulated received signal was computed for 0.14° angular increments, a center frequency of 10 GHz, and processed as an actual signal would be. The intensity of the resulting images is shown in Figure 4.30. Three computed images are shown for angular sectors centered on 0°, 36°, and 90°. The response of the central point represents the point-spread function of the imaging system, and conforms well to the theoretical resolution of 20 cm. The point-spread function exhibits some degradation for object points displaced from center. Images of points near the boundaries of the object space are reduced in intensity by a factor of two, and the spatial response is broadened. The degradation for points with large range offset from the center of rotation is due to the focusing error described in the preceding section. Points with maximum cross-range are also imaged with similar degradation because they traverse more than one range-resolution cell in the course of the angular rotation. This effect is reduced by the windowing process, which decreases the effective angular extent. A slight artifact is noted at the right of the 90° image. This residue is aliased leakage from the response to the left, which is avoided when the object is well contained in the unambiguous image space.

We show the effect of windowing the data prior to Fourier transforming by processing the identical data with a rectangular window replacing the Hann window in both (range and cross-range) transforms. The results, shown in Figure 4.31, differ from previous results in the following aspects: the resolution of the central point is enhanced, sidelobes caused by the abrupt truncation of the rectangular window are evident, and the off-center points are imaged with increased spreading and reduced peak magnitude. This degradation is a result of the increased effective width of the windows. This decreases both the depth of field and the size of the range-resolution cell, thereby causing more smearing than in the previous case. All subsequent images

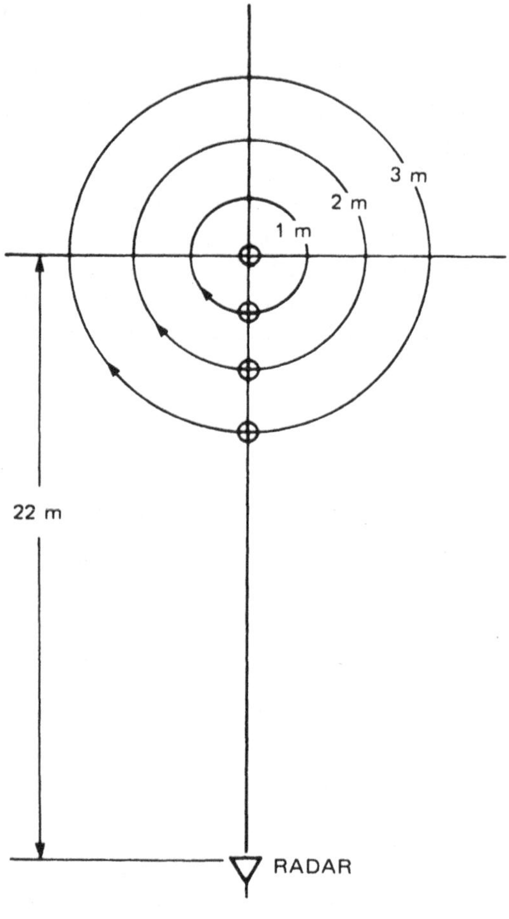

Figure 4.29 Configuration of linear array of four point objects.

are processed using the Hann window. Figure 4.30 represents the performance of the imaging process. Subject to the degradations noted over the 6 m extent of the image field, the method is adequate for two-dimensional imaging, and represents the limit of achievable performance when the efficient FFT algorithm is used.

4.3.4 Examples of Two-Dimensional Images

Several examples of images processed from experimental data are shown in the following figures. Figure 4.32 is a 10-GHz image of two cylinders 10 cm in diameter

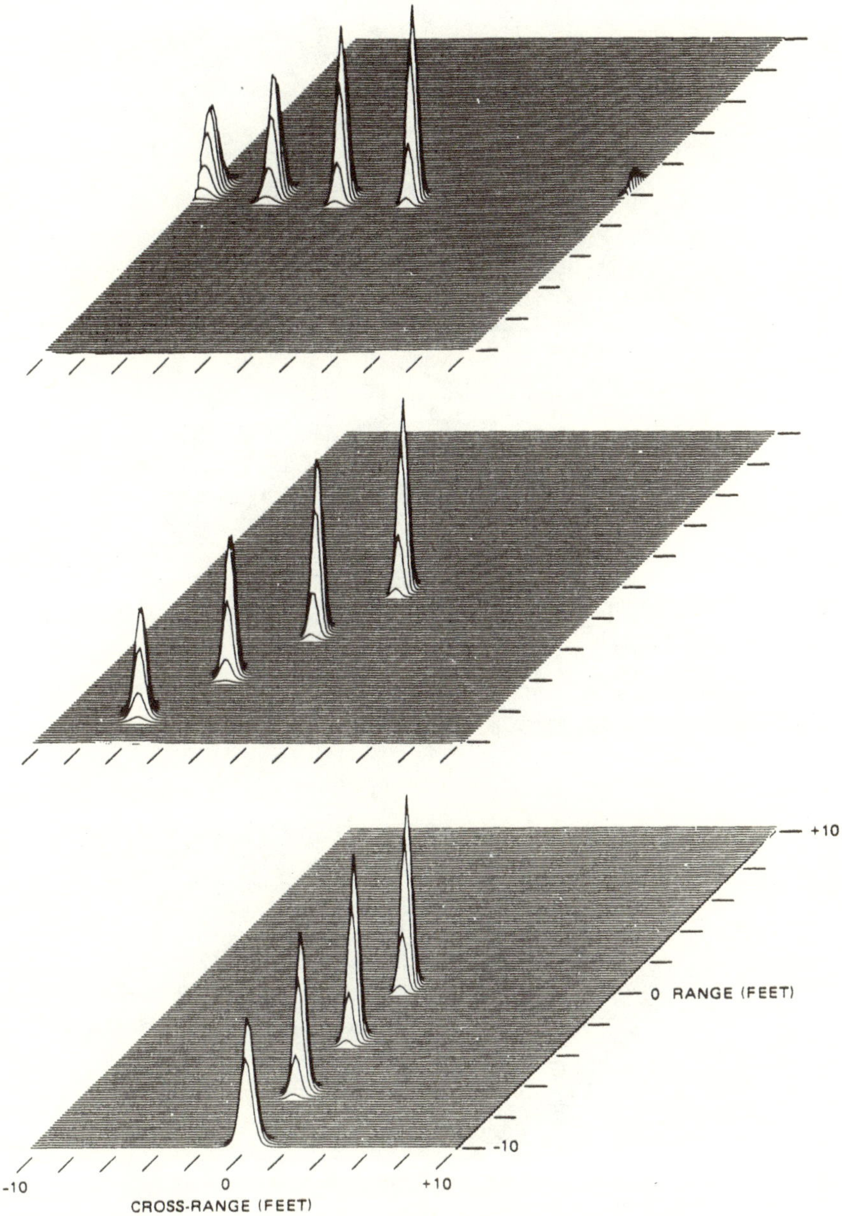

Figure 4.30 Intensity of image processed from simulated data using Hann windows.

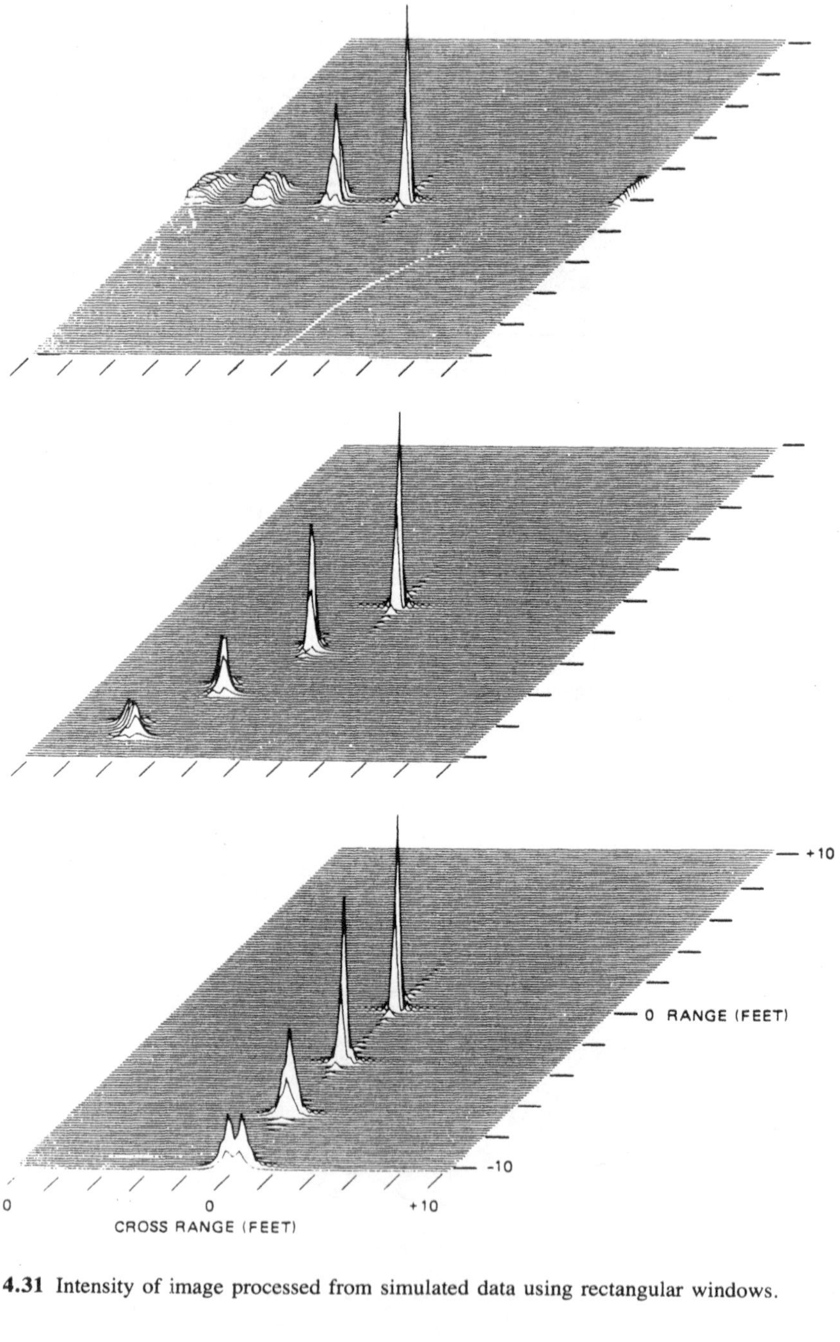

Figure 4.31 Intensity of image processed from simulated data using rectangular windows.

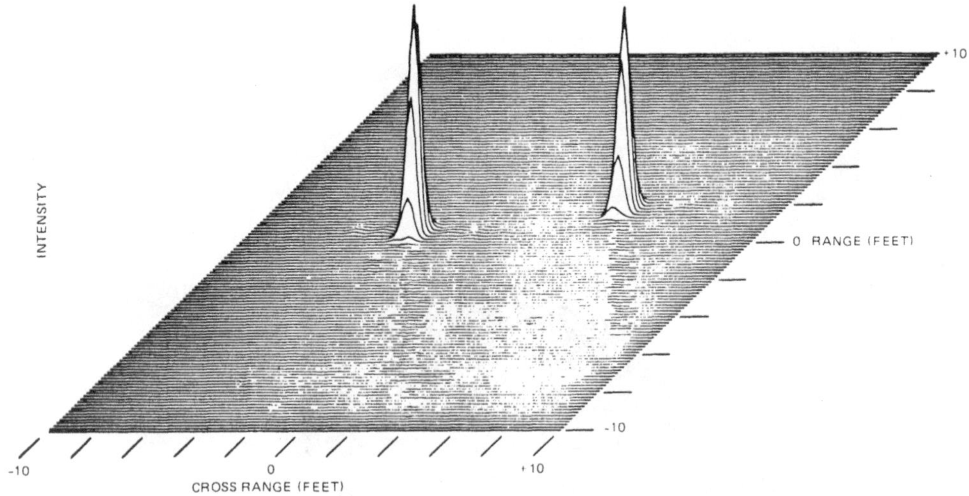

Figure 4.32 Image of two cylinders.

and 15 cm long oriented with their longitudinal axes normal to the line-of-sight and parallel to the axis of rotation. The two cylinders are diametrically opposed 120 cm from the center of rotation. Due to their small physical size, the cylinders behave essentially as point objects. The image field is free from spurious responses, artifacts and noise. The slight difference in peak intensity is caused by errors in vertical alignment of the cylinder axes.

Figure 4.33 shows a 10 GHz image of a missile body superimposed. The image was obtained by processing the range profiles shown in Figure 4.5 for frontal aspects. This example demonstrates the value of two-dimensional imaging: resolution in cross range allows the separation of the two scatterers at the rear of the missile, which would otherwise not be isolated by range processing alone.

Figure 4.34 shows a 10-GHz image of a BQM-34S target drone viewed in a frontal aspect. Scattering from the wingtip pods and tail structures, resolved in cross range, cannot be resolved in range profiles. Multiple responses from the front part of the vehicle are attributed to reverberations from structures internal to a semitransparent fiberglas cowling making up the front section.

Figure 4.35 shows two-dimensional images of a 1:10 scale model aircraft, viewed from a frontal aspect, obtained from 15-GHz wideband data. Several dominant scattering centers coincide with engine and radome locations. The identical image data are plotted in two formats. The right figure is an isometric display similar to the one shown in the preceding example. The vertical scale is logarithmic and spans a 30-dB amplitude range. The left figure is a contour plot showing lines of constant RCS

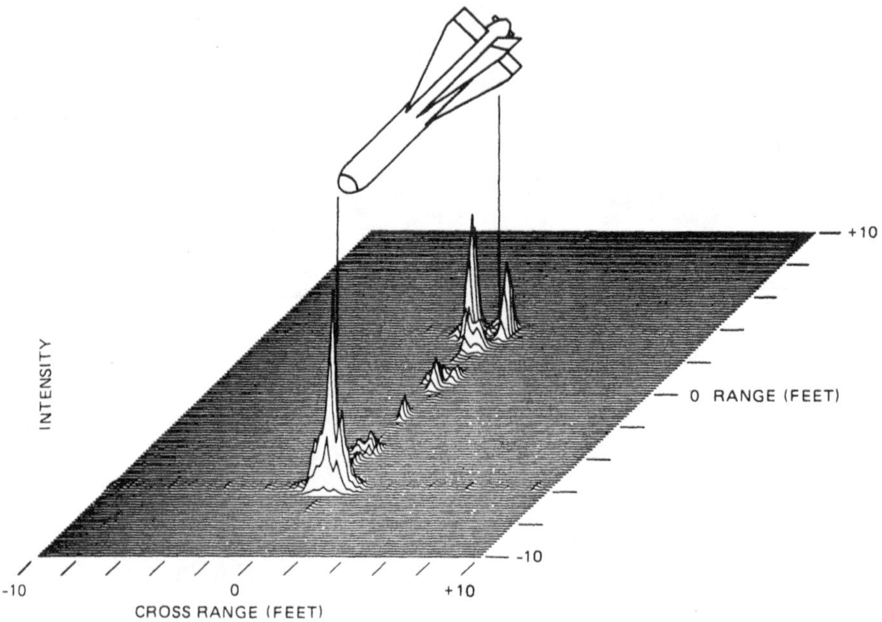

Figure 4.33 Image of missile body in zero-degree position.

in 3-dB increments. While the isometric plot effectively displays the relative magnitude of individual scatterers, the contour plot provides graphic association between the location of scatterers and the physical outline of the vehicle. Figure 4.36 shows similar images for the model aircraft viewed from the rear aspect, with the plotted values spanning a 20-dB range.

Figure 4.37 is a 15 GHz image of a 1:3.5 scale aircraft model viewed from a front quarter aspect. The series of range-extended scattering centers approximately 1.5 feet to the left of center is due to signal reverberations in the inlet duct. The scattering centers appear at the cross-range position of the duct inlet with a down-range displacement caused by the time delay of the reverberations. Figure 4.38 is an image of the same aircraft model viewed from a rear quarter aspect. Discrete scattering centers are evident at the tips of the wing and tail. A range-extended response approximately 4 feet to the left of center is due to reverberations occurring within the exhaust duct.

Figure 4.39 is a two-dimensional image of an open wedge formed by two vertical planes. The wedge measures approximately 2.2 m in length, and 1 m in height

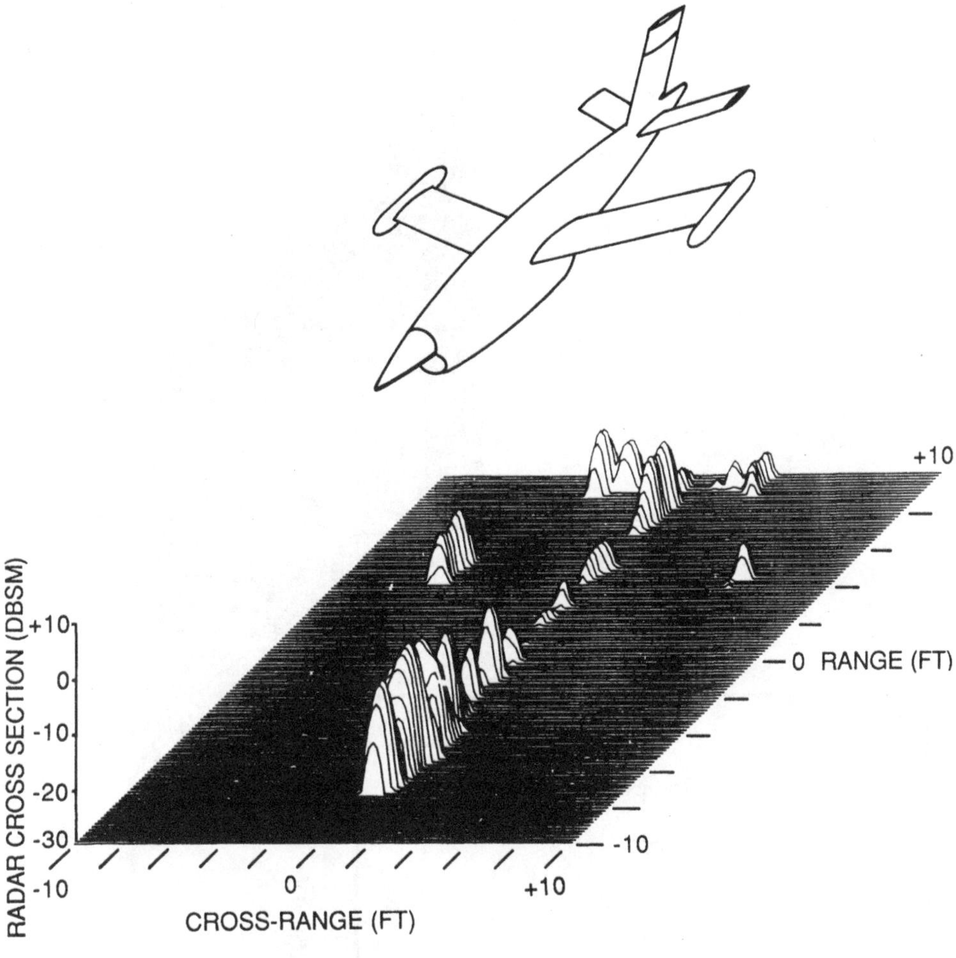

Figure 4.34 10 GHz image of BQM-34S target drone.

and width. The results were obtained from 7.5 GHz wideband data using horizontal polarization. Reflections from point F and diffractions from points A and B are properly correlated with the leading and trailing edges. Scattering centers, labeled D and E, are due to triple diffraction from one trailing edge to the other. Because the received signal is delayed in time by the round-trip distance between trailing edges, the responses D and E are displaced down range by a distance equal to the separation between trailing edges. The response at location C is due to double diffraction from one trailing edge to the other. Because the effective signal path-length remains constant as a function of rotation, the response properly appears at zero cross range with

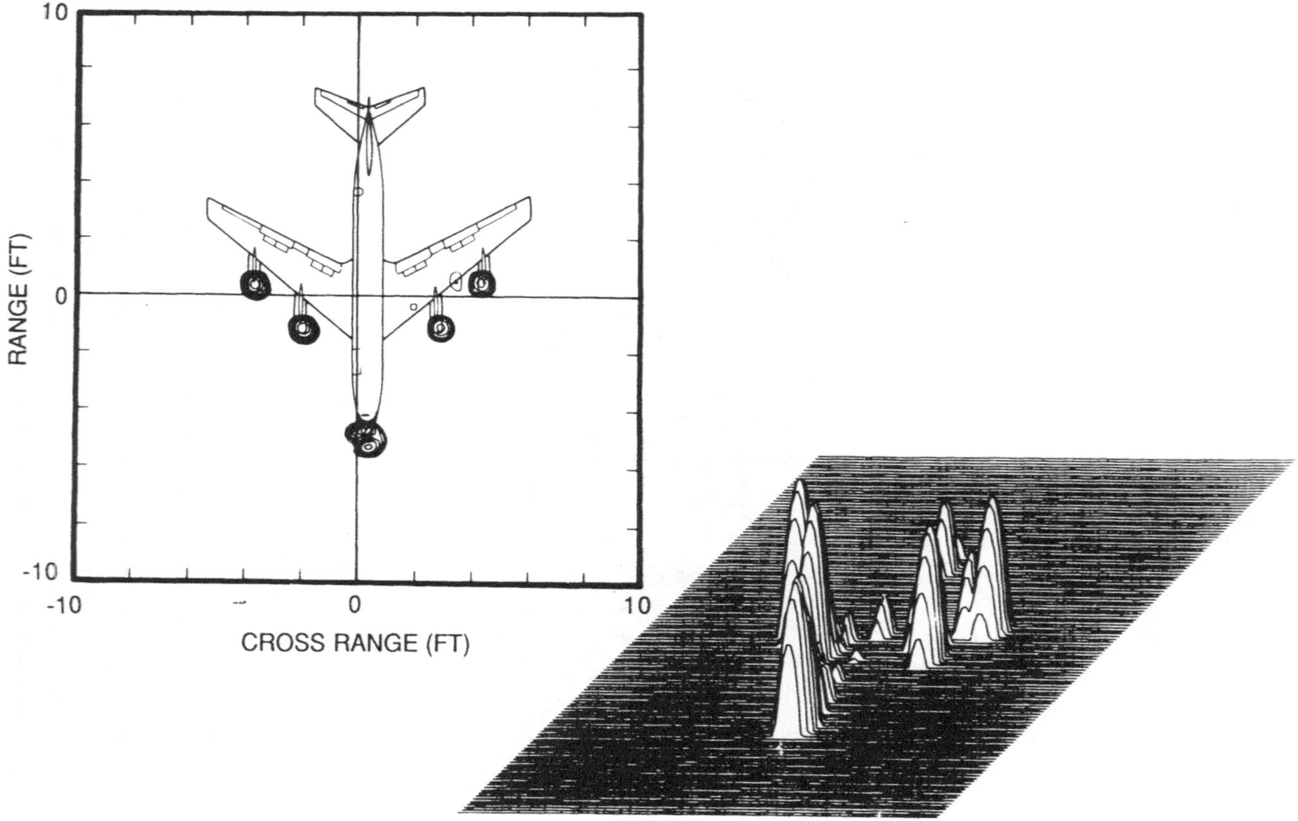

Figure 4.35 15 GHz image at 1:10 scale model viewed from front aspect.

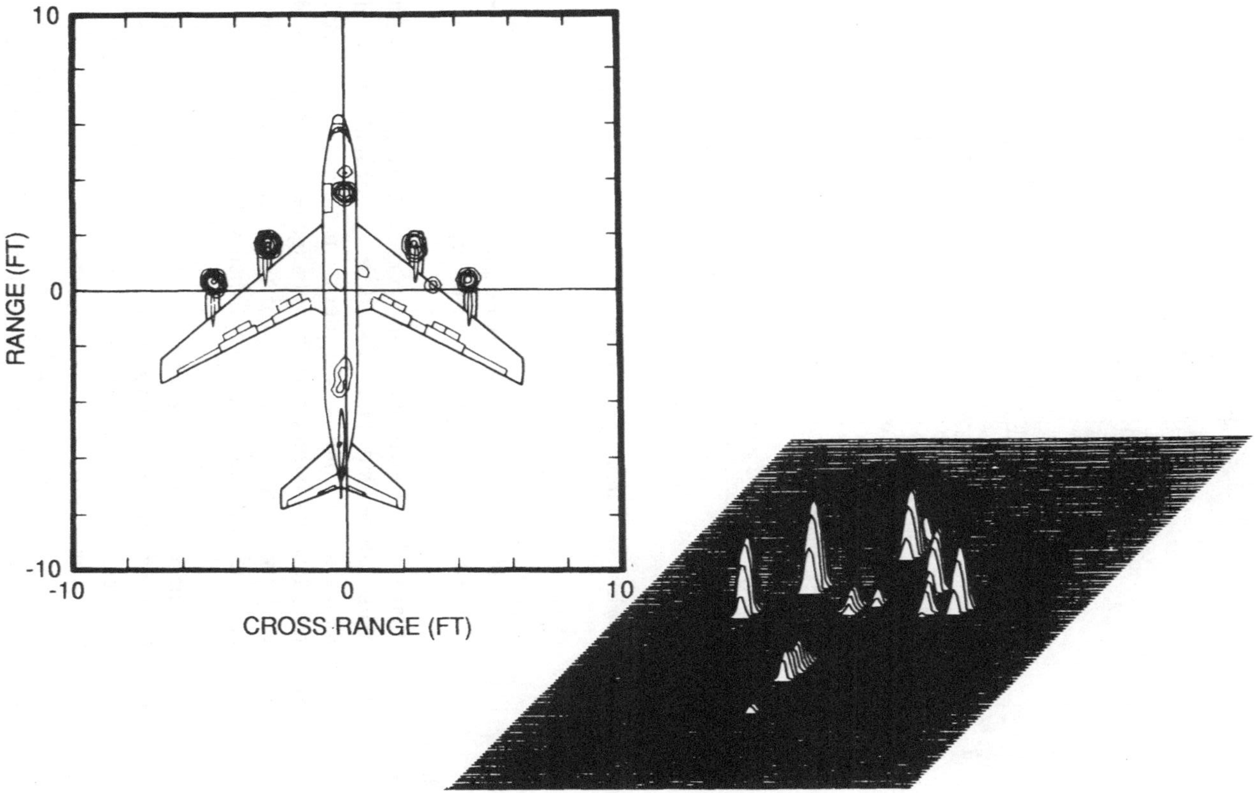

Figure 4.36 15 GHz image at 1:10 scale model viewed from rear aspect.

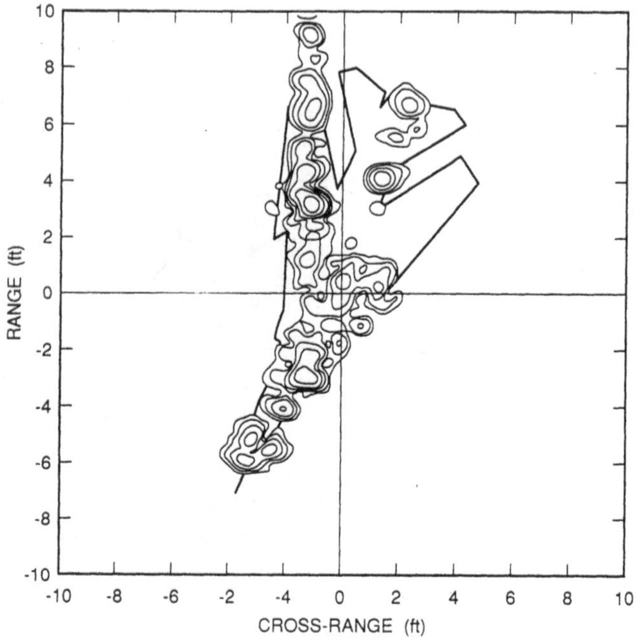

Figure 4.37 Image of aircraft model viewed from front quarter.

a range delay equal to one-half of the separation between trailing edges. The difference in the magnitude of responses D and E is attributed to asymmetries in the panels forming the wedge.

Figure 4.40 is an image of a planar shape showing scattering from various edges and discontinuities with an extended series of scatterers evident along the range axis. The latter features are caused by stationary clutter from the imperfect anechoic enclosure. Because stationary features induce no phase change as a function of object rotation, they are imaged with zero cross range. Their range position, however, is properly registered because range imaging is accomplished without object rotation.

Figure 4.41 is a 15 GHz image of a 1:3.5 scale model of an aircraft plotted with 3-dB contour levels. The unique feature of this image is the range-extended scattering approximately 2 feet left of center. This feature was confirmed to be caused by multiple reflections involving the fuselage and wing pylon, indicated by the bold line. The multiple reflection is registered as a time or range delay.

Figure 4.42 is a 35 GHz image of a 0.6 m high × 1.2 m wide × 2.4 m long block of expanded bead polystyrene foam normally used to fabricate support columns for test objects. The lower threshold is -65 dB/m^2, and the contours are in 5-dB

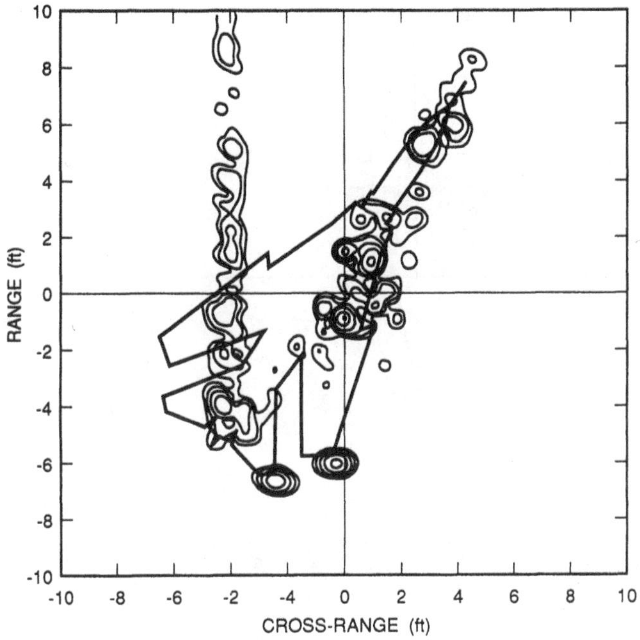

Figure 4.38 Image of aircraft model viewed from rear quarter.

increments. The image shows relatively homogeneous scattering from the interior volume of the block. The minor feature at -2.5 m range is due to stationary clutter in the anechoic chamber.

Figure 4.43 shows two 35 GHz images of a compound conical shape terminating in a blunt tip for vertical and horizontal polarizations. The dominant features of the image are the tip scattering and the scattering from the circumferential discontinuities. The images illustrate the polarization dependence of edge diffractions.

4.4 EXTENSION TO THREE-DIMENSIONAL IMAGING

The preceding sections described a method for two-dimensional imaging that used time-delay sorting to obtain resolution in range and object rotation for resolution in cross range along an axis normal to the axis of rotation of the object. This technique is adequate for imaging three-dimensional objects if resolution is obtained in the second cross-range direction by using fan beams to irradiate selected slices of the object. When the entire object is uniformly irradiated, however, the images represent

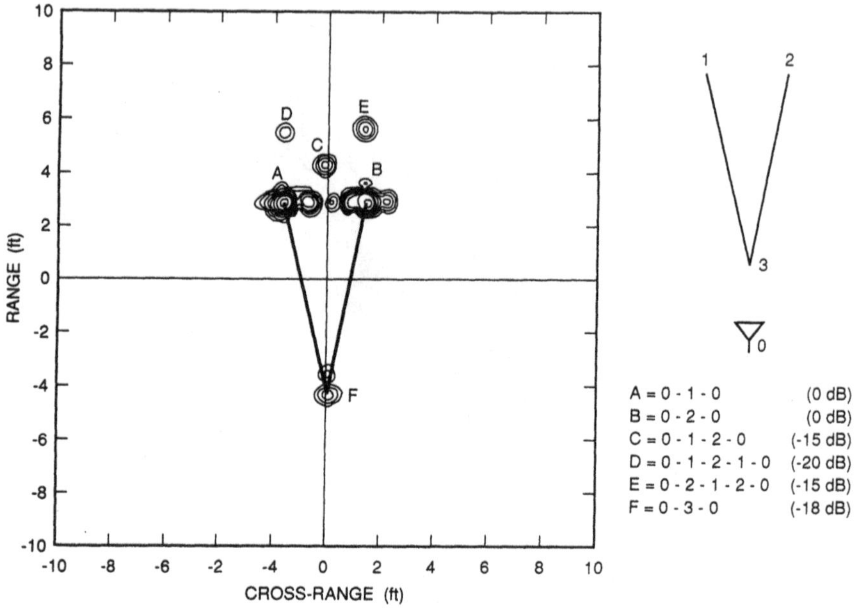

Figure 4.39 Image of open wedge showing multiple diffractions.

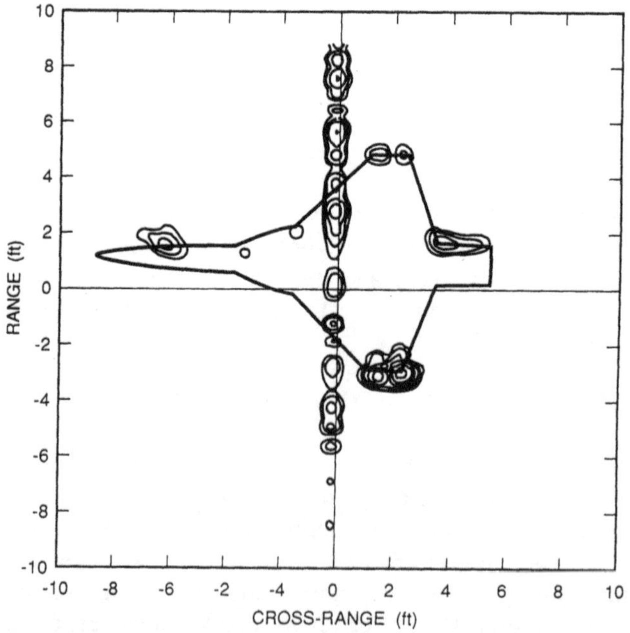

Figure 4.40 Image of planar shape showing stationary clutter.

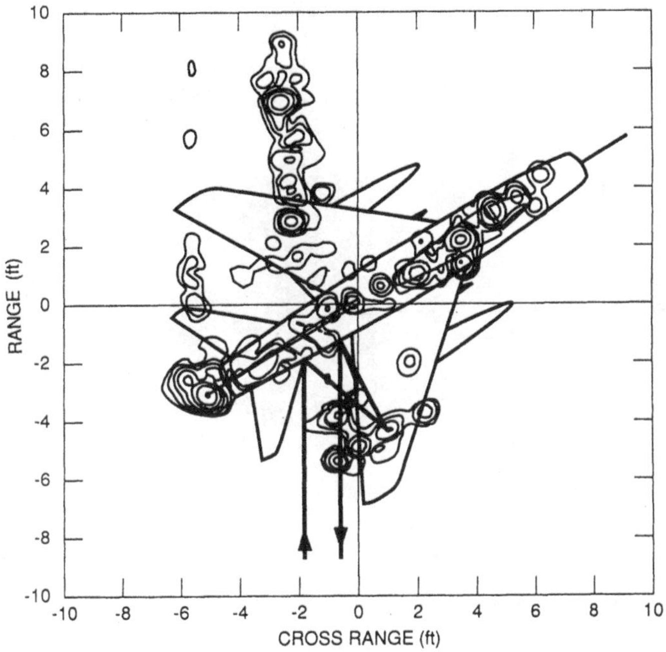

Figure 4.41 15 GHz image showing effects of multiple reflections.

projections of the three-dimensional reflectivity distribution on the two-dimensional image plane. In some applications, such as the locating of scattering centers on known objects, the lack of resolution in the third dimension is not critical. In these cases, the location of reflecting points can be inferred from the two-dimensional image because the object shape is known. Thus, *a priori* information can be used to remove ambiguities not resolved by the imaging process. When imaging complex objects without *a priori* information about the reflectivity distribution, however, three-dimensional resolution may be required to unambiguously define the image.

Resolution in both cross-range directions can be obtained by rotation of the object about two mutually orthogonal axes. The signal content of each range cell, resolved by the same method used in the preceding section, is coherently processed as a function of the two rotation angles. In this case, observing a rotating object with a stationary radar is equivalent to observing a stationary object with a radar moving on a spherical surface. This condition is shown in Figure 4.44, in which the object is rotated by the angles θ and ψ about the y- and x-axes, respectively. The total angular excursions, Θ and Ψ, form a two-dimensional aperture, which is a section of a sphere. For each range cell, the image of the object is found by a two-dimensional DFT of the received signals sampled as a function of the two angles θ,

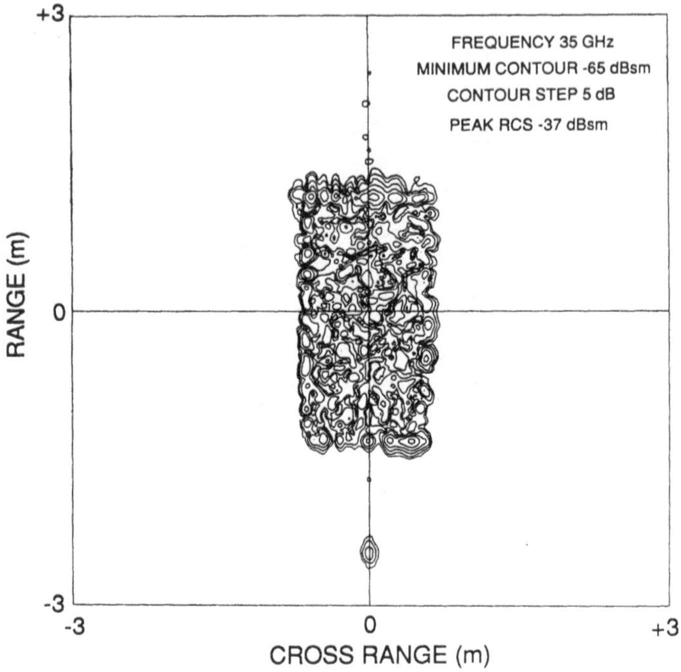

Figure 4.42 35 GHz image of polystyrene foam.

ψ. The result is a sequence of two-dimensional images for adjacent range slices. Because the angles Θ and Ψ are relatively small ($<\pi/6$), the synthesized aperture is nearly rectangular and the point-spread function, $h(x,y)$, is separable in rectangular coordinates and given by the product $h_X(x)\ h_Y(y)$. As a result, each of the relations derived for the one-dimensional synthetic aperture applies to the two-dimensional case. The required signal processing is more extensive than in the preceding case because a one-dimensional DFT is replaced with a two-dimensional DFT to process the signal content of each range cell as a function of rotation angle. Although more stored data and computations are required for the two-dimensional synthetic aperture case than for the one-dimensional case, the types of data and computations are similar. Resolution in the second cross-range dimension can thus be achieved at the cost of increased storage capacity and computation time.

The physical implementation of the two-dimensional synthetic aperture requires means for rotating the object about two orthogonal axes. This can be accomplished by a gimballed support system or by using other mechanisms by which the object can be independently tilted and rotated relative to the radar axis.

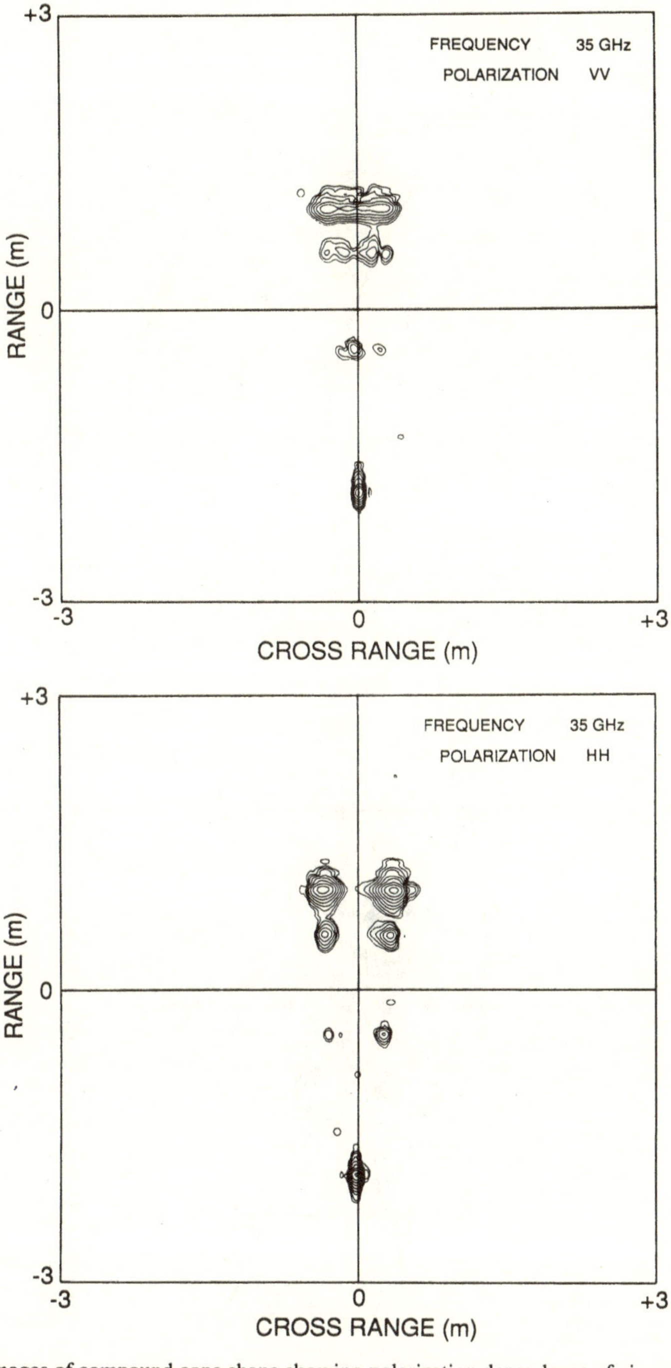

Figure 4.43 Images of compound cone shape showing polarization dependence of circumferential joints.

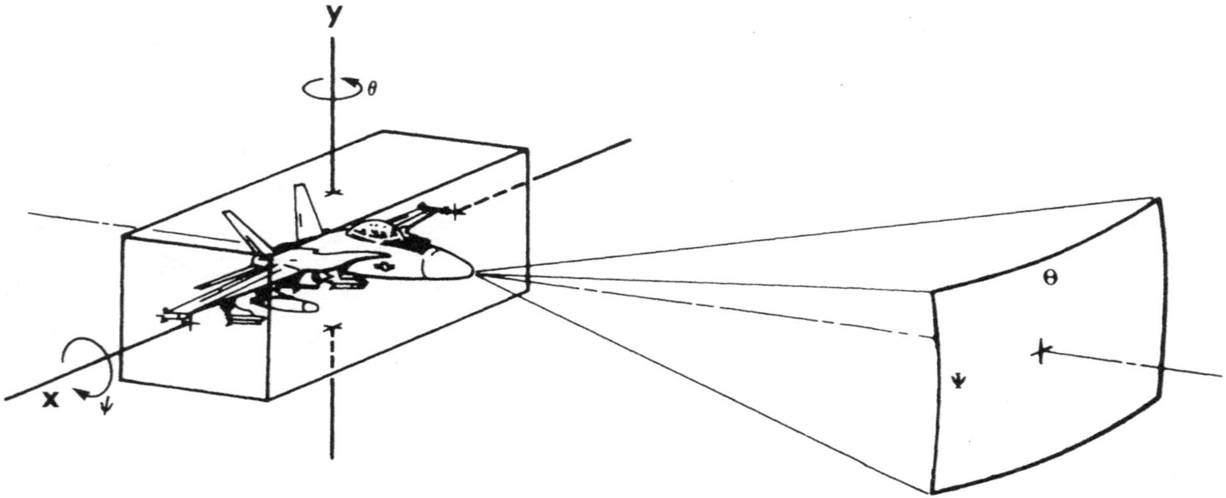

Figure 4.44 Two-dimensional synthetic aperture formed by object rotation.

REFERENCES

1. Harris, F.J., "On the Use of Windows for Harmonic Analyses with the Discrete Fourier Transform," *Proc. IEEE*, Vol. 66, No. 1, January 1978, pp. 51–83.
2. Harger, R.O., *Synthetic Aperture Radar Systems, Theory, and Design*, New York, Academic Press, 1970.
3. Mensa, D.L., "Wideband Radar Cross Section Diagnostic Measurements," *IEEE Trans. Instrumentation and Measurement*, Vol. IM-33, No. 3, September 1984, pp. 206–214.
4. Otnes, R.K., and L.I. Enochson, *Digital Time Series Analysis*, New York, John Wiley and Sons, 1972, pp. 420–424.
5. Halevy, S., G.R. Heidbreder, and D.L. Mensa, "Efficient Range-Doppler Imaging of Rotating Objects," *Proc. IEEE National Radar Conference*, 1984, pp. 53–57.

Chapter 5
Focused Synthetic-Aperture Processing

The preceding chapter described a method for two-dimensional imaging of rotated objects. The resolution in range was shown to be determined by the signal bandwidth and the resolution in cross range by the angular extent of the synthetic aperture. Signal processing consisted almost exclusively of Fourier transforms, which can be implemented by highly efficient FFT algorithms. The Fourier transform is the exact operation required for range processing, and no approximations are involved in its application. However, use of the Fourier transform in synthetic aperture processing constitutes an approximation which limits resolution. The image produced is subject to focusing errors which are predominantly range-dependent; consequently, Fourier transforming signals received from a rotating object results in an imaging system that is space-variant and does not provide a diffraction-limited image throughout the object space. This chapter considers methods for focusing the synthetic aperture in order to obtain diffraction-limited, space-invariant, point-spread functions. As shown in the preceding chapter, focusing errors are phase factors which depend on the wavelength of the irradiation. In the following section, we develop the focusing process for the case of irradiation with a single wavelength. In the subsequent section, we extend the analysis to include wideband irradiation.

5.1 SYNTHETIC APERTURE FOCUSING FOR CW SIGNALS

Consider the problem of obtaining a two-dimensional image of an object by processing reflected signals using a source and receiver located in the cross-sectional plane. The cross section is characterized by a two-dimensional distribution of scattering centers, denoted by $g(x,y)$, which is termed the *reflectivity density function*. The geometry shown in Figure 5.1 depicts an object being uniformly illuminated by a stationary CW source while reflected signals are recorded as a function of the objec

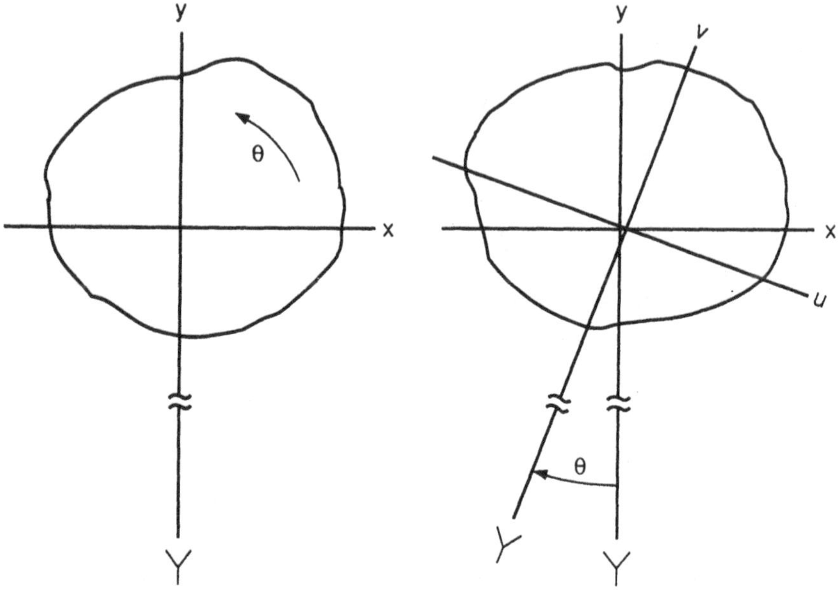

Figure 5.1 Imaging geometry for a planar object.

rotation angle.* The observation distance is assumed to be sufficiently greater than the object extent so that isorange contours can be assumed to be straight lines normal to the line-of-sight. The objective is to obtain a reconstruction of $g(x,y)$ by processing the signals received from various aspect angles. The term "signal" is taken to mean the complex envelope (phase and amplitude) of the received CW carrier.

The formulation is directly applicable to planar objects and to three-dimensional objects when a fan beam is used for selectively illuminating a particular cross section of the object. If a three-dimensional object is uniformly illuminated, $g(x,y)$ is interpreted as the projection of the three-dimensional reflectivity density onto the x-y plane. The imaging process is tomographic in nature because the reconstructed image can be a slice of the object obtained from measurements of signals propagating in the plane of the slice. This is the case when a fan beam is used. The process, however, differs from conventional x-ray tomography [1–4] in which the image is reconstructed using measurements of line integrals or projections of the object density. As we will show in the following analysis, the imaging process being described reconstructs the image from measurements of total reflected signals.

*When imaging complex reflective objects, parts of the object may be shadowed by other parts. In our development, we assume that shadowing does not occur.

Assume that the observations are started with the sensor aligned with the y axis; x is then the cross-range coordinate and y is the range coordinate. When the object is rotated through any angle θ, the range and cross-range coordinates become v and u, respectively. For a fixed θ, the signal from a particular range, v, is given by the reflectivity density integrated over u for a fixed v. This assumes that all reflectors from a given range add in-phase or, equivalently, that the reflection coefficients are of equal phase:

$$p(v;\theta) \propto \int_{-\infty}^{+\infty} g_\theta(u,v)\, du \tag{5.1}$$

where

$$g_\theta(u,v) = g(x,y)$$

$$u = x \cos\theta - y \sin\theta$$

$$v = x \sin\theta + y \cos\theta$$

The function $p(v;\theta)$ is a line integral consisting of a projection of the reflectivity density onto the v axis. This function is one-dimensional with respect to the single variable v with the second variable, θ, interpreted as a parameter to denote the orientation of a projection.

The total received signal is the integral of signals along the projection modified by a phase factor determined by the round-trip phase. If the propagation velocity in the medium is constant, the signal is expressed within a generally complex constant by

$$\begin{aligned} G(\theta) &= \int_{-\infty}^{+\infty} p(v;\theta) \exp(-j4\pi v/\lambda)\, dv \\ &= \int_{-\infty}^{+\infty}\int_{-\infty}^{+\infty} g_\theta(u,v) \exp(-j4\pi v/\lambda)\, du\, dv \end{aligned} \tag{5.2}$$

The integrals can be expressed with infinite limits because the integrand is zero outside the bounded object.

The relation $v = y \cos\theta + x \sin\theta$ allows (5.2) to be expressed as

$$G(\theta) = \int_{-\infty}^{+\infty}\int_{-\infty}^{+\infty} g(x,y) \exp[-j4\pi/\lambda(y \cos\theta + x \sin\theta)]\, dx\, dy \tag{5.3}$$

If the variables f_x and f_y are defined as $f_x = 2(\sin\theta)/\lambda$, $f_y = 2(\cos\theta)/\lambda$, then (5.3) can be written as

$$G(f_x,f_y) = \int_{-\infty}^{+\infty}\int_{-\infty}^{+\infty} g(x,y) \exp[j2\pi(f_x x + f_y y)]\, dx\, dy \tag{5.4}$$

Equation (5.4) has been expressed as a two-dimensional Fourier transformation. If $G(f_x,f_y)$ were known for f_x and f_y running from $-\infty$ to $+\infty$, $g(x,y)$ could be readily determined by an inverse Fourier transformation, resulting in

$$g(x,y) = \int_{-\infty}^{+\infty}\int_{-\infty}^{+\infty} G(f_x,f_y) \exp[-j2\pi(f_x x + f_y y)]\, df_x\, df_y \tag{5.5}$$

Equations (5.4) and (5.5) would then form a two-dimensional transform pair symbolized by

$$g(x,y) \leftrightarrow G(f_x,f_y) \tag{5.6}$$

where $g(x,y)$ is the spatial function; $G(f_x,f_y)$ is the corresponding spatial spectrum; and f_x and f_y are planar wave spatial frequency components in the x and y directions, respectively.

Equation (5.5) shows that $g(x,y)$ could be determined by Fourier transforming $G(f_x,f_y)$, which is related to the observed data. Because $g(x,y)$ is space-limited, $G(f_x,f_y)$ extends over the entire f_x-f_y plane. However, the constraints $f_x = 2(\sin\theta)/\lambda$ and $f_y = 2(\cos\theta)/\lambda$ show that the actual data are limited to values of $G(f_x,f_y)$ falling along a circle in the f_x-f_y plane with radius $2/\lambda$. In the following analysis, we determine how the limited sample of the spatial spectrum afforded by the measured values $G(\theta)$ affects the reconstructed image.

Let $\hat{g}(x,y)$ be the reconstructed reflectivity density obtained by setting $G(f_x,f_y) = 0$ everywhere except on a circle of radius $2/\lambda$. The radial symmetry allows simplification by converting to polar coordinates, defined in Figure 5.2. Here, $g_R(x,y)$ and $g_P(r,\psi)$ are spatial descriptions of the object in rectangular and polar coordinates; $G_R(f_x,f_y)$ and $G_P(\rho,\theta)$ are the corresponding spectral functions expressed in rectangular and polar coordinates. The expression for the reconstruction in polar coordinates is

$$\begin{aligned}\hat{g}_P(r,\psi) &= \int_0^{\infty}\int_0^{2\pi} G_P(\rho,\theta)\delta(\rho - 2/\lambda) \exp[-j2\pi\rho r \cos(\theta - \psi)]\, d\theta\, \rho\, d\rho \\ &= 2/\lambda \int_0^{2\pi} G_P(\theta) \exp[-j4\pi r\lambda^{-1} \cos(\theta - \psi)]\, d\theta \end{aligned}\tag{5.7}$$

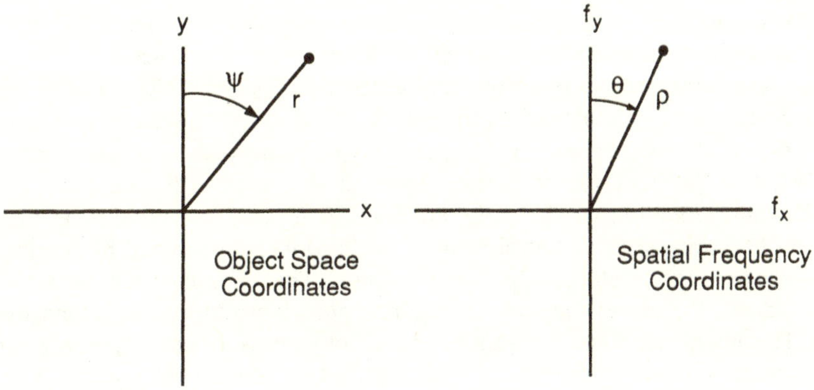

Figure 5.2 Space and frequency domain coordinates.

The values $G_p(\rho,\theta)$ are precisely the observed data $G(\theta)$ as described in (5.3).

Equation (5.7) expresses the response of a circular line aperture which surrounds the object and is focused to each point in the object space. The exponential term is the phase correction required to bring signals received from an object point at r,ψ to an in-phase condition. Although (5.7) is expressed as a single integral as a result of the radial symmetry, it constitutes a two-dimensional Fourier transform. The two-dimensional operation is required to focus the aperture to all points in the object plane.

The point-spread function of the imaging system can be obtained by employing the transform relation:

$$\hat{g}_P(r,\psi) = \mathcal{F}\{[G_P(\rho,\theta)][\delta(\rho - 2/\lambda)]\} \tag{5.8}$$

Because $\hat{g}_P(r,\psi) = \mathcal{F}[(G_P(\rho,\theta)]$, (5.8) becomes

$$\hat{g}_P(r,\psi) = g_P(r,\psi) \underset{2}{\star} h(r,\psi) \tag{5.9}$$

where $h(r,\psi) = \mathcal{F}[\delta(\rho - 2/\lambda)]$ and the symbol $\underset{2}{\star}$ denotes two-dimensional convolution.

The calculation of $h(r,\psi)$ is performed by a Fourier integral in polar coordinates:

$$\begin{aligned} h(r,\psi) &= \int_0^\infty \int_0^{2\pi} \delta(\rho - 2/\lambda) \exp[-j2\pi\rho r \cos(\theta - \psi)]\, d\theta\, \rho\, d\rho \\ h(r) &= (2/\lambda) \int_0^{2\pi} \exp[-j2\pi r(2/\lambda) \cos(\theta - \psi)]\, d\theta = (4\pi/\lambda) J_0(4\pi r/\lambda) \end{aligned} \tag{5.10}$$

where $J_0(\cdot)$ is the zero-order Bessel function.

The reconstruction $\hat{g}(r,\psi)$ is a convolution of the object reflectivity distribution, $g(r,\psi)$, with the function $h(r,\psi)$, which is the point-spread function of the imaging process. The point-spread function, expressed by (5.10), exhibits radial symmetry with a central lobe width (radial distance between zeros adjacent to the peak) of 0.4λ; the zero-crossing spacing is not precisely constant with r, and the first sidelobe is 8 dB below the peak. Its magnitude squared is plotted in Figure 5.3.

The interpretation of the point-spread function of the process is that object points will be imaged as a central peak of width 0.4λ, surrounded by monotonically decreasing circular sidelobe ridges and valleys. The imaging process is linear and shift invariant; therefore, the form of the response is uniform throughout the object space. The image response to multiple-point objects is the superposition of point-spread functions, each weighted by the amplitude of the object point and shifted to the spatial coordinates of the object point. Because the first zero of the point-spread function occurs at a radial distance of 0.2λ from the peak, two point objects of comparable amplitude, separated by more than 0.2λ, will produce an image in which they are resolved.

Figure 5.4 shows the intensity of the image reconstructed from data computed to simulate a signal received from a point object. The simulated wavelength was 0.03 m, and the image field was calculated for an area of 0.5×0.5 m and consisted of 128×128 pixels. The object point was placed at a radial distance of 0.15 m and 300 uniform angular increments were used for a complete rotation. The reconstructed image conforms to the theoretical point-spread function and, not being centered, confirms the shift invariance of the imaging process. Small artifacts result for plotting the radially symmetric response on a rectangular format.

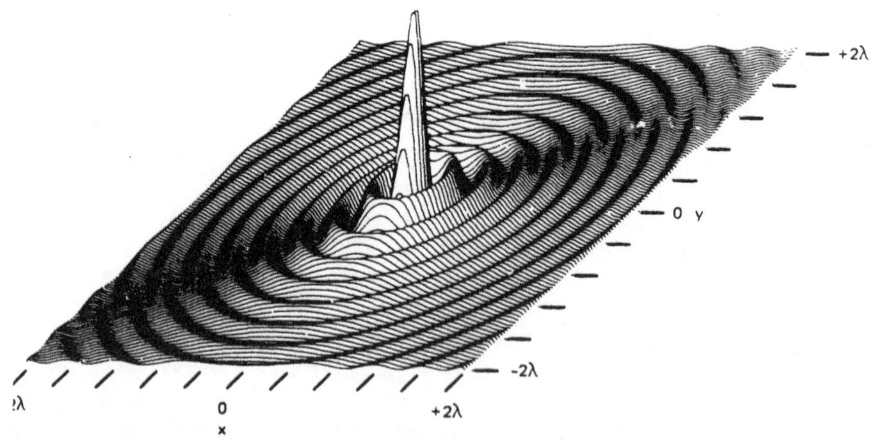

Figure 5.3 Intensity of the point-spread function for the imaging process.

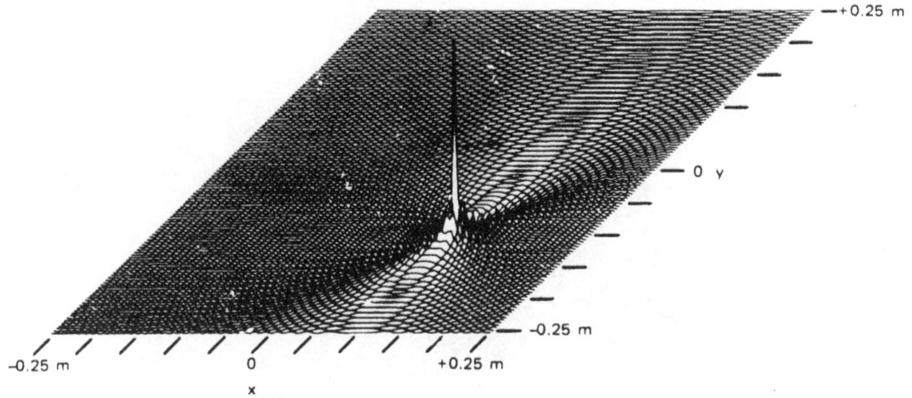

Figure 5.4 Intensity of reconstructed image of a line object ($\lambda = 0.03$ m).

To confirm the practical performance of the imaging process, a microwave measurement of a dual point object was conducted and the image was reconstructed from the recorded data. Synthetic data corresponding to parameters duplicating those of the test were also generated and processed. The test object was a pair of identical rods 0.64 cm in diameter and 34 cm in length. The rods, separated by 17 cm and mounted vertically on a rotating support, were irradiated by a radar with a wavelength of 0.06 m from a distance of approximately 20 m. The rotation axis was parallel to the rods and normal to the radar line-of-sight. The test object is essentially a pair of line objects because the rod diameter is small relative to the wavelength; the projection of the line objects onto the x-y plane is a pair of points. The reconstructed image field is a 128 × 128 cell array spanning 0.5 × 0.5 m. Figures 5.5 and 5.6, displaying the intensity of the images reconstructed from real and synthetic data, show excellent correlation. The image artifacts are induced by interference between the periodic sidelobe structures of the two object responses.

The imaging process can be considered from a spatial spectrum viewpoint with the reconstructed image interpreted as a synthesis of elemental plane-wave components. Consider the reconstruction of a single point-object of unit amplitude with coordinates (x_0,y_0):

$$G_R(f_x,f_y) = \int_{-\infty}^{+\infty}\int g_R(x,y) \exp[j2\pi(f_x x + f_y y)]\, dx\, dy \tag{5.11}$$

If $g_R(x,y) = \delta(x - x_0, y - y_0)$, (5.11) becomes

$$G_R(f_x,f_y) = \exp[j2\pi(f_x x_0 + f_y y_0)] \tag{5.12}$$

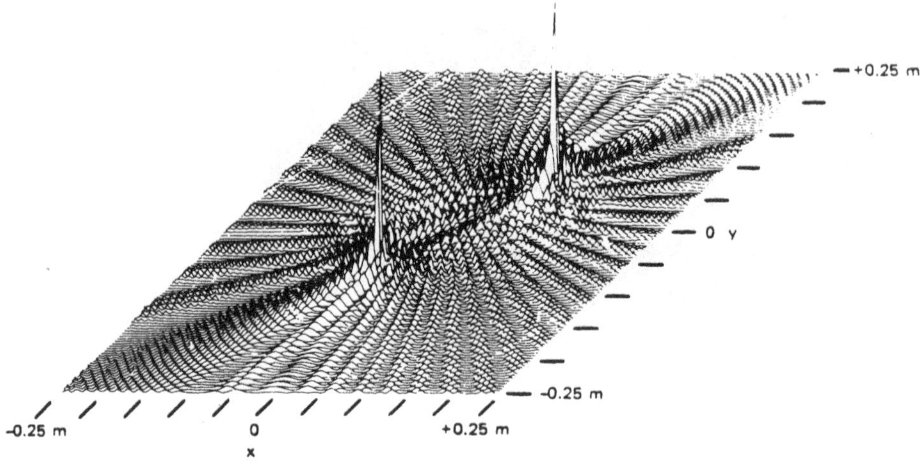

Figure 5.5 Intensity of image reconstructed from measured data for two-line objects (λ = 0.06 m).

The spatial spectrum of (5.12) has constant magnitude and linear phase; that is, the spectrum contains all spatial frequencies and has a phase linear with frequency and proportional to the displacement of the object. The reconstruction of the spatial function is obtained by evaluating (5.7) for $G(\theta)$ corresponding to $g_R(x,y) = \delta(x - x_0, y - y_0)$. Algebraic manipulations after converting the result to rectangular coordinates yield:

$$\hat{g}_R(x,y) = (2/\lambda) \int_0^{2\pi} \exp\{-j2\pi[2/\lambda \sin\theta(x - x_0) + 2/\lambda \cos\theta(y - y_0)]\}\, d\theta \tag{5.13}$$

Equation (5.13) indicates that the reconstruction is formed by a spectrum of planar wave components each with spatial frequency of $2/\lambda$, directed over the complete range of angles from 0 to 2π. The reconstruction can be visualized as a superposition of sinusoidally corrugated infinite sheets of fixed corrugation spacing $\lambda/2$ and orientations varying from 0 to 2π. The amplitude of the corrugation is defined by $|G(\theta)|$; the displacement normal to the corrugations is defined by the phase Arg$[G(\theta)]$, and the orientation is $\theta = \tan^{-1}(f_x/f_y)$. Figure 5.7 shows the real part of an elemental function of the reconstruction process. The superposition of such elemental functions, represented in (5.13), yields a central peak whose width is less than one-half the corrugation spacing, $\lambda/4$, with the peak occurring at the point where the corrugations add in phase.

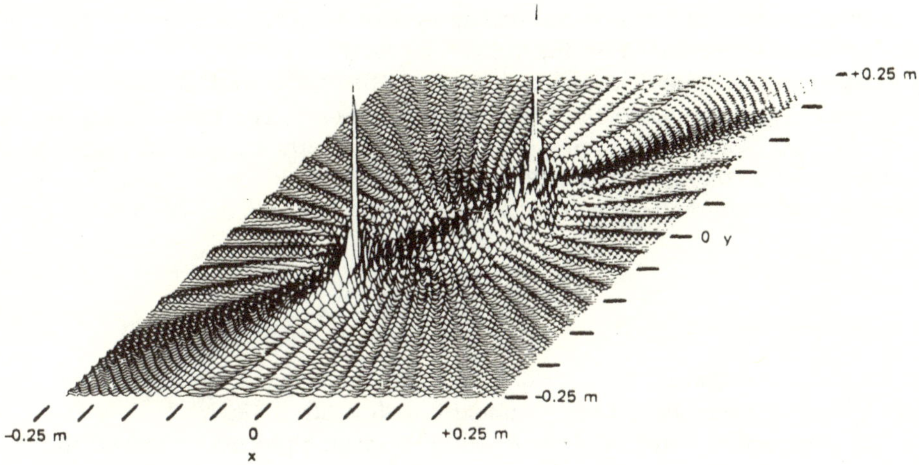

Figure 5.6 Intensity of the image reconstructed from synthesized data for two-line objects (λ = 0.06 m).

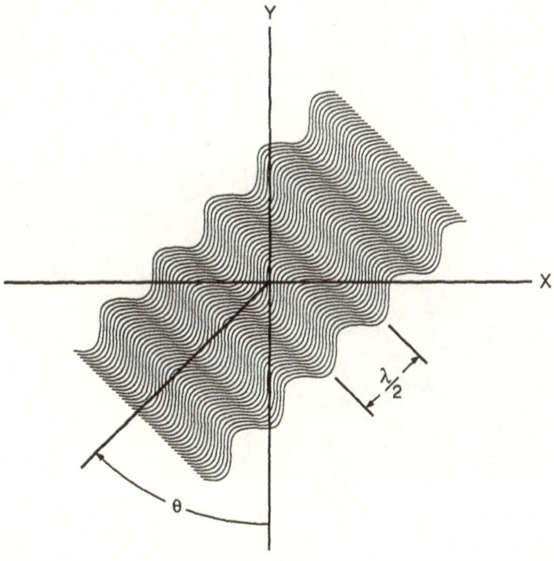

Figure 5.7 Elemental function of reconstruction process.

Point-spread functions corresponding to partial circular apertures of 180°, 90°, 60°, 30°, and 10° are shown in Figures 5.8 through 5.12. The plots were generated by carrying out a numerical integration of (5.10) over a limited range of θ. The range axis (y) is normal to the center of the aperture. The point-spread function broadens most rapidly in the range direction as the aperture angle decreases. As the angular extent of the aperture decreases, the widths of the point-spread function in the range and cross-range directions approach the limits derived in Chapter 4:

$$\Delta x \approx \frac{\lambda}{2\Theta}; \quad \Delta y \approx \frac{\lambda}{\Theta^2} \tag{5.14}$$

As the angular extent of the aperture increases to surround the object, the point-spread function becomes radially symmetric with width 0.4λ.

The operations required to implement the imaging process consist of obtaining received data samples corresponding to uniformly spaced positions on a circular ring in the spatial frequency domain and performing a two-dimensional Fourier transform. This operation inherently focuses the aperture throughout the object field and produces a point-spread function which is diffraction-limited and shift-invariant.

The reconstructed image is a good approximation to the reflectivity density of the objects considered in view of the very limited sampling of the spatial spectrum. The relatively large sidelobes in the point-spread function result from the discontinuous nature of the annulus in the spectral domain.

Although the imaging process can be analyzed from a spatial viewpoint, additional insight is obtained by considering the process from a spatial frequency viewpoint. Because the object image and its spatial spectrum are related by a Fourier

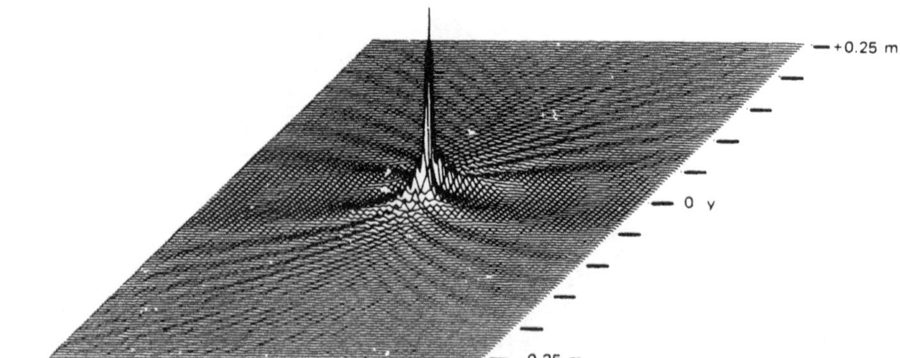

Figure 5.8 Point-spread function for 180-degree partial aperture (λ = 0.03 m).

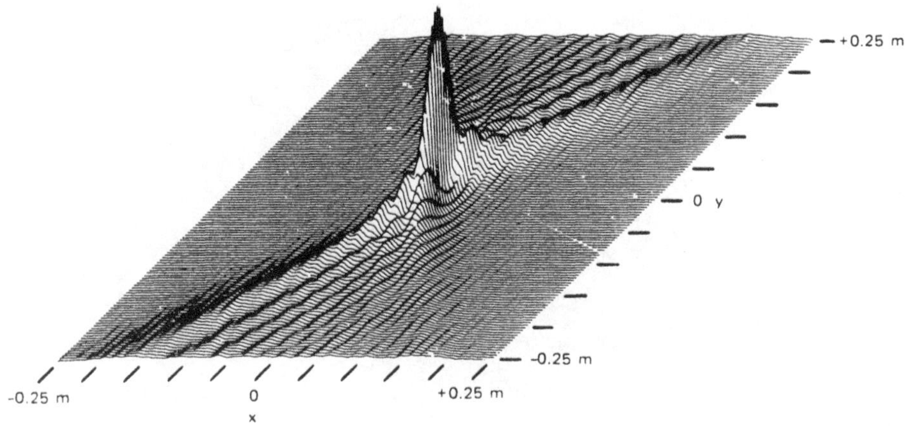

Figure 5.9 Point-spread function for 90-degree partial aperture ($\lambda = 0.03$ m).

transform, a description of the spectrum over the entire frequency plane is equivalent to complete knowledge of the object in the spatial domain. Such knowledge, therefore, would provide an exact reconstruction. The data obtained from object rotation using a single wavelength constitute samples of the spectrum along a circular arc of radius $2/\lambda$, where the angular extent of the arc is equal to the angular rotation of the object. Figure 5.13 illustrates the spatial frequency domain representation. For small angular apertures, the circular arc is suitably approximated by a straight-line segment, and the spatial response is adequately approximated by a one-dimensional

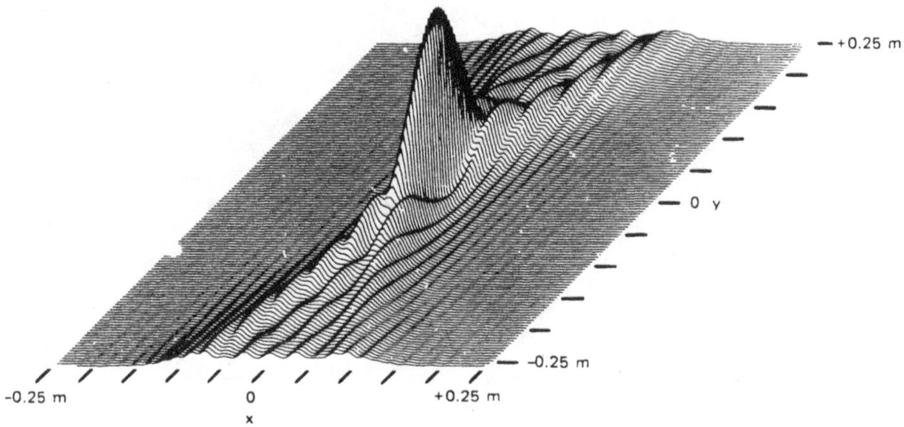

Figure 5.10 Point-spread function for 60-degree partial aperture ($\lambda = 0.03$ m).

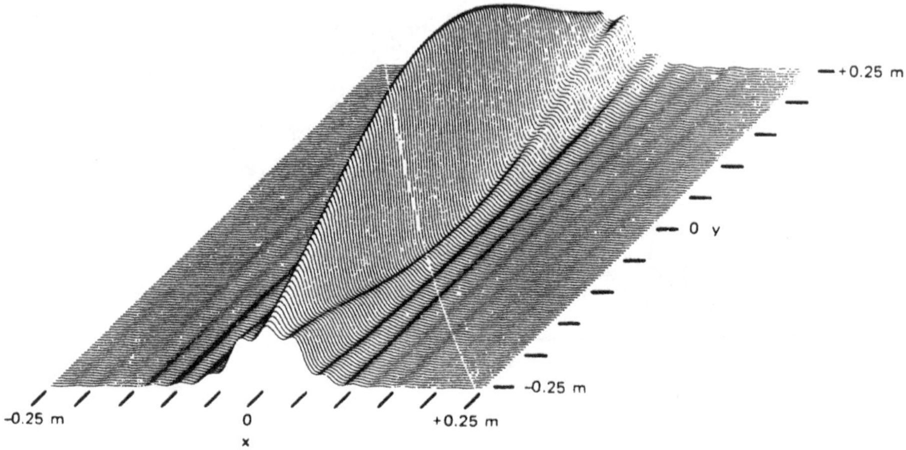

Figure 5.11 Point-spread function for 30-degree partial aperture (λ = 0.03 m).

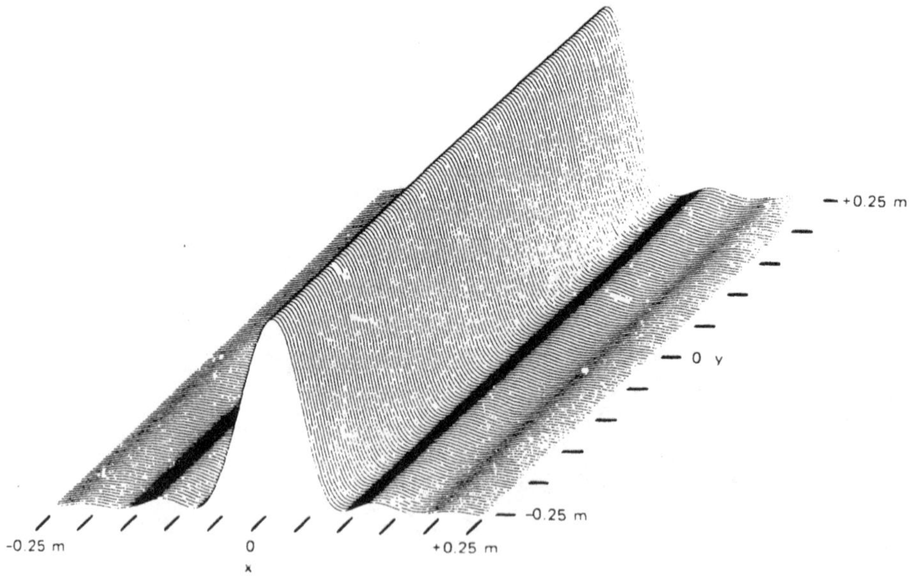

Figure 5.12 Point-spread function for 10-degree partial aperture (λ = 0.03 m).

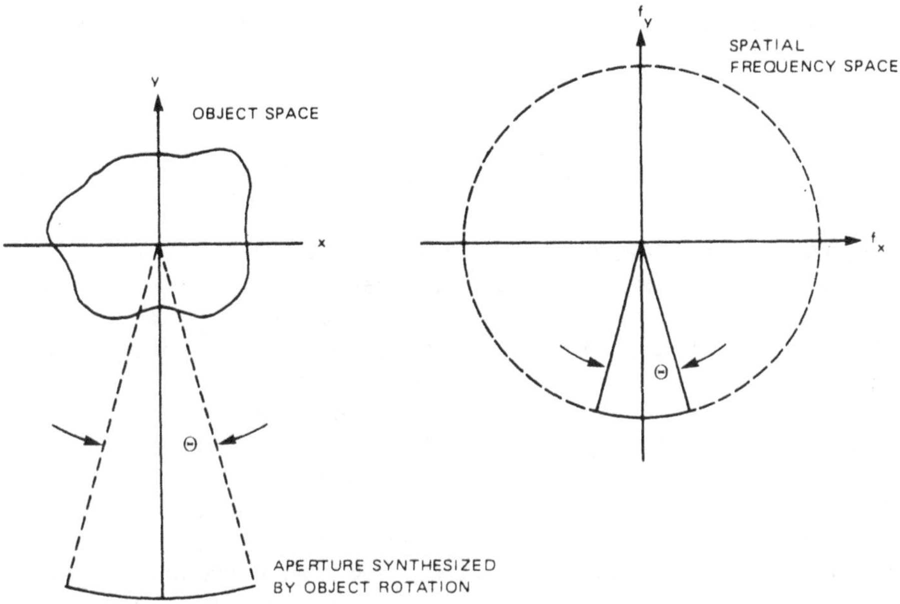

Figure 5.13 Spatial frequency sampling obtained from object rotation.

Fourier transform of the observed data. From this consideration, the nature of the approximation incurred by using a one-dimensional Fourier transform for synthetic aperture processing can be visualized. Unfocused synthetic aperture processing is equivalent to approximating a circular segment of the spatial spectrum as a linear segment. The processing required to focus the aperture is the evaluation of the two-dimensional Fourier transform along a circular arc.

The principal advantage of focused synthetic-aperture processing by using CW signals is the high degree of two-dimensional spatial resolution achievable without the use of wideband signals. In contrast, the conventional method of obtaining range resolution by time-delay sorting requires wideband signals. The disadvantages of the CW method are the high sidelobes in the point-spread function, which limit the dynamic range of the image and the degradation in image resolution if object points cannot be viewed over large angles ($\geqslant\pi$ radians). The method, therefore, allows high-resolution imaging of sparse arrays of object points which are small compared to a wavelength and are of nearly equal magnitude.

5.2 SYNTHETIC APERTURE FOCUSING FOR WIDEBAND SIGNALS

The preceding chapter explored the limits of range and cross-range resolution for wideband ISAR images obtained by processing signals recorded as a function of frequency and angular rotation. The range and cross-range resolutions achievable from data collected over a frequency band B centered on f_0 and an angular rotation Θ are

$$\Delta r = \frac{c}{2B} \quad \text{and} \quad \Delta x = \frac{c}{2f_0 \sin\Theta} = \frac{\lambda}{2 \sin\Theta} \tag{5.15}$$

The processing was implemented by a set of one-dimensional Fourier transforms applied to frequency responses to obtain range resolution, and by a subsequent set of Fourier transforms on the content of each range cell to obtain cross-range resolution. This procedure is mathematically equivalent to a two-dimensional Fourier transform of the two-dimensional array of recorded data.

For a fixed aspect angle, the Fourier transform is the exact computation necessary to obtain range resolution; however, as the object rotates, the changing range causes a migration in the apparent location of object points. As a consequence of this migration, the Fourier transform is only an approximation to the exact computation required to obtain optimum cross-range resolution. The approximation ignores the coupling between range and cross-range induced by the rotational motion of the object, and causes a blurring of the image, approximated by the extent of circular arcs described by the motion of individual scatterers in the course of the angular rotation. For a given angular rotation, the extent of the blur in the image of a point scatterer is proportional to its radial distance from the center of rotation. A point located at the center of rotation experiences no blurring, and is imaged with resolution defined by (5.15), while points offset from the center of rotation are imaged with degraded resolution. This type of error is undesirable in imaging systems because images of object points depend on their location in the target space. The latter constitutes a shift-variant system, in contrast to a shift-invariant system in which the form of the image of an object point is independent of its location in the object space. As indicated in (5.15), the rotation angle required for a given cross-range resolution is directly proportional to wavelength; consequently, image blurring is most severe when attempting high-resolution imaging of large objects at low frequencies. This section explores computational techniques for achieving theoretical limits of resolution throughout the object space.

We can gain some insight into the source and nature of focusing errors in wideband ISAR imaging by considering the response of a single point object. Figure 5.14 shows a planar object irradiated by a monochromatic signal of frequency from a distance R_0 and angle θ. The variables r and ψ locate a point on the coordinate plane fixed to the object. Assuming $R_0 \gg r$ establishes the isorange lines as parallel and normal to the line-of sight. The signal reflected from an object point is expressed by:

$$v_R(t) = R_e\{\exp(j2\pi ft)\sigma \exp[-j4\pi fc^{-1}[R_0 + r\cos(\theta - \psi)]]\} \tag{5.16}$$

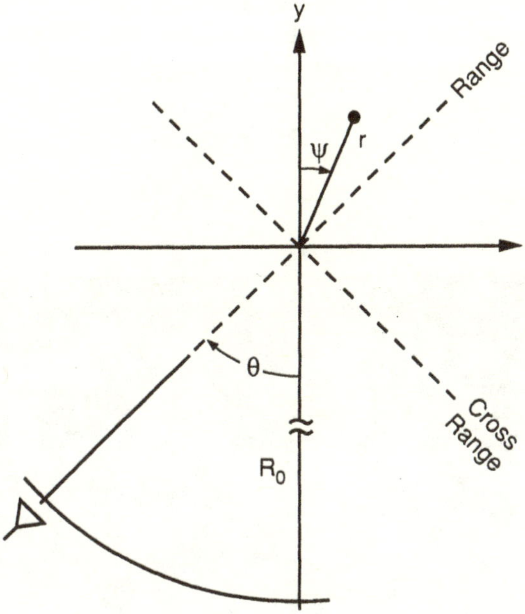

Figure 5.14 Planar object imaging geometry.

The first bracketed quantity expresses the temporal variations due to the transmitted signal carrier; the second bracketed quantity is the complex envelope, which represents the modulation imposed on the carrier by the propagation delay and object properties; σ is a complex constant, which includes the reflectivity of the object and the attenuation due to the propagation distance. The term $\exp(-j4\pi fR_0/c)$ in the complex envelope expression is a constant phase factor, which plays no significant role; for convenience, the phase term can be included in σ. The remaining term determines the phase variations in the received signal, and is the significant part of the complex envelope, given by

$$G(f,\theta) = \exp[j4\pi frc^{-1}\cos(\theta - \psi)] \tag{5.17}$$

The general behavior of the complex envelope can be shown as a function of f and θ by displaying the isophase lines, which satisfy the equation:

$$2\pi n = 4\pi frc^{-1}\cos(\theta - \psi) \tag{5.18}$$

Integer values of n map loci of points in the f-θ plane, for which the phase is uniform and equal to $2\pi n$. Figure 5.15 shows plots of (5.18) for r = 1.5 m, n ranging from

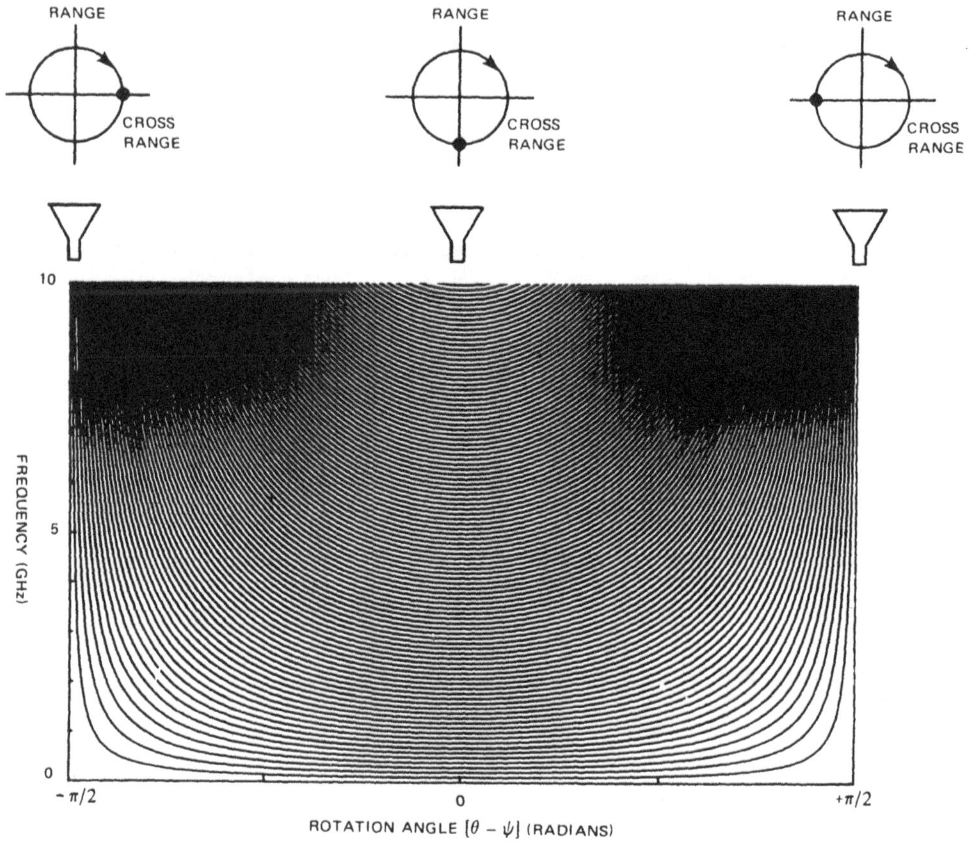

Figure 5.15 Isophase contours of complex envelope for rotating point-object.

0 to 100, f from 0 to 10 GHz, and $(\theta - \psi)$ from $-\pi/2$ to $+\pi/2$. The isophase lines indicate the cyclic variations of the complex signal envelope as a function of f and θ. Examination of Figure 5.15 indicates that the local spacing of the isophase lines in the vertical (f) and horizontal (θ) directions are inversely proportional to the range and cross-range coordinates of the object, respectively. This can be verified by differentiating (5.18):

$$\left.\frac{dn}{df}\right|_{\theta=\text{constant}} = 2c^{-1}[r\cos(\theta - \psi)] \tag{5.19}$$

$$\left.\frac{dn}{d\theta}\right|_{f=\text{constant}} = 2fc^{-1}[r\sin(\theta - \psi)] \tag{5.20}$$

In Figure 5.14, the bracketed quantities in (5.19) and (5.20) are the coordinates of the object point measured along, and normal to the line-of-sight; these distances are the range and cross-range coordinates of the point relative to the sensor. For a fixed angle, the spacing of the isophase lines is uniform, and varies inversely with range; for a fixed temporal frequency, the spacing of the isophase lines varies inversely with cross range.

Over limited regions of the f-θ plane, the isophase lines are approximately linear and uniformly spaced; over such regions, therefore, the variations in the complex envelope are uniquely described by the horizontally and vertically directed spatial frequency components of the isophase line plot. A two-dimensional Fourier transform of the complex signal over the limited region will produce an image of the point properly located in the range, cross-range space with the resolution determined by the size of the f-θ region. For an object consisting of an array of point reflectors, the f-θ map of the complex envelope is a superposition of responses for single reflectors, properly shifted in position and amplitude in accordance with the location and magnitude of each point. The superposition properties of the Fourier transform provide the decomposition required to image individual points. If the processing is limited to a restricted region, the two-dimensional imaging is represented by a scaled Fourier transform pair between (x,y) and (f,θ), as illustrated symbolically in Figure 5.16. Fourier transforming data from a larger region in order to increase resolution results in degraded imagery, as shown in Chapter 4.

The method described here has been used to process microwave data optically [5], but imaging can be performed by processing the sampled received signal using a computer [6]. If optical processing is used, the recording of the fringe pattern represented by Figure 5.15 is, by necessity, real and the Fourier transform will exhibit conjugate images and background illumination due to undiffracted light. If the processing is performed digitally on a record of the complex envelope, a single image is obtained.

Early workers in optical processing of ISAR data recognized [7] that resolution could be improved by using lenses as matched filters to compensate for local curvature of the fringes in the f-θ plane. Following this notion, we can note from (5.18)

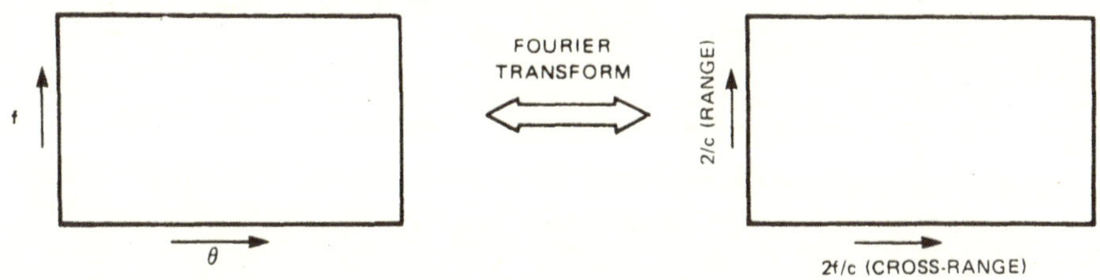

Figure 5.16 Fourier transform relation between complex envelope and object point coordinates.

that the lines of constant phase correspond to a constant product $f\cos(\theta - \psi)$ which, by expanding in a trigonometric identity, can be written as:

$$f\cos(\theta - \psi) = f\sin\theta\,\sin\psi + f\cos\theta\,\cos\psi \tag{5.21}$$

Equation (5.21) represents a constant expressed in terms of a set of mutually orthogonal coordinates, $f\sin\theta$ and $f\cos\theta$, which have been rotated through an angle θ. It is thus possible to determine a transformation which maps the curved lines of Figure 5.15 into a set of straight parallel lines. The conformal mapping of $G(f,\theta)$ into a new function, $G'(X,Y)$, is accomplished by the following relations, illustrated in Figure 5.17:

$$\begin{aligned} X &= f\sin\theta \\ Y &= f\cos\theta \end{aligned} \tag{5.22}$$

The function $G(f,\theta) = \exp[j4\pi rc^{-1}\cos(\theta - \psi)]$ of (5.17), therefore, maps onto the function:

$$G'(X,Y) = \exp[j4\pi rc^{-1}(X\sin\psi + Y\cos\psi)] \tag{5.23}$$

Substituting the rectangular coordinates of the point object shown in Figure 5.14 into (5.23) yields:

$$G'(X,Y) = \exp[j4\pi c^{-1}(Xx + Yy)] \tag{5.24}$$

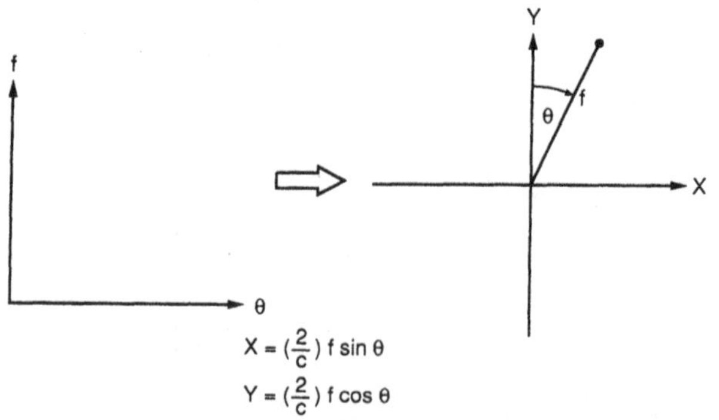

Figure 5.17 Transformation for mapping $G(f,\theta)$ into $G'(X,Y)$.

If the isophase lines of the complex envelope expressed in (5.23) and (5.24) are plotted in the X,Y space, they will consist of a set of parallel lines rotated from the Y axis by the angle ψ as shown in Figure 5.18. We can make the following observations about the fringe pattern plotted in the X-Y space:

1. The fringes for a point object are parallel lines with orientation determined by the angular position, ψ, of the point object.
2. The fringe spacing is inversely proportional to the radial coordinate, r, of the object, and is given by $c/2r$.
3. The fringe pattern for a point located on the y-axis is a set of horizontal lines; that for a point on the x-axis is a set of vertical lines. A point at the origin ($r = 0$) produces no fringes because the received signal phase is constant.
4. The radial limit of the fringe pattern is determined by the maximum temporal frequency of the irradiating signal; if the irradiating signal frequency is limited to a range $f_0 \pm B/2$, the fringe pattern will be an annular section of radial extent equal to B.

The transformation of (5.22) precisely corrects the phase curvature, and therefore constitutes the required focusing correction. Subsequent to this correction, a two-dimensional Fourier transform can be applied to all the available data. By this method, spatial frequencies can be separated to produce images of object points properly located in a two-dimensional space. The realization that the polar transformation

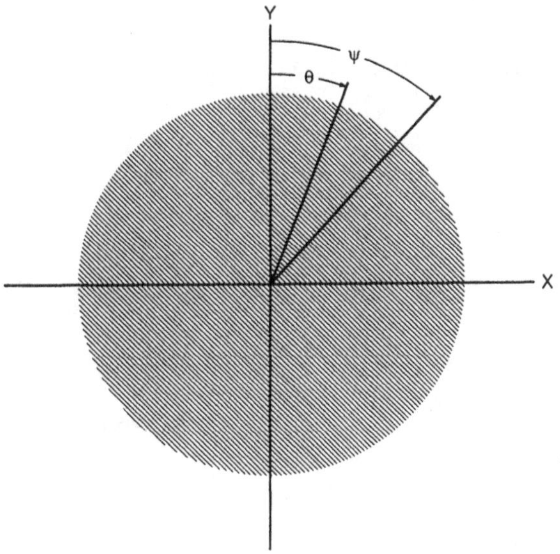

Figure 5.18 Isophase contours of complex envelope for rotating point object in transformed coordinates.

provides the necessary correction for focusing is attributed to Walker [8], who applied the technique to optically record data directly in a polar format for subsequent Fourier transformation.

The polar formatting necessary to obtain focused images can be readily derived as follows. Let $G(f,\theta)$ denote the response of the system to an object with reflectivity distribution $g(x,y)$, where y,x are the range and cross-range coordinates of object points as shown in Figure 5.19, and f,θ are the frequency and rotation angle. The measured response to an array of points is expressed by

$$G(f,\theta) = \sum_x \sum_y g(x,y) \exp[j4\pi c^{-1}(xf \sin\theta + yf \cos\theta)] \tag{5.25}$$

where the argument of the exponential represents the round-trip phase corresponding to the distance z in Figure 5.19. By defining the variables $X = 2f \sin\theta/c$ and $Y = 2f \cos\theta/c$, (5.25) can be expressed as

$$G(f,\theta) = \sum_x \sum_y g(x,y) \exp[j2\pi(Xx + Yy)] \tag{5.26}$$

The preceding is recognized as a two-dimensional Fourier transform which can be inverted to obtain an estimate of the image:

$$\hat{g}(x,y) = \sum_f \sum_\theta G(f,\theta) \exp[-j2\pi(Xx + Yy)] \tag{5.27}$$

Except for a constant, (5.26) and (5.27) constitute a two-dimensional Fourier transform pair between the variables:

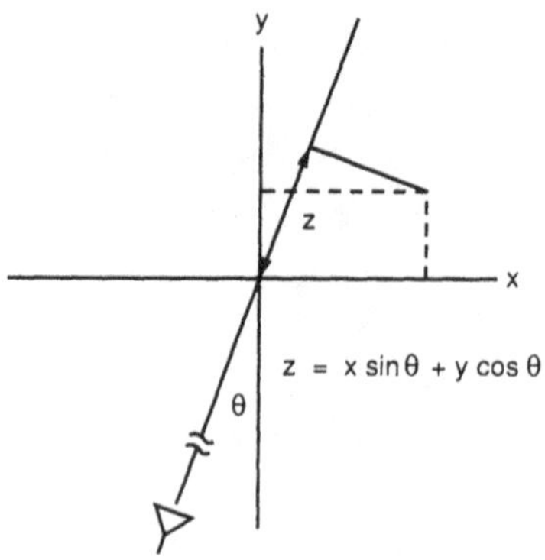

Figure 5.19 Coordinate system for rotating target measurement.

$$x \leftrightarrow X = 2f\sin\theta/c$$

$$y \leftrightarrow Y = 2f\cos\theta/c$$

The image is, therefore, obtained from a two-dimensional Fourier transform of the data sampled on uniform increments of X and Y. These samples differ from those obtained from the measurements which are disposed on uniform increments of f and θ. Processing the latter with a two-dimensional FFT implies the assumptions that: $\cos\theta \approx 1$, $\sin\theta \approx \theta$, and variations with f and θ are independent. Under such assumptions, (5.27) becomes:

$$\hat{g}'(x,y) = \sum_f \sum_\theta G(f,\theta) \exp[-j2\pi(2c^{-1}f_0\theta x + 2c^{-1}fy)] \qquad (5.28)$$

which is a Fourier transform relation between the variables:

$$x \leftrightarrow 2f_0\theta/c$$

$$y \leftrightarrow 2f/c$$

The quantity $\hat{g}(x,y)$ expressed by (5.27) represents the processing for focused imaging; $\hat{g}'(x,y)$, expressed by (5.28), represents the processing for unfocused imaging used for the examples in Chapter 4.

The upper part of Figure 5.20 shows a qualitative diagram of the response and image of a single point object. The two-sided arrow indicates the two-dimensional Fourier transform relation linking the two quantities. The response $G(f,\theta)$ consists of a complex exponential function in f and θ, represented by the curved lines which are the loci of constant phase. A two-dimensional Fourier transform of $G(f,\theta)$ produces the unfocused image; one-dimensional Fourier transforms along vertical lines (θ = constant) yield range responses for fixed rotation angles; one-dimensional Fourier transforms along horizontal lines (f = constant) yield angular responses for fixed frequencies. The curvature of the isophase lines is responsible for blurring the image. Focusing the image is accomplished by applying to the function $G(f,\theta)$ a transformation consisting of a mapping to polar coordinates, defined by $X = f\sin\theta/c$ and $Y = f\cos\theta/c$, as indicated in the lower part of Figure 5.20. The polar transformation maps a rectangular section of the f-θ plane into an annular sector in the X-Y plane. The mapping rectifies the curvature of the isophase lines so that a subsequent two-dimensional Fourier transform results in a focused image.

Practical imaging implementations use FFTs to process rows and columns of the two-dimensional data arrays. For unfocused imaging, the FFTs are performed directly on the two-dimensional data array consisting of object responses sampled in uniform increments of frequency and angle. Unfocused image processing using FFTs

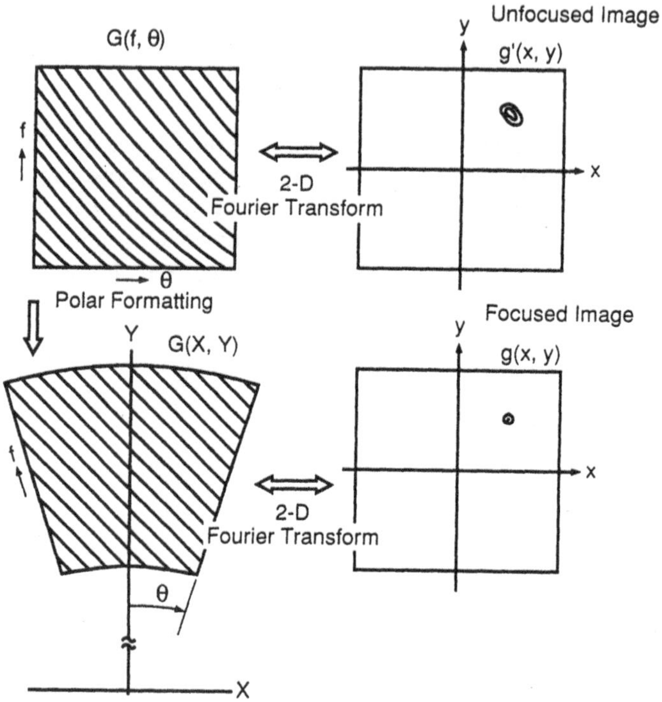

Figure 5.20 Qualitative measured responses and images of a point target.

combines samples of object responses obtained as a function of frequency while the rotation angle is held constant. These procedures uncouple the frequency-angle interdependence by assuming that range and cross-range responses depend only on frequency and angle responses, respectively. For focused imaging, the Fourier transform is performed on the two-dimensional data array after it is formatted in polar coordinates.

The requirement to compute two-dimensional Fourier transforms of data sampled on a polar raster precludes the use of the highly efficient FFT algorithms, which can only be applied to a rectangular raster. To retain the computational efficiency, the two-dimensional function can be resampled by interpolating the known samples disposed on a polar raster to obtain new samples disposed on a rectangular raster. The interpolation of the complex function can be performed by a number of numerical schemes including zero-padded FFTs, curve-fitting polynomials, or splines. While zero-padded FFTs can perform a complex interpolation and thus treat the two signal components simultaneously, numerical interpolations must be applied to each of the two signal components. A conceptual alternative to the resampling process is directly sampling the target response on uniform samples of $f\sin\theta$ and $f\cos\theta$, thus avoiding the interpolation process at the cost of increased complexity in the sampling scheme.

Figure 5.21 shows the rasters of available and required sample locations. The interpolation can be performed first along radii to obtain samples along parallel horizontal lines, and then along horizontal lines to obtain samples on a rectangular grid. For systems using continuous angular scanning rather than discrete angular steps, the initial samples are disposed along spiraling radial lines; in this case, an interpolation along lines of constant radius is first performed to resample the function to a polar raster, followed by the interpolations described previously.

Figure 5.22 shows examples of unfocused and focused images obtained from synthetic data simulating four point scatterers rotated over 18° using a center frequency of 10 GHz, a bandwidth of 3.15 GHz, and Hann windowing to achieve a maximum resolution of approximately 10 cm. The contours are 5 dB levels spanning a range of 50 dB. The focusing process reduces the significant blurring shown in the unfocused image, and demonstrates the shift-invariant nature of the focused image.

Figure 5.23 shows focused and unfocused images of a small aircraft viewed from a frontal aspect using vertical polarization, a center frequency and bandwidth of 3.4 and 2.1 GHz, respectively. The 5-dB contours span a 50-dB amplitude range. Because the rotation angle is varied over 35°, the unfocused image is subject to considerable blurring. The focused image eliminates the blurring caused by the target rotation, and produces a significant improvement in the image quality.

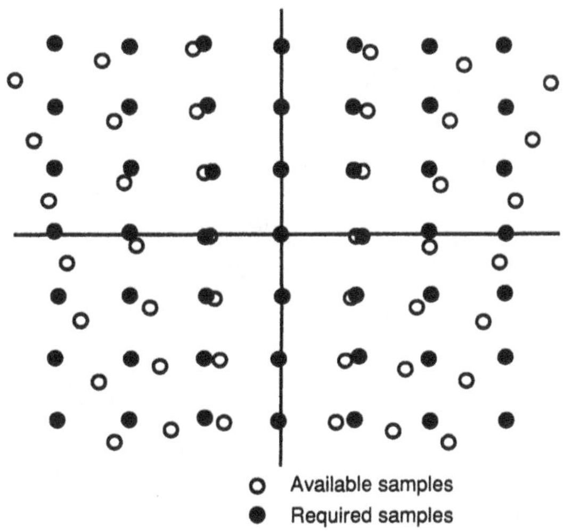

Figure 5.21 Raster of polar and rectangular sample locations for image focusing.

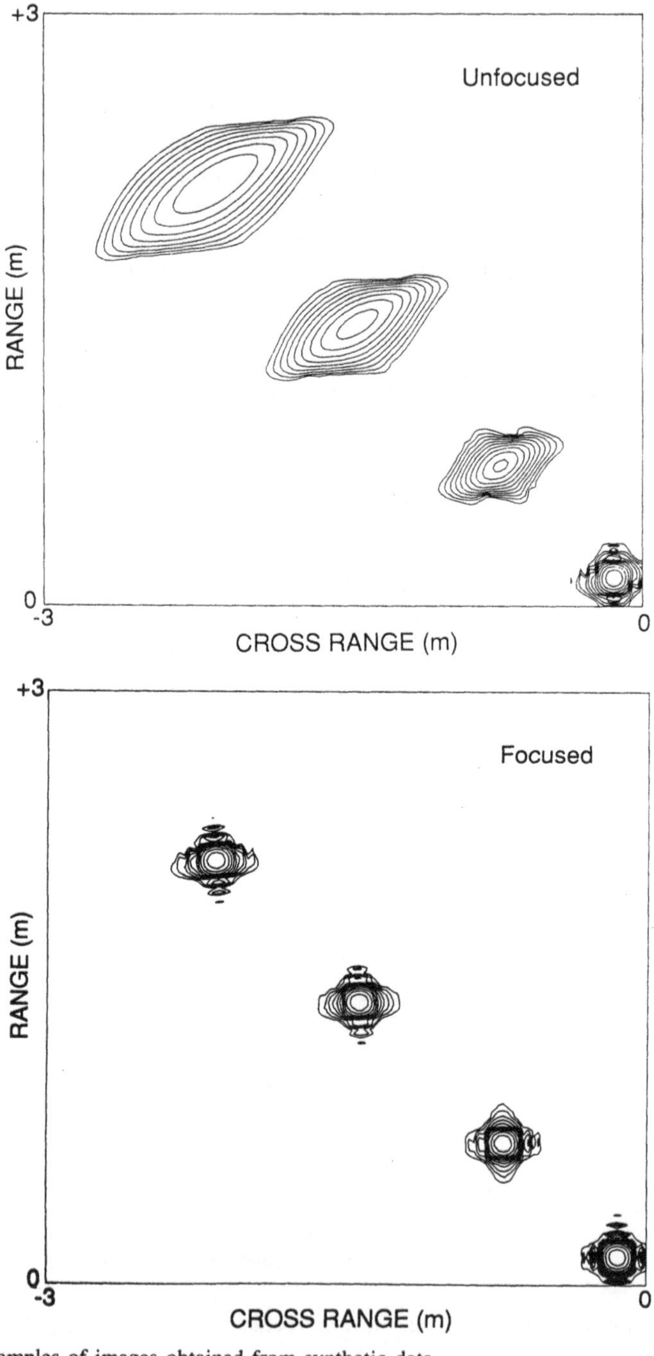

Figure 5.22 Examples of images obtained from synthetic data.

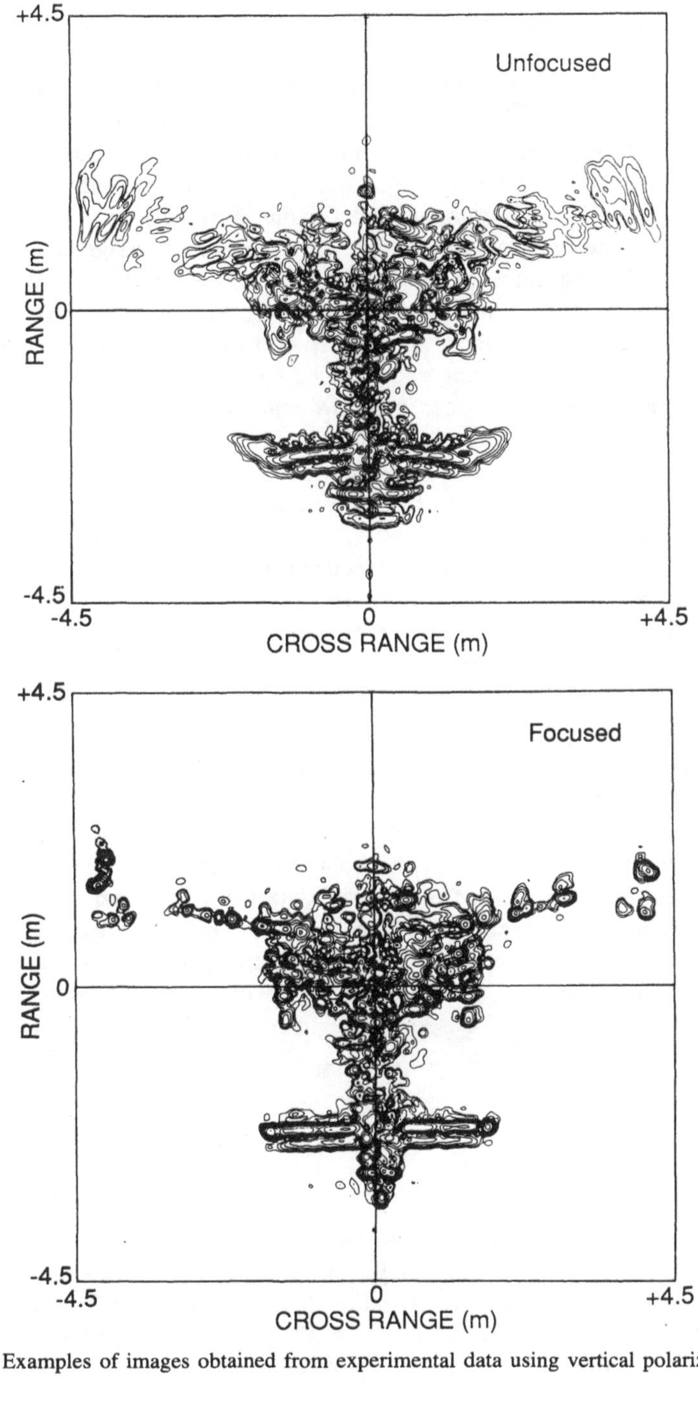

Figure 5.23 Examples of images obtained from experimental data using vertical polarization.

Figure 5.24 shows similar results for the same vehicle using horizontal polarization. The focusing sharpens the features of the image, with the exception of the scatterer on the downrange axis at the rear of the vehicle. This portion of the image, expanded in Figure 5.25, shows the scatterer blurred by the focusing process. The scatterer on the downrange axis represents a stationary echo due to the anechoic chamber. Because the clutter echo is stationary, it is imaged with zero cross range. The focusing process, which compensates for the blurring caused by target motion, assumes that all scatterers are rotating. The rotating point's image is sharpened, thus increasing its peak magnitude to its true value; the image of a stationary scatterer having a sharp peak in the unfocused image, however, is blurred by the focusing process because the scatterer lacks the presumed rotation. Consequently, stationary scatterers are imaged with blurring and reduced peak magnitude. The images of Figures 5.23 through 5.25 were processed from near-field data measured at a distance of 22 m. Under these conditions, lines of constant range and cross range describe a set of radial lines and concentric circles centered on the antenna. The resulting keystone geometric distortion has been rectified in the plotting process by the procedure described in Appendix A.

Figure 5.26 shows focused and unfocused images obtained from data simulating three rotating points and one stationary point located at maximum range. The unfocused image shows blurring in the moving scatterers, and no blurring in the stationary one, while the reverse condition applies to the focused image.

5.3 WIDE-ANGLE, WIDEBAND FOCUSED SYNTHETIC APERTURE PROCESSING

The focusing correction provided by the polar formatting described in the preceding section removes the restriction of processing data from a limited angular sector. As the angular extent of the aperture is increased, the resolution can also increase to the limit corresponding to a complete rotation of the object. This feature, explored for CW signals in Section 5.1, can be similarly developed for wideband signals. The expression for obtaining the focused image in terms of the recorded frequency or angle response, given in discrete form by (5.27), is repeated as an integral in

$$g(x,y) = \int_{-\infty}^{+\infty} \int_{-\infty}^{+\infty} G(X,Y) \exp[-j2\pi(Xx + Yy)]\, dX\, dY \tag{5.29}$$

where $G(X,Y)$ represents the measured data and $g(x,y)$ is the computed image.

When a single temporal frequency f_0 is used in a CW measurement, $G(X,Y)$ is measured only on a circular ring of radius $2f_0/c$, as shown in Figure 5.27. In this case $G(X,Y) = \delta[2c^{-1}(f - f_0)]$, where $f^2 = X^2 + Y^2$; this is precisely the result of (5.7) derived in Section 5.1. In order to perfectly reconstruct the object reflectivity

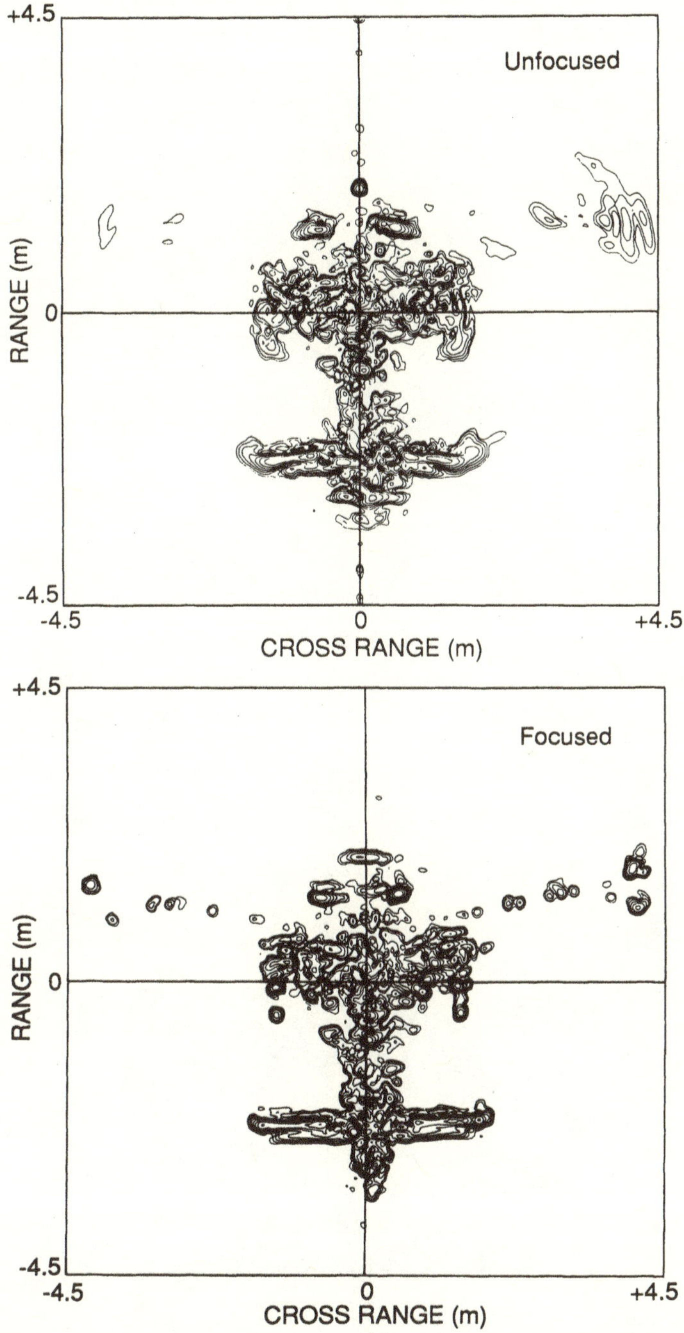

Figure 5.24 Examples of images obtained from experimental data using horizontal polarization.

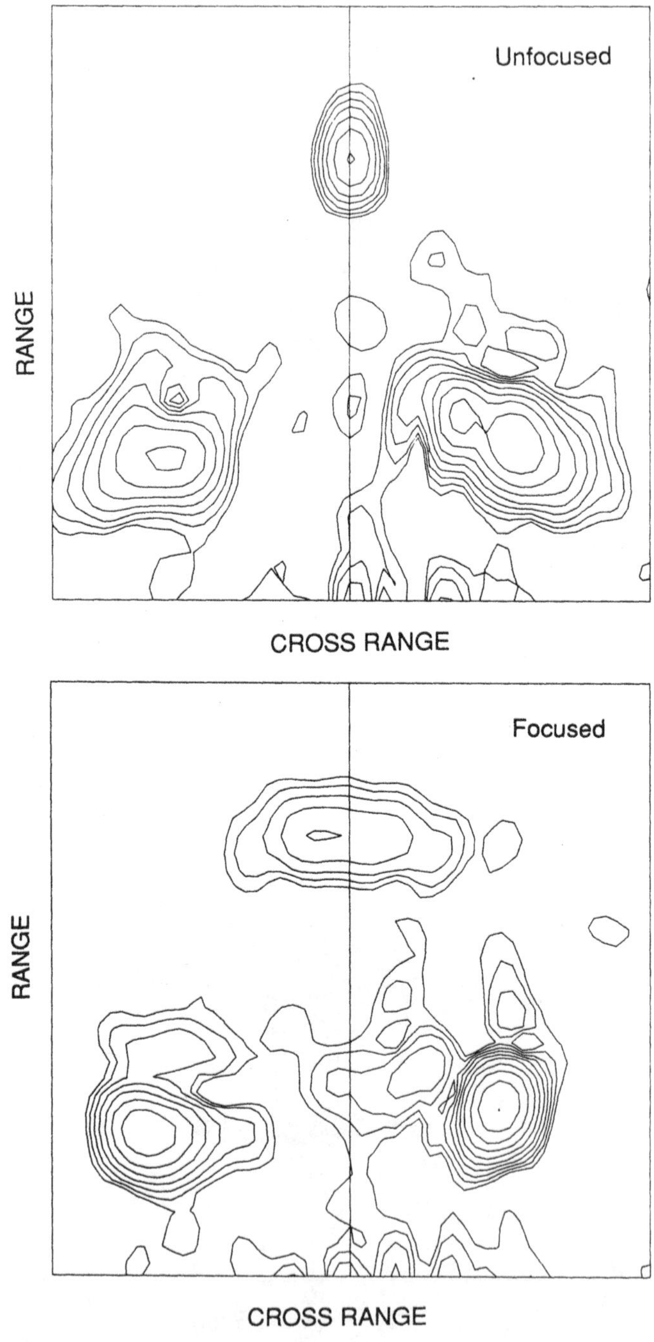

Figure 5.25 Expanded images of rear portion of Figure 5.24 (3 dB contours).

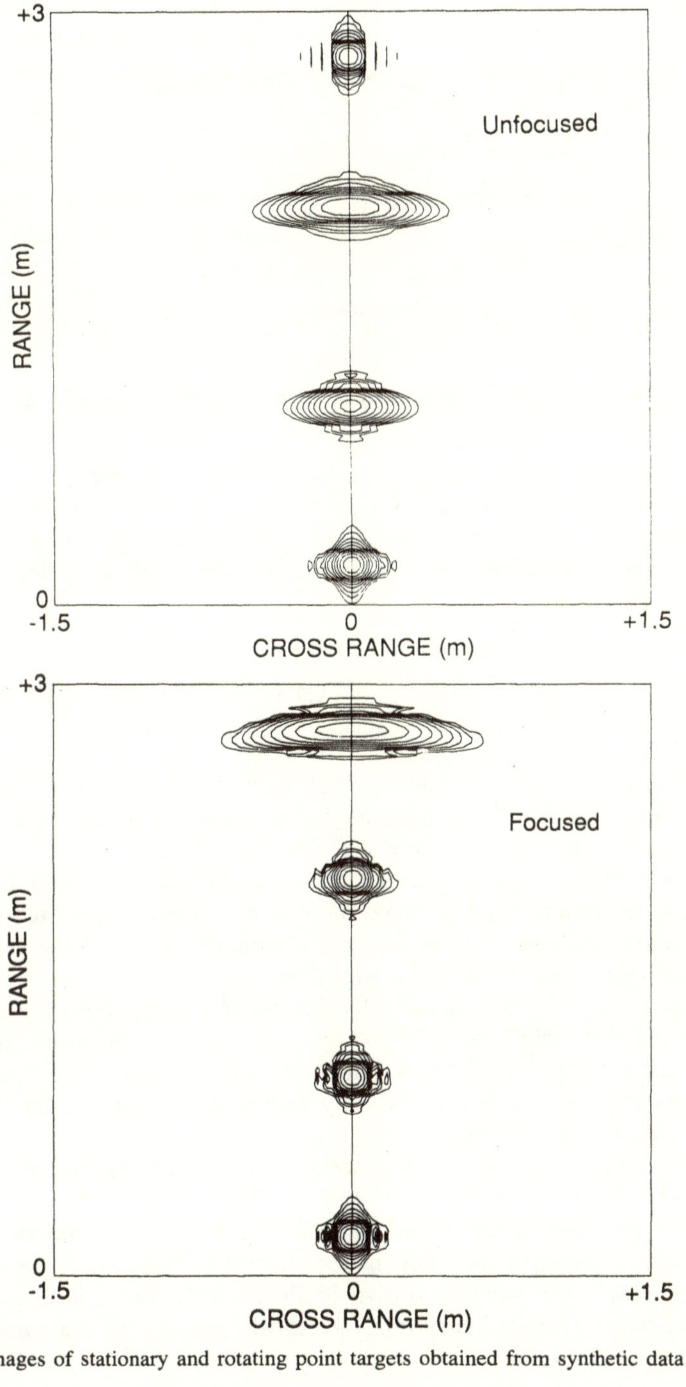

Figure 5.26 Images of stationary and rotating point targets obtained from synthetic data.

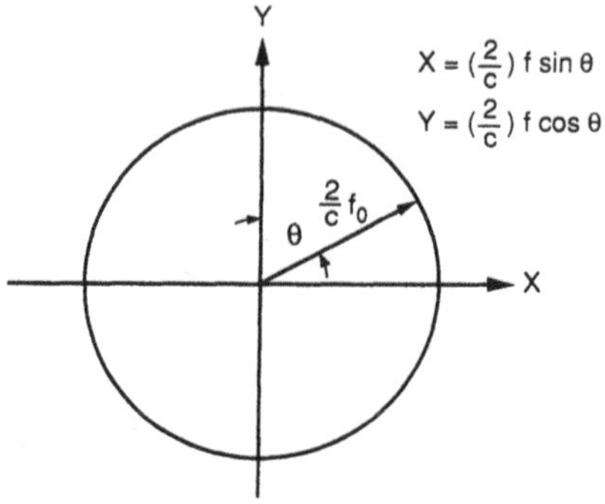

Figure 5.27 Spatial spectrum region sampled by a CW measurement at frequency f_0.

density, the spatial spectrum must be known over the entire spatial frequency plane; this corresponds to an impulse point-spread function. In all practical cases, however, the spatial spectrum is measurable only over a limited region, and the point-spread function is spatially extended. The image reconstructed from a limited region of the spectrum is

$$\hat{g}(x,y) = \int_{-\infty}^{+\infty} \int_{-\infty}^{+\infty} H(X,Y)G(X,Y) \exp[-j2\pi(Xx + Yy)] \, \mathrm{d}X \, \mathrm{d}Y \quad (5.30)$$

where $H(X,Y)$ is a window function, which accounts for the spectral bound. The point-spread function, $h(x,y)$, of the imaging process is determined by the inverse Fourier transform of the window function. Several radially symmetric functions and corresponding point-spread functions are shown in Figures 5.28 through 5.32. These are determined analytically using two-dimensional transform pairs tabulated in [9]. A notable feature of these plots is the nearly identical width of the central portion of all the point-spread functions. Evidently, the ability to discriminate between two closely spaced point objects is principally determined by the high spatial frequency components. As the radial extent of the spectral window decreases, however, the sidelobes of the point-spread functions increase considerably. The sidelobes constitute image ambiguities which are not fully removed when a limited region of the spectrum is used for reconstruction.

A heuristic interpretation of the process for reconstructing the image of a unit-amplitude point-object can be made as follows. Each point in the spatial spectrum represents a complex planar wave component, and each pair of diametrically opposed points represents a cosinusoidally corrugated sheet with orientation corresponding to that of the spectral point-pair, as shown in Figure 5.7. The spatial reconstruction corresponding to an extended spectrum, therefore, consists of the superposition of

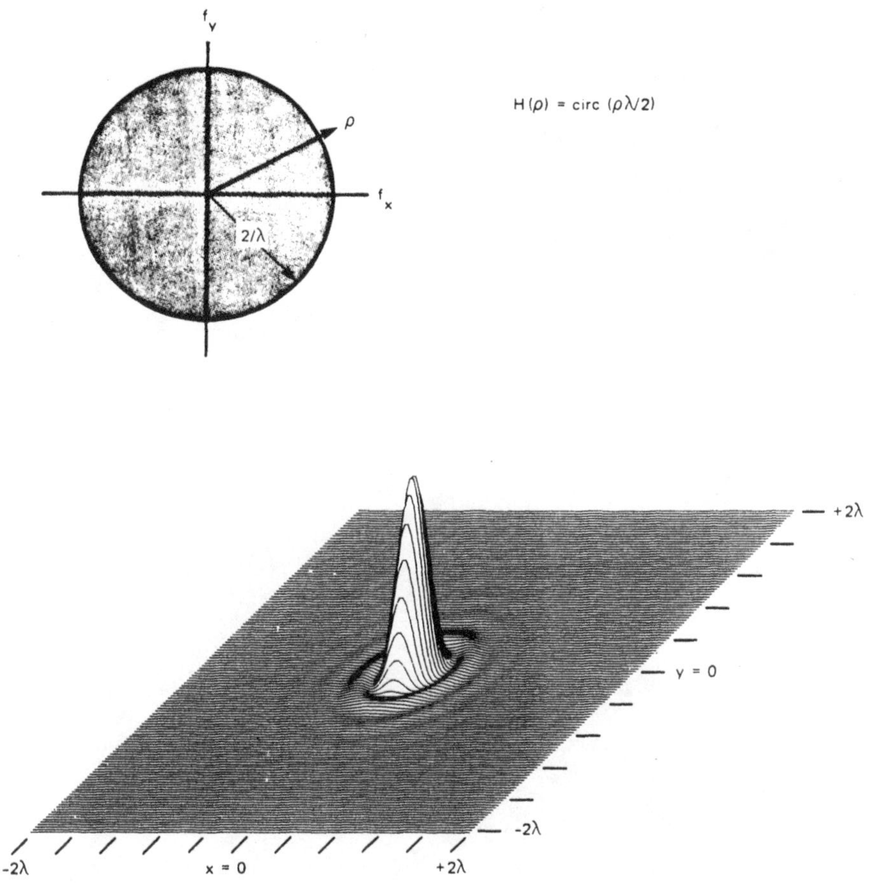

Figure 5.28 Intensity of point-spread function corresponding to a disk spectral window.

such corrugated sheets. A spectrum consisting of a circular ring, as shown in Figure 5.27, represents a continuum of sheets of equal corrugation spacing and differing orientations. Upon superposition, these corrugated sheets have a common maximum; for all other spatial locations, some cancellation occurs. The incompleteness of this cancellation gives rise to residual sidelobes. As other spectral components are included in the superposition, sheets with different corrugation spacings are included, and the cancellation is thus more complete. Finally, if all spectral components are employed, the complete cancellation results in a reconstruction that is maximum at the location of the point-object and zero elsewhere.

The presence of large sidelobes in the point-spread function is significant because they limit the ability to distinguish images of objects with small magnitude in

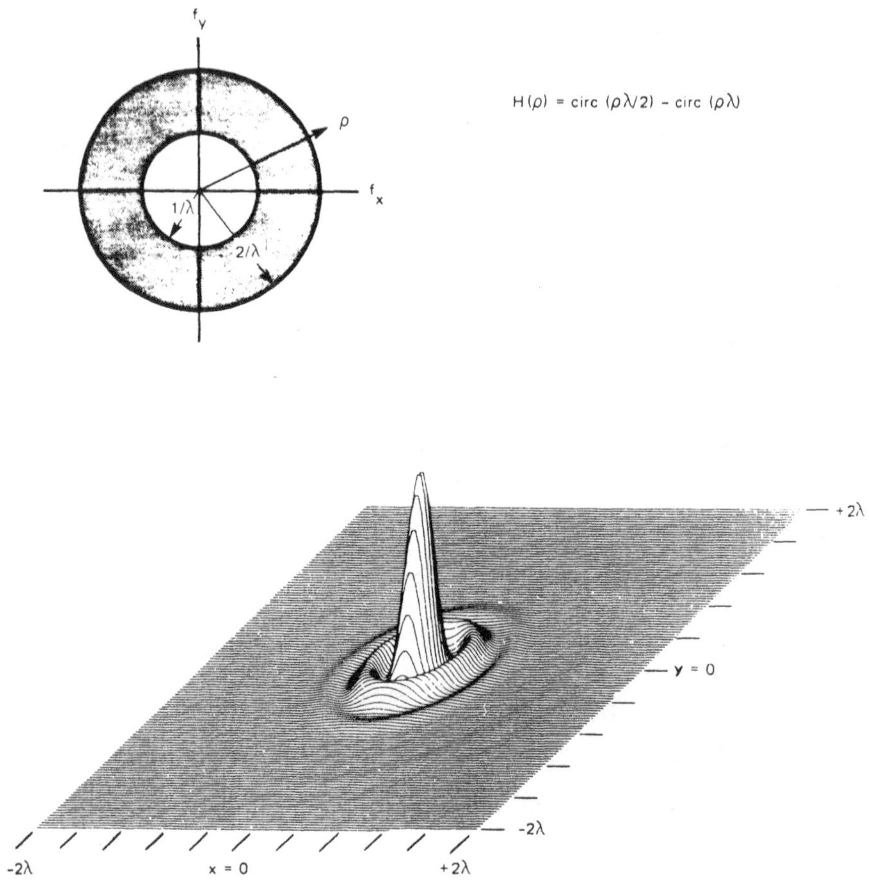

Figure 5.29 Intensity of point-spread function corresponding to a wide annulus spectral window.

proximity to those with large magnitude. The use of tapered windows may be considered for the reduction of sidelobes (analogous to the use of time windows) to reduce spectral leakage associated with finite observation intervals [10]. Figure 5.32 shows the point-spread function resulting from a radially tapered window, which is an annulus with cosinusoidal cross section. The result, observed by comparing Figure 5.29 and Figure 5.32, is that the sidelobes of the point-spread function for the tapered window are increased over those for a uniform window of equal radial extent. We will give a more complete analysis of the use of tapered two-dimensional windows applied to annular spectra for smoothing the sidelobes in the point-spread function in Appendix B.

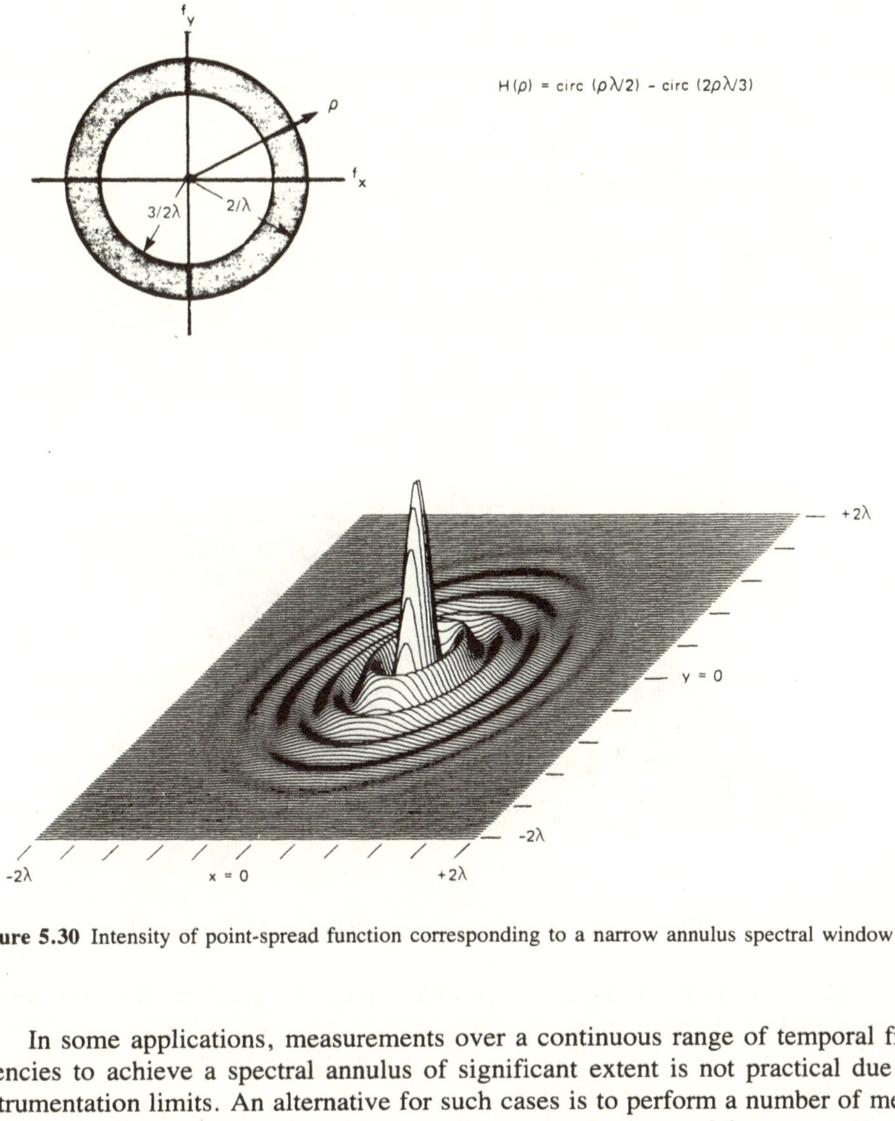

Figure 5.30 Intensity of point-spread function corresponding to a narrow annulus spectral window.

In some applications, measurements over a continuous range of temporal frequencies to achieve a spectral annulus of significant extent is not practical due to instrumentation limits. An alternative for such cases is to perform a number of measurements using discrete temporal frequencies to sample the spatial spectrum over a number of concentric rings. In the limit of diminishing ring spacing, the continuous and discrete frequency cases become identical. Of practical interest is the imaging performance achievable with relatively few (<10) discrete frequencies. The periodic sampling of the spatial spectrum over discrete rings results in image artifacts due to ambiguities associated with the discrete spectral sampling. Examples of analogous ambiguities in multifrequency acoustic holography are shown in [11].

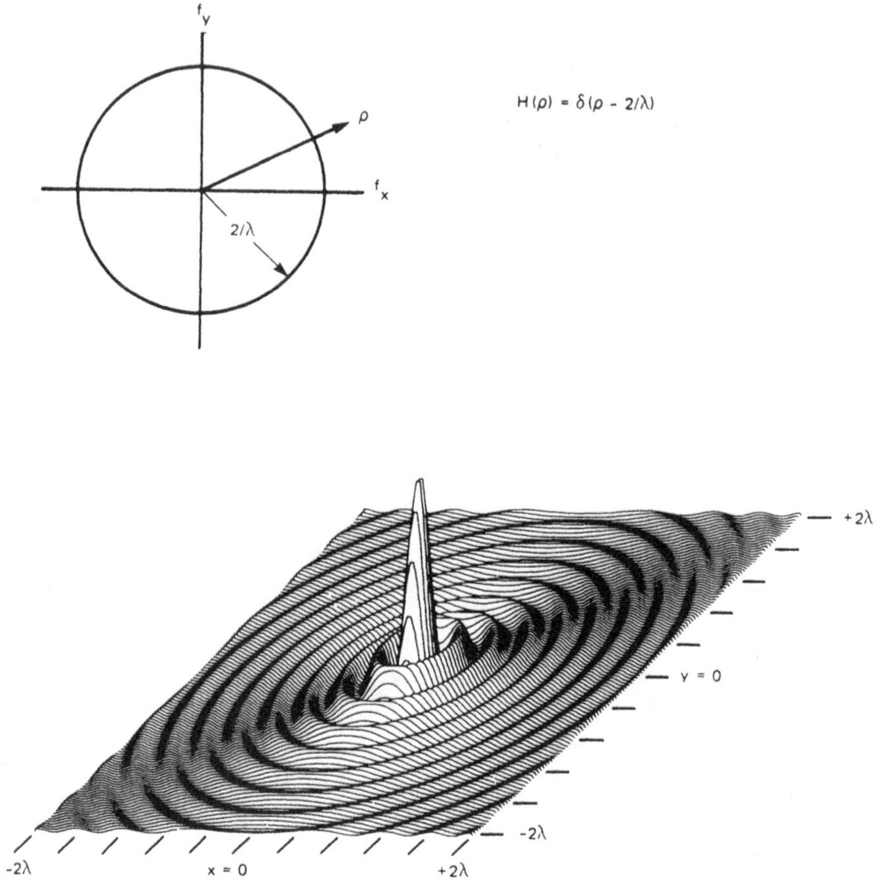

Figure 5.31 Intensity of point-spread function corresponding to a ring spectral window.

Owing to the linearity properties of the Fourier transform, the point-spread function and the image can be obtained by the superposition of responses corresponding to individual rings. Thus, if $2/\lambda_n$ is the radius of a discrete spectral ring, the point-spread function for N rings is

$$h(r) = \sum_{n=1}^{N} \frac{4\pi}{\lambda_n} J_0(4\pi r/\lambda_n) \tag{5.31}$$

Figures 5.33 and 5.34 show point-spread functions for window functions consisting of five and three spectral rings which are uniformly spaced between $1/\lambda$ and $2/\lambda$. The central part of the point-spread functions is very similar to the correspond-

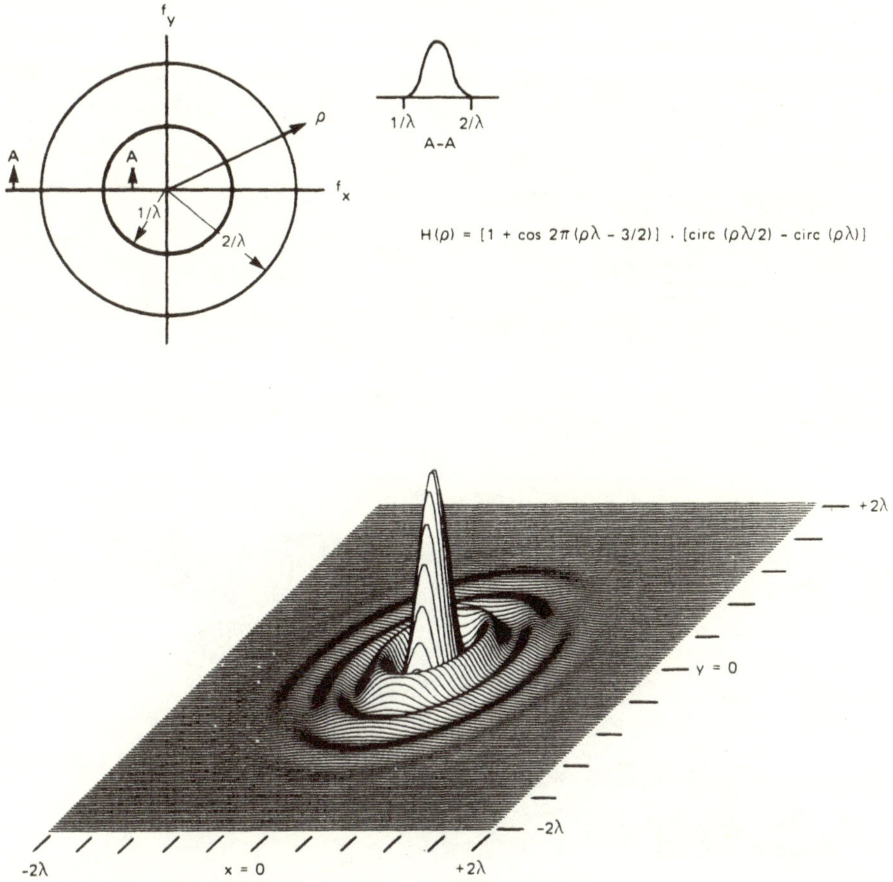

Figure 5.32 Intensity of point-spread function corresponding to tapered annulus spectral window.

ing continuous annulus of Figure 5.29. Figure 5.34, the three-ring response, shows a circular artifact with radius 2λ, which is due to periodic spectral sampling. The ambiguity artifact is not visible in Figure 5.33 because it occurs outside the plotted range.

Figures 5.35 and 5.36 are expanded quadrants of the point-spread functions that illustrate the presence of periodic circular artifacts. A uniform spectral separation between rings, $\Delta\rho$, corresponds to a spatial period in the space domain, Δr, given by

$$\Delta r = \frac{1}{\Delta\rho} \tag{5.32}$$

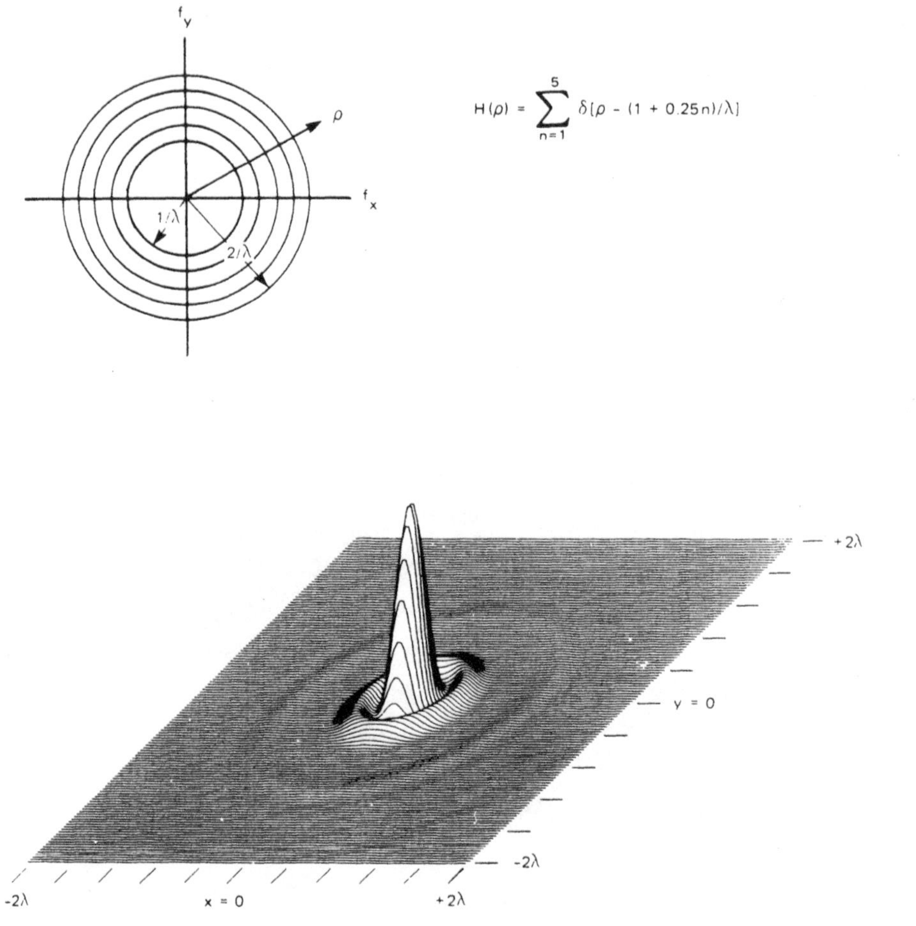

Figure 5.33 Intensity of point-spread function corresponding to a five-ring spectral window.

Substituting the coordinate relations shown in Figure 5.27 leads to the equivalent expressions:

$$\Delta r = \frac{c}{2\Delta f} \tag{5.33}$$

and

$$\frac{\Delta r}{\lambda} = \frac{f}{2\Delta f} \tag{5.34}$$

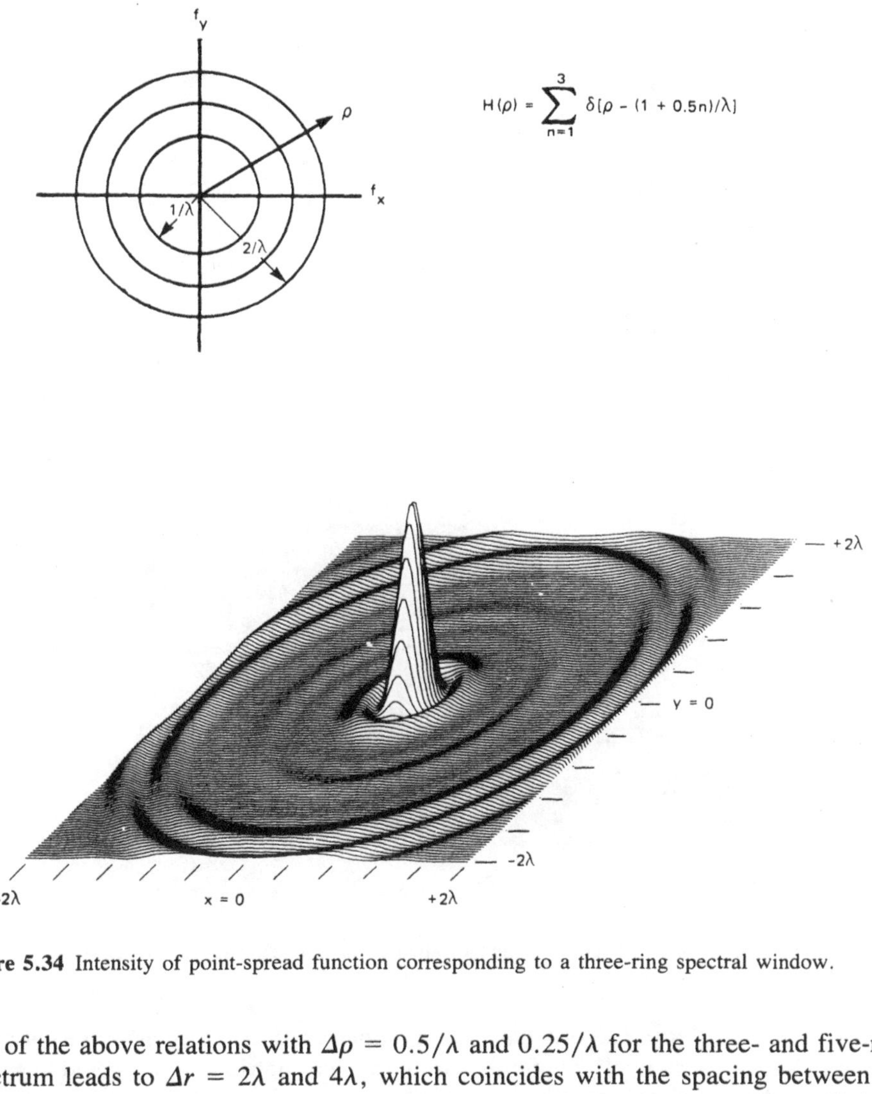

Figure 5.34 Intensity of point-spread function corresponding to a three-ring spectral window.

Use of the above relations with $\Delta\rho = 0.5/\lambda$ and $0.25/\lambda$ for the three- and five-ring spectrum leads to $\Delta r = 2\lambda$ and 4λ, which coincides with the spacing between the ambiguity artifacts in Figures 5.35 and 5.36. Note that the circular artifacts diminish in magnitude as their radius increases. This inverse relation means that close spacing of the spectral rings causes low-level artifacts at large distances from the central region of the point-spread function; of course, the small number of closely spaced rings is equivalent to a narrow spectral annulus with attendant sidelobes.

The nature of the periodic ambiguities can also be established by the following development. The point-spread function for a single spectral ring of wavelength λ is $J_0(4\pi r/\lambda)$; the point-spread function for several rings is the superposition of $J_0(\cdot)$

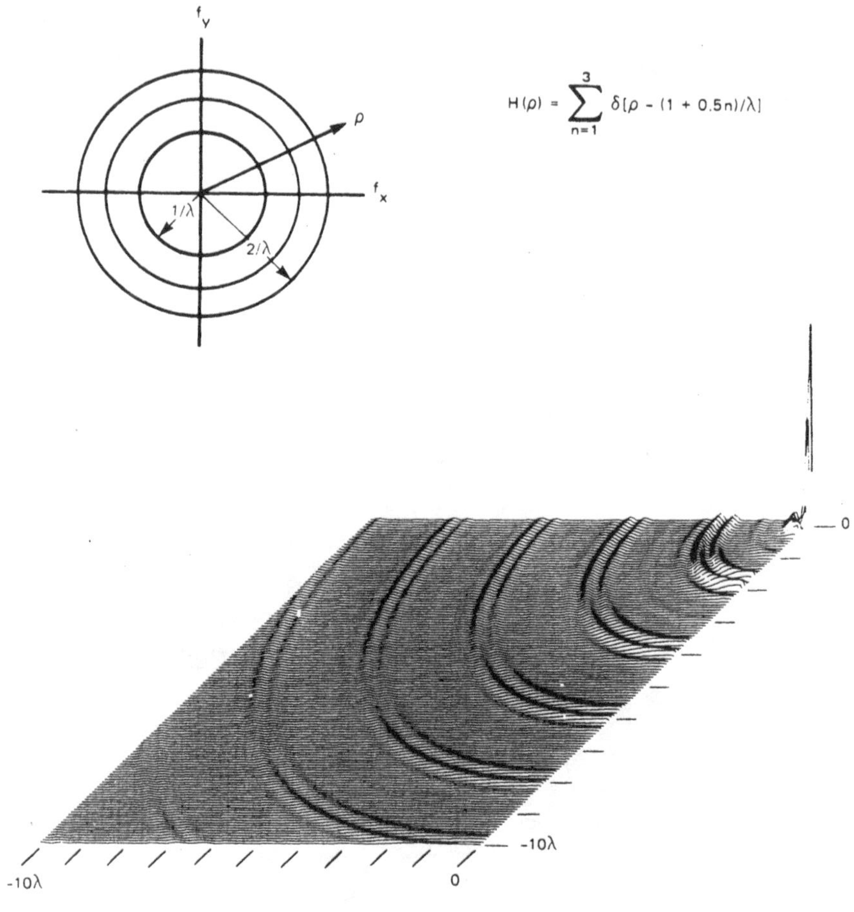

Figure 5.35 Intensity of expanded point-spread function for three-ring spectral window.

functions with differing arguments. In the superposition, the central peaks of the functions add constructively, while the circularly symmetric sidelobes tend to cancel. At some unique radial distances, the sidelobes reinforce, and the superposition increases by the same factor as the central region. Because the magnitude of the sidelobes decreases monotonically with increasing distance from center, the magnitude of the periodic ambiguities decreases similarly with increasing radius. To a good approximation, beyond the first zero, the following approximation can be made [12]:

$$J_0(4\pi r/\lambda) \cong \left(\frac{2\lambda}{\pi^2 r}\right)^{1/2} \cos(4\pi r/\lambda - \pi/4) \tag{5.35}$$

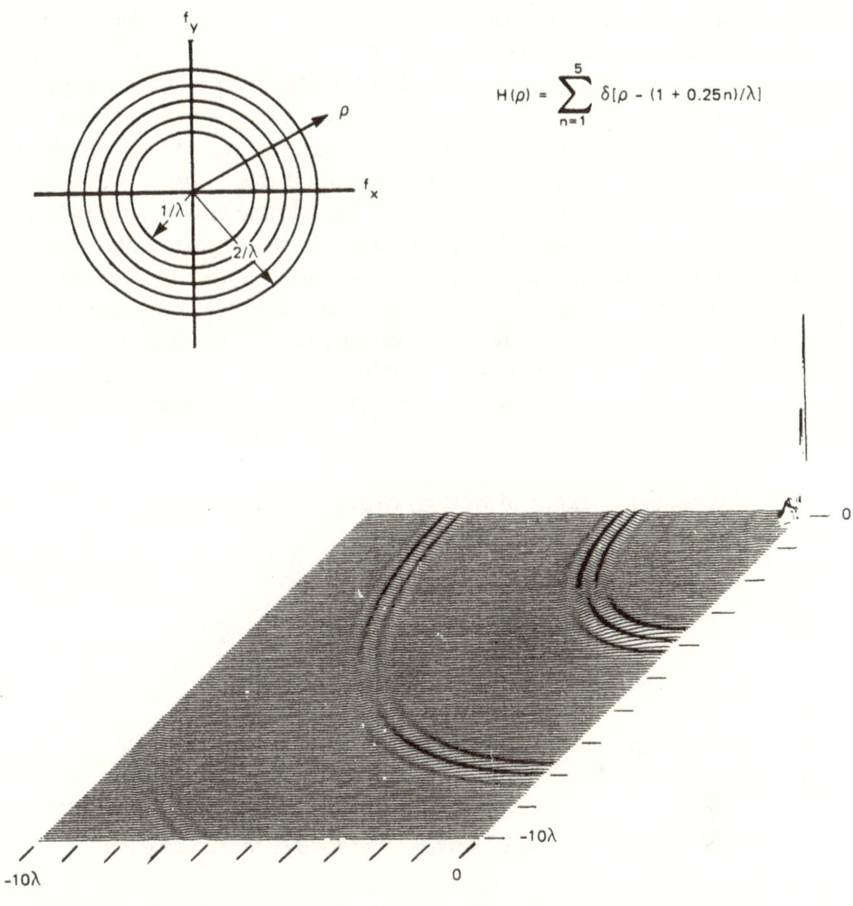

Figure 5.36 Intensity of expanded point-spread function for five-ring spectral window.

The magnitude of the oscillations decreases as $(r)^{-1/2}$; the intensity of the periodic ambiguities therefore varies inversely with r. This behavior is verified in Figures 5.35 and 5.36.

5.4 SPECTRAL SAMPLING FROM BISTATIC MEASUREMENTS

The preceding section considered image reconstruction using data from discrete spectral rings. Because the radius of a particular ring in the spatial frequency domain is $2/\lambda$, data for concentric rings are obtained from measurements at multiple temporal frequencies. In some applications, however, the available instrumentation will not permit changing the frequency of the transmitted signal. This section considers bistatic measurements as a means of sampling the spatial spectrum over discrete concentric rings while retaining the convenience of using a single irradiating frequency.

The transmitter and receiver of a bistatic radar are spatially separated, in contrast to a monostatic condition in which the two are colocated. The angle of separation measured relative to the object center is called the *bistatic angle*. Consider a point object with coordinates r, ϕ, observed by monostatic and bistatic radars located at large equal distances R_0, and angles θ relative to the object center, as shown in Figure 5.37. The elements of the bistatic radar are symmetrically displaced from the monostatic radar axis by angles $\beta/2$, where β is the bistatic angle which lies in the plane of the object. As the angle θ is varied, the phase of signals received by the radars varies due to the changing round-trip distance. The complex envelopes of signals received as a function of the angle for the monostatic and bistatic cases are, respectively,

$$G_M(\theta) = \exp(-j4\pi R_0\lambda^{-1}) \exp\{j2\pi\lambda^{-1}[2r\cos(\theta - \phi)]\} \tag{5.36}$$

and

$$\begin{aligned} G_B(\theta) = \exp(-j4\pi R_0\lambda^{-1}) \exp\{j2\pi\lambda^{-1}[&r\cos(\theta - \phi + \beta/2) \\ &+ r\cos(\theta - \phi - \beta/2)]\} \end{aligned} \tag{5.37}$$

Application of the trigonometric identity $\cos(A + B) + \cos(A - B) = 2\cos A \cos B$ to (5.36) results in

$$G_B(\theta) = \exp(-j4\pi R_0\lambda^{-1}) \exp\{j2\pi[(\cos\beta/2)\lambda^{-1}2r\cos(\theta - \phi)]\} \tag{5.38}$$

Comparison of (5.36) and (5.38) shows that the variable parts of the complex envelope have identical forms, with the exception of the factor $\cos(\beta/2)$. This indicates that results of measurements with a bistatic angle β are equivalent to results that would be obtained with a monostatic geometry using a wavelength $\lambda\sec(\beta/2)$. This fact has been recognized in radar work, and is known as the *bistatic equivalence theorem* [13]. The increase in effective wavelength resulting from the bistatic geometry allows measurements made with bistatic angles and a single temporal frequency to simulate measurements with multiple temporal frequencies.

To test the practicality of the bistatic method, experimental data from a set of measurements performed on a two-point object using different bistatic angles were used to reconstruct an image. The test object was a pair of thin vertical rods separated by 34 cm, and irradiated by a wavelength of 6 cm from a distance of approximately 5 m. Separate measurements were made using bistatic angles of 0°, 58°, 83°, 102°, and 120°. This provided five spectral rings with radii uniformly spaced between

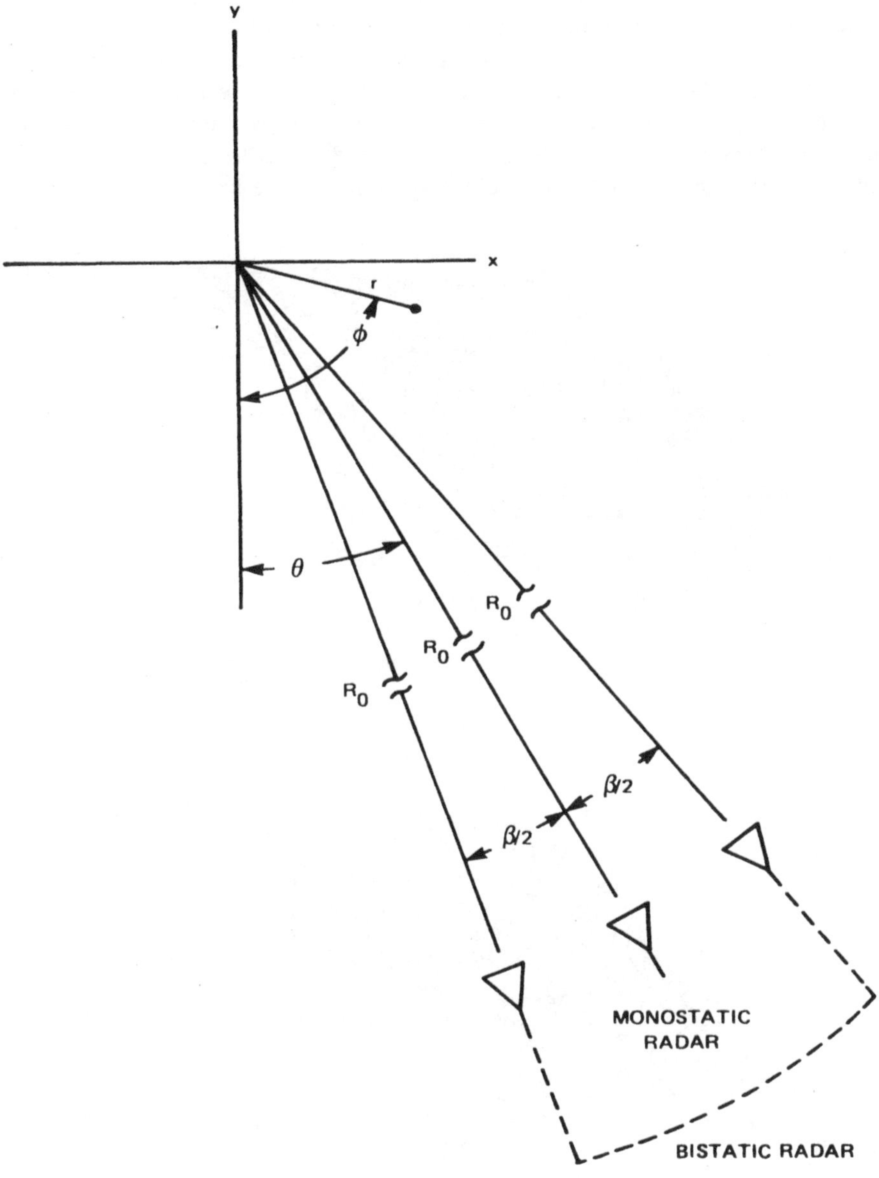

Figure 5.37 Monostatic and bistatic radar configurations.

$1/\lambda$ and $2/\lambda$, simulating the window function shown in Figure 5.33. The images reconstructed from data obtained from measurements at each bistatic angle are shown in Figure 5.38. The results illustrate the decreasing resolution as the bistatic angle is increased, and the effective wavelength is increased accordingly. The combined image, formed by superimposing the complex amplitudes of the individual images, is shown in Figure 5.39. The results show excellent correlation with the theoretical point-spread function of Figure 5.33; the magnitudes and locations of the near side-lobes and of the ring artifacts are in close agreement. The image reconstructed from

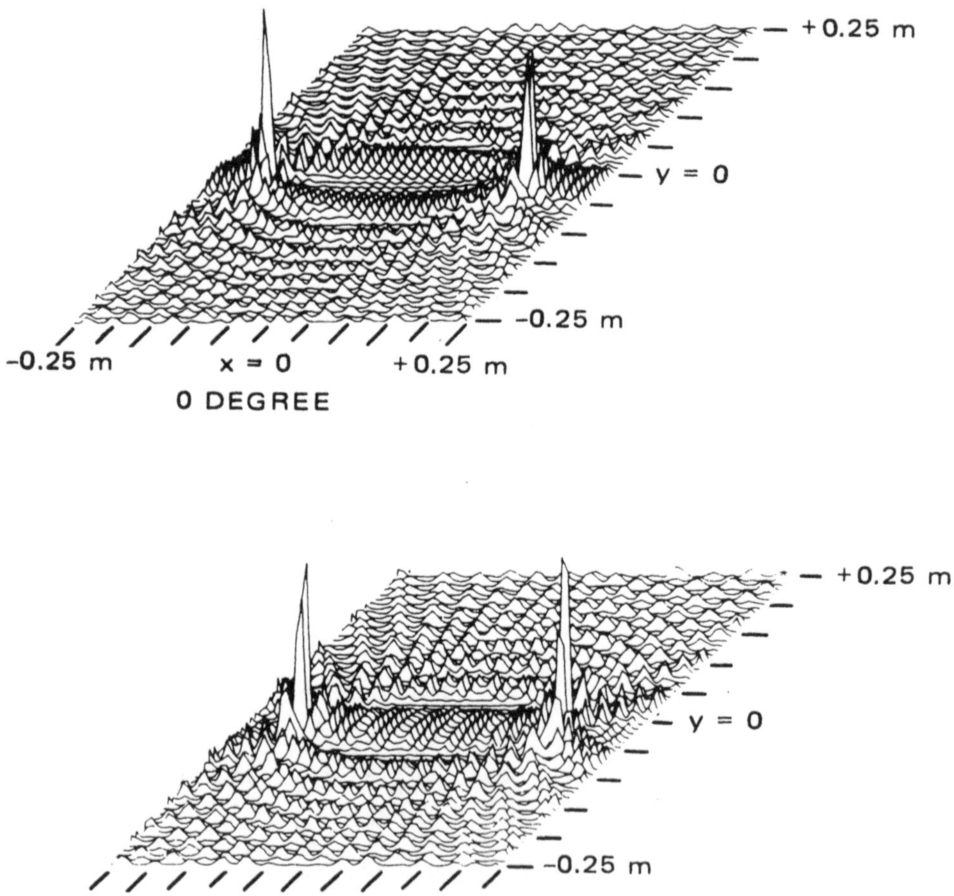

Figure 5.38 Intensity of images of two point objects reconstructed from experimental data for bistatic angles of 0, 58, 83, 102, and 120 degrees.

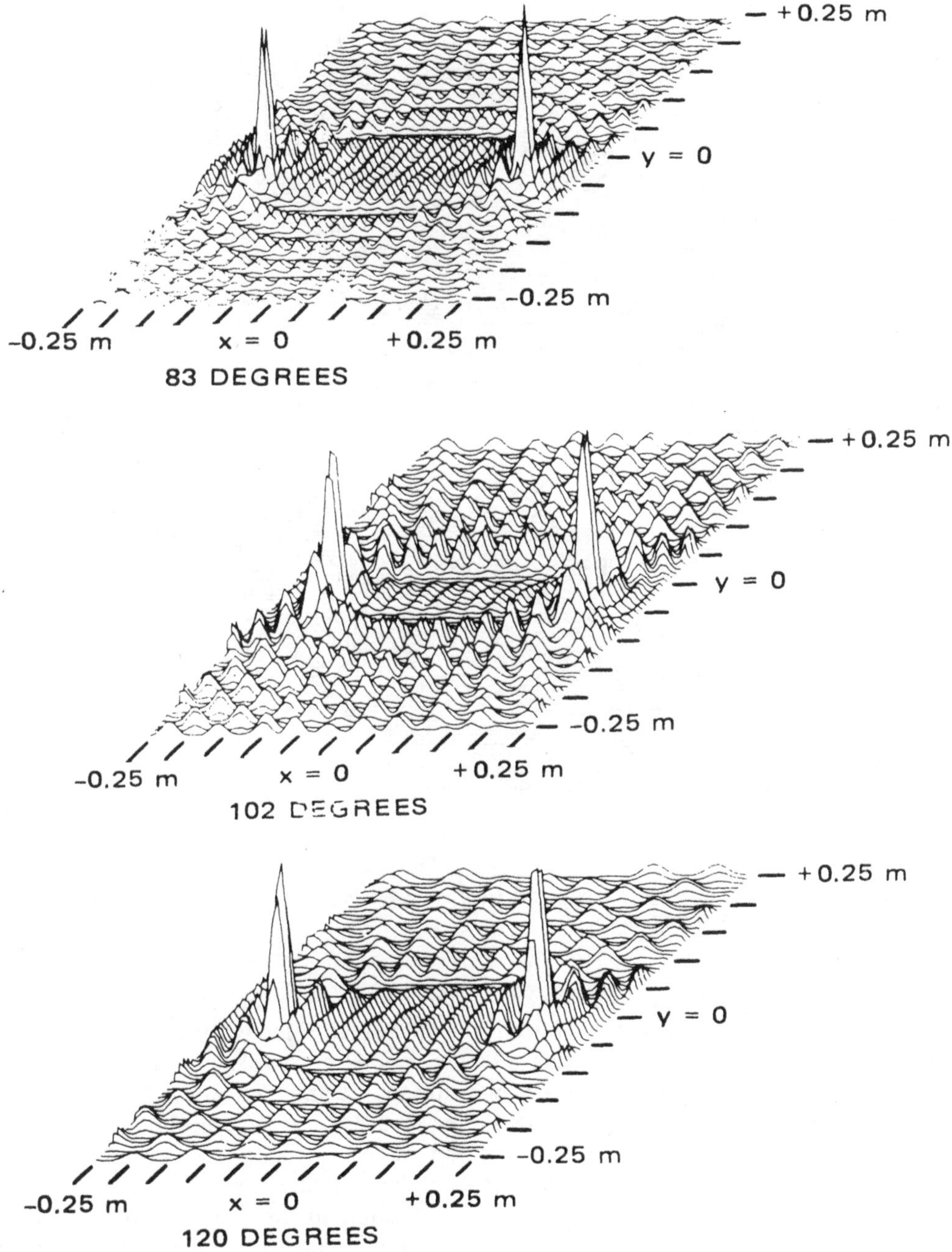

Figure 5.38 (continued).

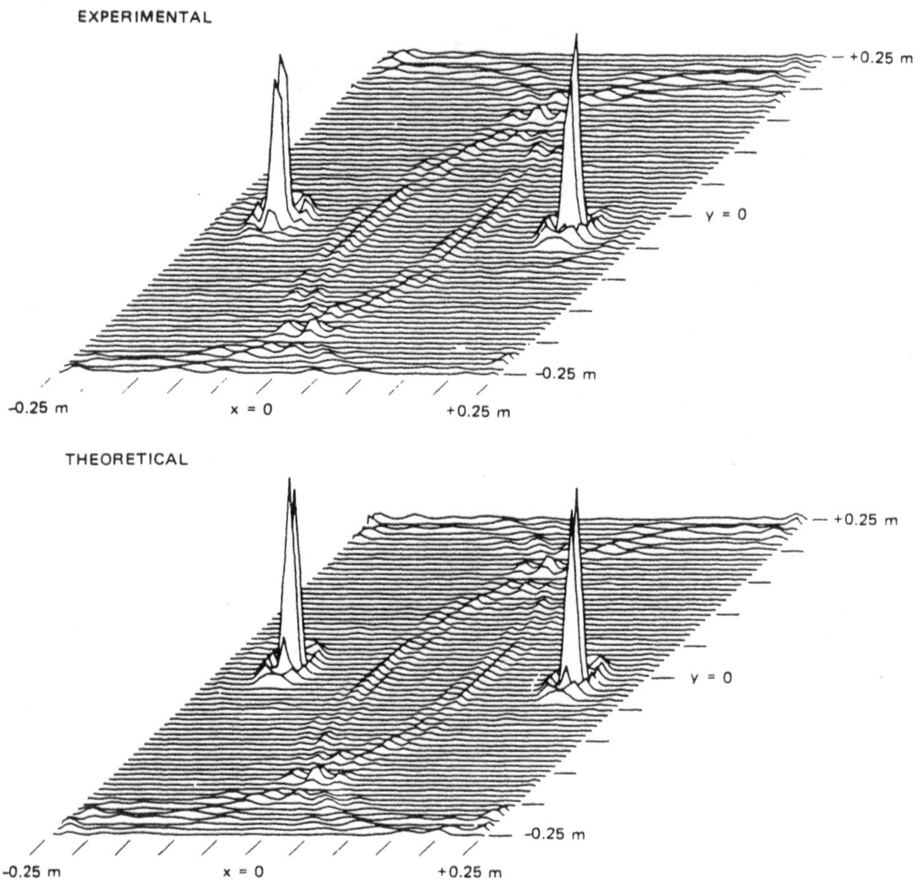

Figure 5.39 Intensity of image reconstructed from complex amplitude superposition of five bistatic images ($\lambda = 0.06$ m).

the five spectral rings shows a marked improvement over that from a single ring. The results demonstrate that the bistatic method is an effective way of simulating multifrequency imaging while using a single irradiating frequency. A different reconstruction can be obtained by superimposing the magnitudes rather than the complex amplitude of the individual images. The result of this superposition, shown in Figure 5.40, indicates a lesser improvement in the reconstruction than in the complex amplitude superposition. In this case, the sidelobe structure is averaged while in the complex amplitude superposition, some cancellation in the sidelobes occurs. We hasten to caution that the validity of the bistatic equivalence theorem requires that the amplitudes and location of the scattering centers not be altered by the bistatic

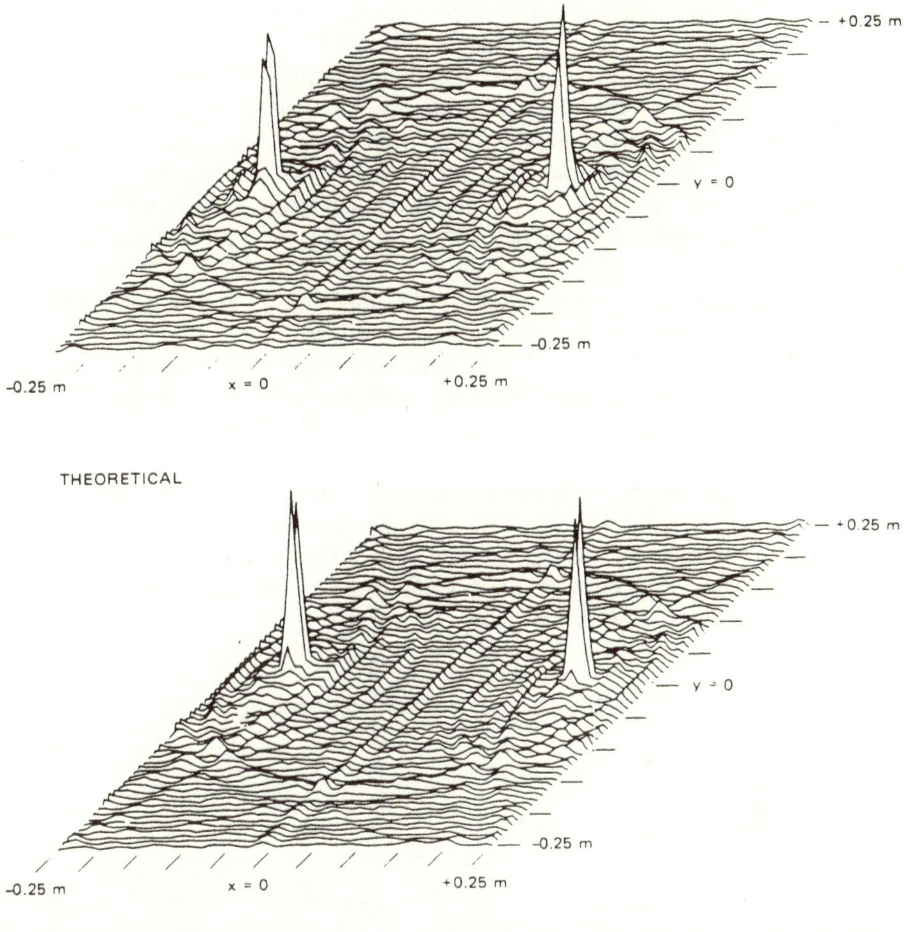

Figure 5.40 Intensity of image reconstructed from magnitude superposition of five bistatic images (λ = 0.06 m).

geometry. This situation prevails for point objects and simple shapes; however, we expect that when complex objects are observed at large bistatic angles, the bistatic equivalence may not be applicable, especially if corner reflectors are present or if the bistatic geometry alters the shadowing of some reflectors by others.

5.5 GENERALIZED FOCUSED IMAGING

We can obtain physical insight into the focusing process by interpreting the polar formatting described in the preceding sections in terms of synthetic aperture focusing (described in Chapter 3). In polar format processing, object responses measured as a function of frequency and angle provide samples of the spatial frequency spectrum disposed on a polar raster from which the focused image is obtained by a two-dimensional Fourier transform. Focused imaging can also be approached from the spatial viewpoint developed in Chapter 3, in which samples of object responses collected along the synthetic aperture are modified by prescribed sets of phase corrections and subsequently summed. The corrections are phase factors which equalize the phases of signals observed across the aperture for each location in the image space.

The two viewpoints are represented by the diagrams in Figure 5.41. The right diagram indicates data samples in the spatial frequency domain, defined by $X = 2f \sin(\theta)/c$ and $Y = 2f \cos(\theta)/c$. The left diagram indicates the summation of phase-corrected data sampled along the aperture for particular angles and frequencies. The result of the summation represents the value of a particular image pixel. The expression below the right diagram is a two-dimensional, polar Fourier transform [9] of the measured data $G(f,\theta)$; the expression below the left diagram is a summation of the same measured data modified by phase factors corresponding to the round-trip distance, z, between each element and each pixel. By applying the trigonometric identity $\sin(A + B) = \sin A \cos B + \cos A \sin B$ to the right expression, we see that it is equivalent to the left expression. This equivalence illustrates that Fourier trans-

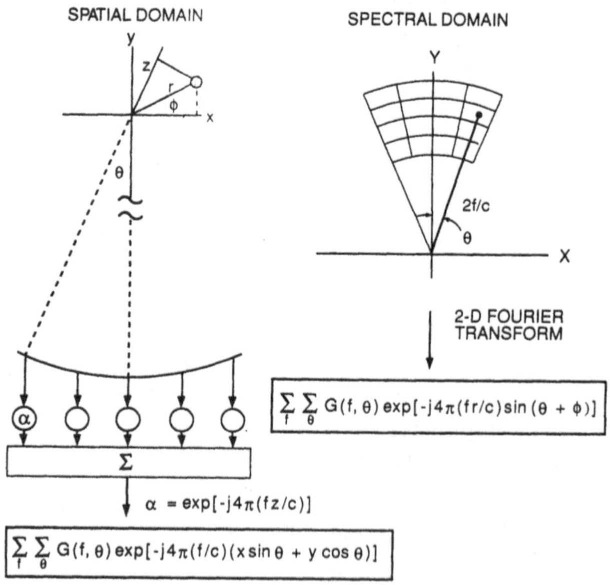

Figure 5.41 Representation of focusing process in spatial and spectral domains.

forming the polar-formatted data is identical to summing phase-corrected samples along the synthetic aperture. Evidently, the Fourier transform kernel provides the precise phase correction necessary to focus the image.

The spatial viewpoint of the focused imaging process leads directly to the following generalized formulation, applicable to all forms of synthetic aperture processing. Signals observed by imaging radars consist of the linear superposition of components associated with individual scattering centers, each exhibiting unique phase histories as a function of frequency and angle determined by their spatial relation to the radar. The task of the imaging algorithm is to decompose the total signal into elemental components while retaining a measure of their scattering strength. Mathematically, the total signal can be decomposed into a set of basis functions of unknown amplitude, each associated with the phase history of a particular scattering center. The imaging algorithm implements a set of matched filters, which are used to test the correlation between the measured data and each of the basis functions corresponding to the location of each pixel. The matched filter or correlation process yields the unknown amplitude coefficients, which then form the image. This process is analogous to the decomposition of a complex waveform into its harmonics, in which case the harmonics are the basis functions and their determined coefficients describe the amplitude associated with each of them. The mathematical decomposition is optimum if the basis functions form an orthogonal set, meaning that the integrated product of any two different basis functions is zero, while the integrated product of any function times itself is a maximum. The most widely used orthogonal set consists of the sinusoidal functions which form the basis for Fourier series expansions. A formal derivation of the least-square error (LSE) approximation to arbitrary functions by a series of orthogonal function can be found in a number of references on linear systems analysis [14–16].

Conceptually, the imaging algorithm for any spatial and frequency diversity can be structured in the preceding framework, as indicated in Figure 5.42, which shows a generalized geometry for an imaging system. The instrumentation radar provides frequency diversity by varying the irradiating frequency and spatial diversity by varying the relative orientation between object and radar. Let $g(x,y)$ be the image to be determined from the complex object response measured as a function of frequency and a spatial or angular coordinate θ. The image is computed by the expression:

$$g(x,y) = \sum_f \sum_\theta G(f,\theta)\Phi(f,\theta;x,y) \tag{5.39}$$

$$\Phi(f,\theta;x,y) = \exp(-j4\pi rf/c) \tag{5.40}$$

This terminology indicates that Φ is a function of f and θ, unique to each x,y, and the dependence on θ, x, and y is implicit in r.

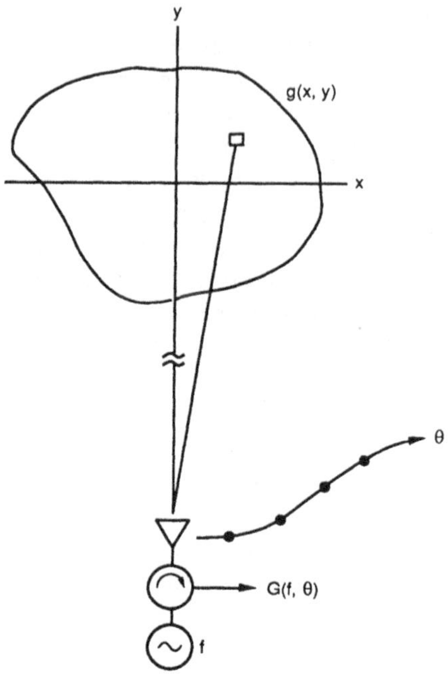

Figure 5.42 Generalized geometry for imaging by frequency and angular diversity.

Equation (5.40) represents the set of basis functions which are the complex conjugate of the response to a unit-strength scatterer located at a particular (x,y). When the object consists of a single scatterer located at (x,y), the measured data and the basis function for that coordinate have a conjugate phase relation, and therefore sum to a maximum in (5.39). All other basis functions produce a lesser value, ideally zero. When (5.39) consists of a Fourier transform, the basis functions form an orthogonal set.

Computational considerations aside, the imaging algorithm for any form of frequency and spatial diversity is eminently simple: the image is formed by the integrated product of the measured data multiplied by the conjugate phase history postulated for a point located at each pixel in the image space. This process can accommodate all forms of frequency and spatial diversities, including the various types of synthetic apertures, motion compensations, and near-field geometries. The quality of the image is thus determined by the accuracy to which the diversities, intentional or accidental, are known.

Practical aspects of RCS imaging involve the application of efficient algorithms to numerically evaluate (5.39). Nearly every imaging scheme in current use formulates the basis functions in terms of two-dimensional Fourier transforms that can be implemented using the FFT algorithm and thus reap the ensuing computational efficiency.

The basis functions for focused imaging represent the phase histories of isotropic, frequency-independent point scatterers moving on circular paths and observed from large distances. The basis functions for unfocused imaging represent similar conditions with the additional assumption that circularity in the motion of the scatterer is negligible.

Results provided by RCS images must be interpreted in light of the correspondence of the basis functions with the behavior of the scattering features to be imaged. If one of the basis functions provides a conjugate match with the phase history of a particular scatterer, the latter will be imaged with a high degree of fidelity; conversely, an inadequate phase match results in aberrations and degradation of image quality.

The ideal imaging process, therefore, would be adaptive in using basis functions unique to every feature of the image. Although this goal may not be practically achievable, it contrasts the arbitrary, nearly universal use of Fourier transform basis functions, driven by addiction to the FFT algorithm.

In previous discussions, we showed that the spatial configuration of the image is driven by the phase of recorded signals and basis functions. To illustrate this notion, images were obtained from a set of measured data using both phase and magnitude, phase only, and magnitude only. The results are shown in Figure 5.43; in each case, the total signal energy has been normalized to the same value. While the image obtained by using only the phase retains the spatial relation of the object features, the image obtained by using only the magnitude is obliterated. This example supports the assertion that the phase information determines the spatial quality of the image [17], while the magnitude is necessary to obtain a measure of the strength of scattering sources.

The following figures show examples of images interpreted in terms of the basis functions used in the imaging process. Figure 5.44(a,b) shows unfocused and focused 10 GHz images of a 20-inch diameter polystyrene foam column used for target support in an anechoic chamber. The column, consisting essentially of discrete point scatterers, is imaged in the proper spatial configuration indicated by the circular central region. The chamber clutter echoes, consisting of stationary scatterers having phase invariant with aspect angle, however, are imaged as streaks along the range axis. The focusing process blurs the clutter image because the algorithm incorrectly presumes the scatterers to follow a circular motion. The focused image illustrates the effects of a mismatch between the basis functions used for the imaging process and the actual phase behavior of scattering centers. The stationary clutter can be eliminated by a filtering process which rejects contributions from all scatterers having rotation-invariant phase. By this simple filtering operation, the contribution of stationary scatterers can be removed from the measured data prior to processing the image. Figure 5.44(c) shows the image of the polystyrene column with the stationary clutter suppressed by the filtering process.

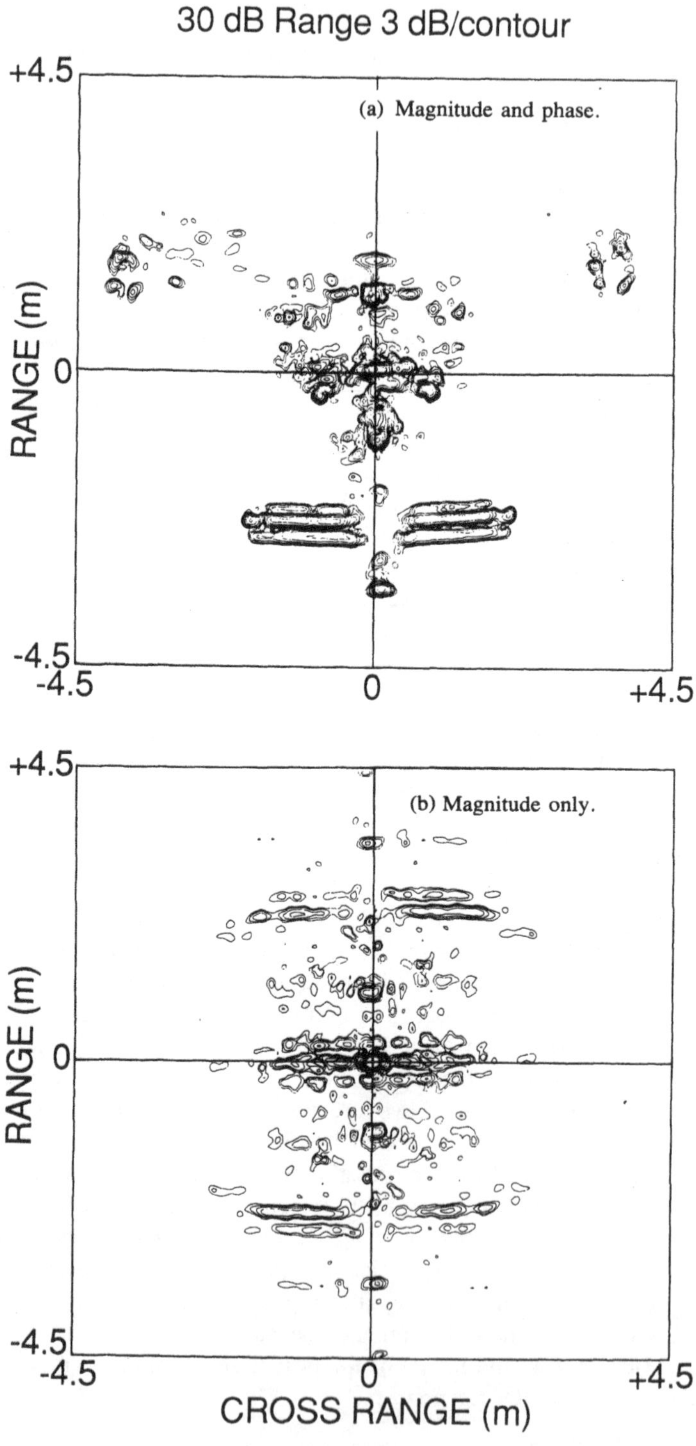

Figure 5.43 Two-dimensional RCS images derived from phase and magnitude data.

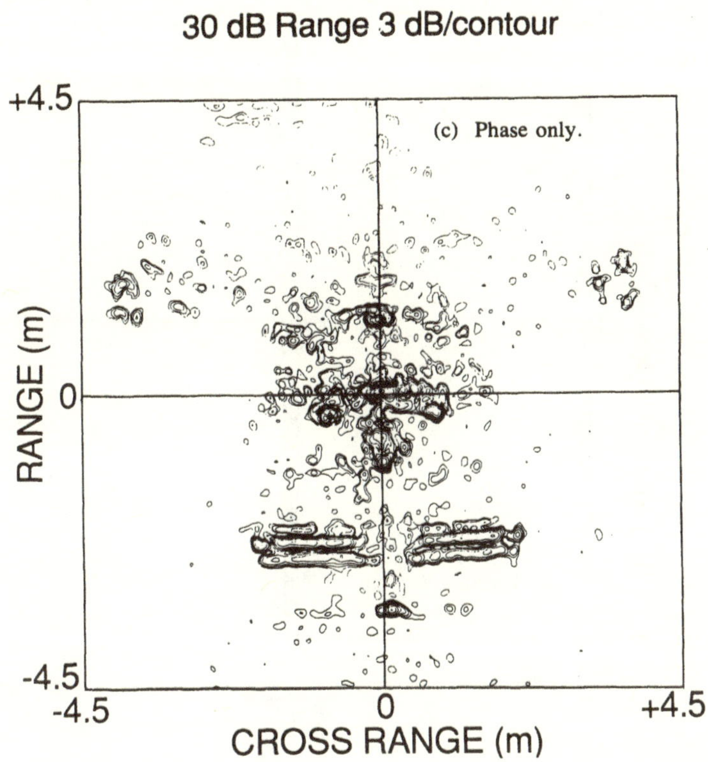

Figure 5.43 (continued).

Figure 5.45 shows examples of images obtained by processing simulated responses of a 20-inch diameter sphere using a center frequency of 1.6 GHz and a bandwidth of 3 GHz. The frontal feature of the unfocused image in Figure 5.45(a) represents the specular scattering from the surface of the sphere; the trailing response represents the creeping wave echo that travels around the rear surface of the sphere and is delayed accordingly. Because the sphere is rotationally symmetric, the apparent scattering centers are aspect-invariant and the unfocused image is sharply resolved. Although the sphere is rotated, the scattering center appears to "slip" along the surface in synchronism with the rotation, thus emulating a stationary scatterer. Figure 5.45(b), the focused image, shows considerable blurring resulting from the improper treatment of stationary scatterers presumed to rotate by the focusing algorithm.

The preceding examples demonstrate that imaging algorithms can produce highly resolved images providing that the basis functions match the actual phase history of individual scatterers. For example, the unfocused imaging algorithm is optimum for stationary scatterers, while the focused algorithm is optimum for scatterers exhibiting

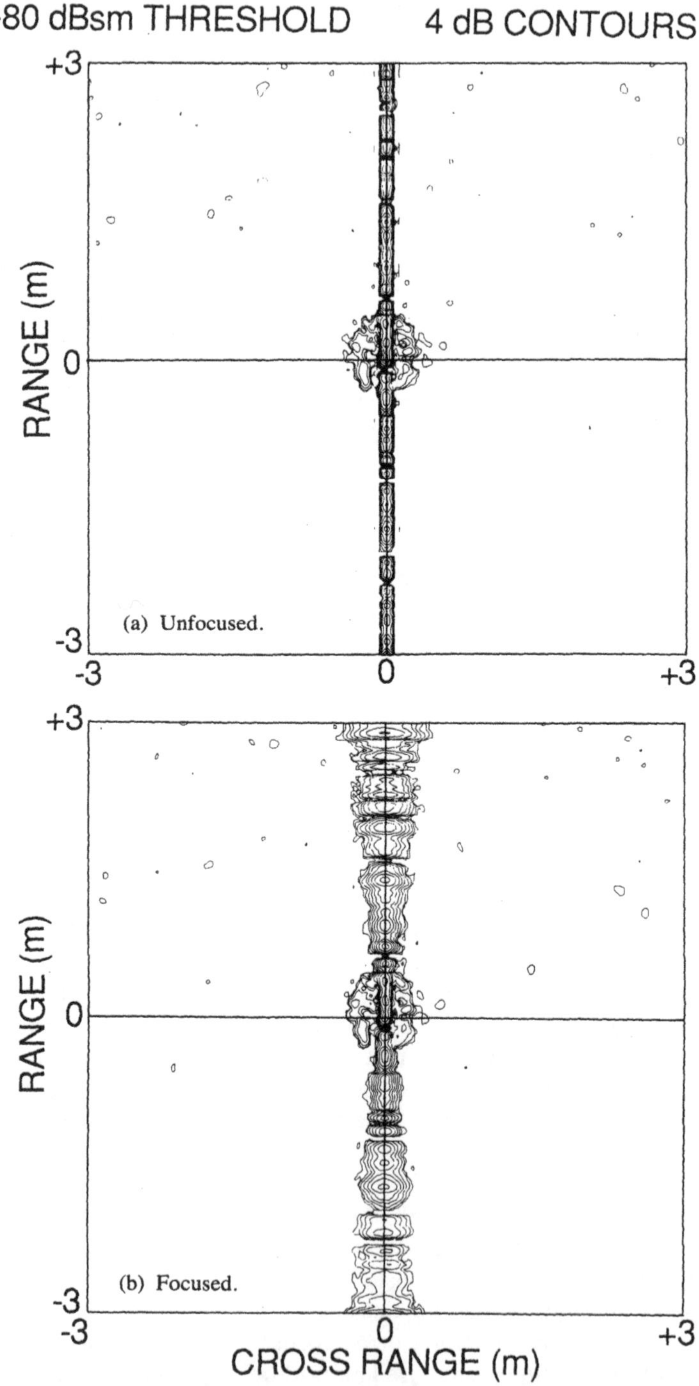

Figure 5.44 Focused and unfocused images of polystyrene foam support column.

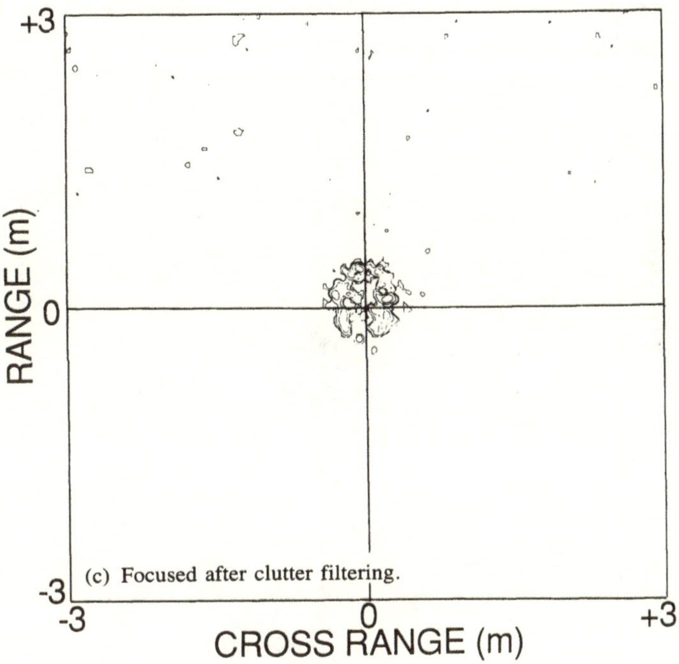

(c) Focused after clutter filtering.

Figure 5.44 (continued).

circular motion. For this reason, a focused algorithm does not consistently produce superior images. This conclusion would suggest that a number of processing algorithm options should be available, each tailored to the behavior of specific classes of scattering features.

Both the focused and unfocused algorithms are derived from the implicit assumptions that the signals are observed from infinite distance. When objects modeled as arrays of isotropic point scatterers are imaged in the near field, the basis functions provided by the Fourier transforms used for far-field imaging are no longer strictly correct. In this case, the basis functions are not complex sinusoids and the FFT algorithms cannot be applied. The image, however, can be formed by a numerical evaluation of (5.39) with the proper basis functions defined in Figure 5.46. A set of wideband data was used to produce a single image derived from a full 360° rotation. The data set, spanning a 2 GHz bandwidth centered on 3.4 GHz, was obtained by measuring the 8.5 m object from a distance of 22 m. The result, shown in Figure 5.47, is a single image which represents the global behavior of the object. Although the phase histories of individual scatterers are highly distorted from the equivalent

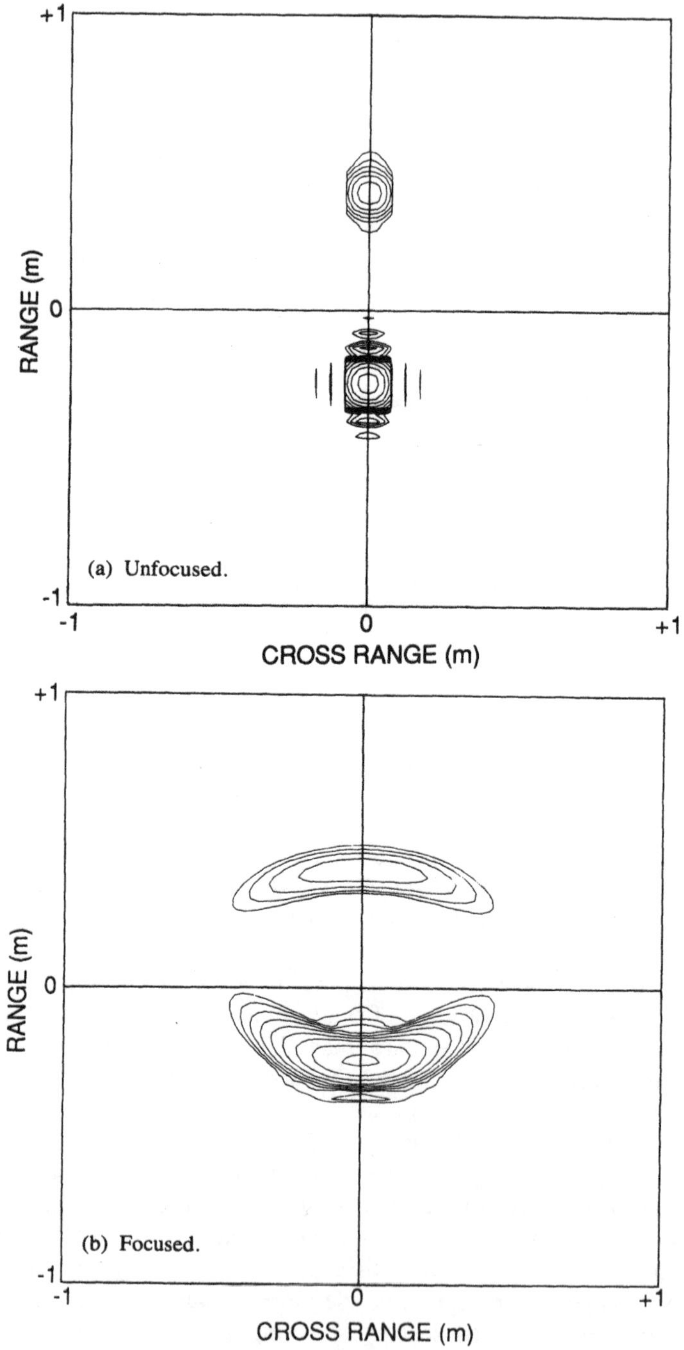

Figure 5.45 Images of 20-inch diameter sphere.

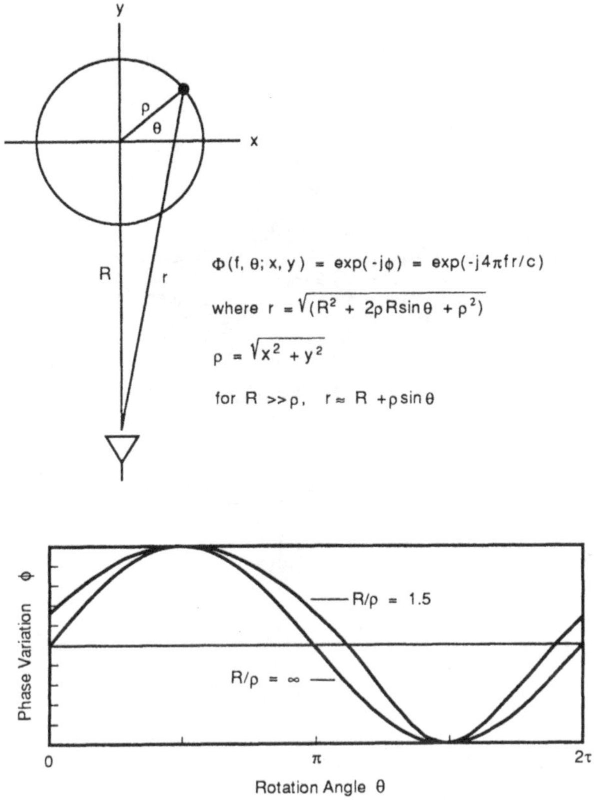

Figure 5.46 Definition of basis functions used for imaging.

far-field case, the image is properly registered and provides a high degree of resolution. The image can be considered as the coherent superposition of subimages obtained over smaller angular sectors, rotated to account for the viewing angle prior to superposition.

Figure 5.48 shows a final example of the utility of the basis function concept in generalized imaging. The image was obtained from a linear SAR configuration in which an anechoic chamber was irradiated by a wideband RCS system mounted on a linear positioner moved horizontally, normal to the line-of-sight. The empty anechoic chamber was irradiated at a center frequency of 10 GHz with a bandwidth of 1.1 GHz by widebeam antennas linearly scanned over 1.26 m horizontal distance in the plane of the chamber aperture. The image, formed by (5.39), in which the basis functions were computed from the known frequency and geometry of the measurement setup, is well focused and shows good correlation with the photograph shown in Figure 5.49.

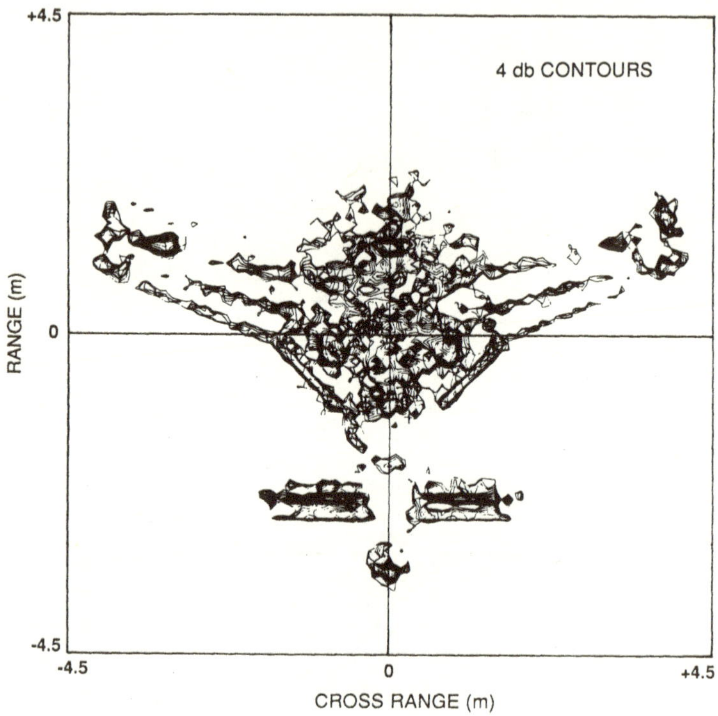

Figure 5.47 Focused, near-field image obtained from complete object rotation.

The preceding considerations lead to a generalized viewpoint of the synthetic imaging process, which can be extended to arbitrary frequency diversities and synthetic aperture geometries. The upper three diagrams of Figure 5.50 indicate conventional synthetic aperture radars involving linear SAR, spotlight SAR, and inverse SAR. The lower diagram indicates a generalized SAR geometry in which either or both radar and object can move relative to one another. As diagrammed in Figure 5.51, the imaging process consists of two distinct steps: the data collection in which object responses to frequency and angle diversities are recorded, and the image reconstruction in which the observed data are processed to obtain the image.

Each of the synthetic imaging processes described in preceding sections can be summarized by the hierarchy indicated in Table 5.1, which cites the consequences of assumptions for the phase behavior of object points.

The top entry assumes the phase of an object point to be linear as a function of frequency and angle, and independent of range. The imaging algorithm is a two-dimensional FFT applied directly to an $N \times N$ array of data measured on uniform

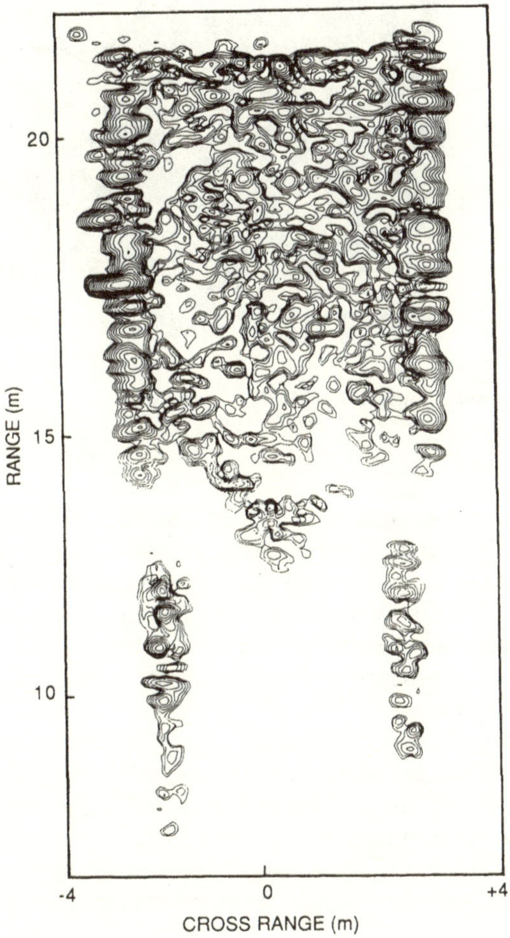

Figure 5.48 SAR image of anechoic chamber: F_0 = 10 GHz, BW = 1.1 GHz, SCAN = 1.26 m, Min. level = -55 dB/m^2, 2 dB contour.

Figure 5.49 Anechoic chamber imaged with SAR system.

increments of angle and frequency, which can be accomplished in $2N^2 \log_2 N$ multiplications for an $N \times N$ image. To obtain acceptable image quality, the total angle and the bandwidth are restricted to small values, and far-field conditions are required. This process results in an unfocused image with range-dependent distortion.

The second entry assumes the phase of an object point to vary linearly with frequency, sinusoidally with angle, and be independent of range. The imaging algorithm is a two-dimensional FFT applied to polar-formatted data resampled to a rectangular grid; the algorithm requires $2N^2 \log_2 N$ multiplications in addition to the computations required for the interpolation process. The algorithm produces focused images for far-field conditions and range-dependent image distortions.

The third entry assumes only arbitrary but known phase variations as a function of frequency, angle, and range. The imaging process is a conjugate-phase matched filter requiring N^4 computations. There are no restrictions, and focused images result for any observation range.

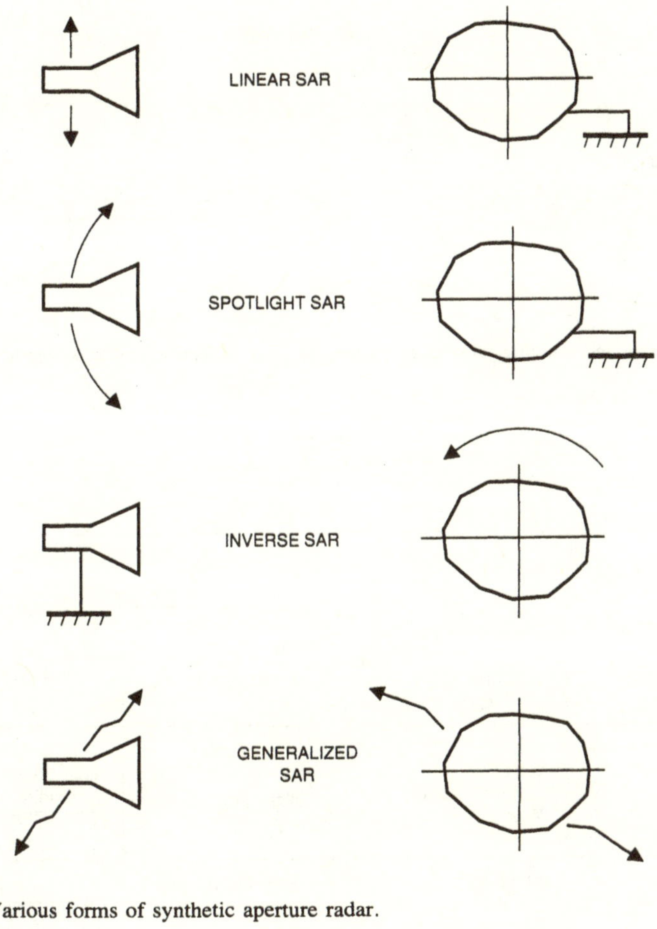

Figure 5.50 Various forms of synthetic aperture radar.

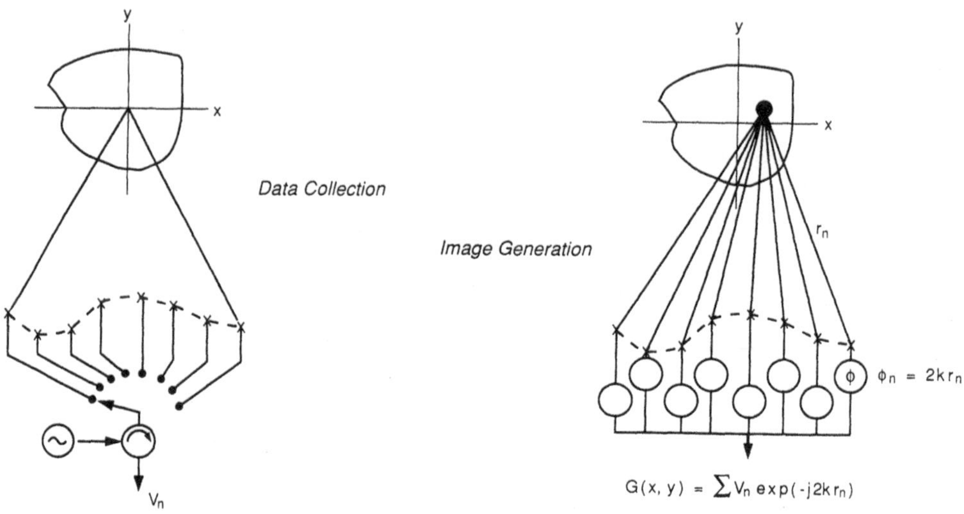

Figure 5.51 Generalized imaging process consisting of data collection and image reconstruction.

Table 5.1
Hierarchy of Image Processing

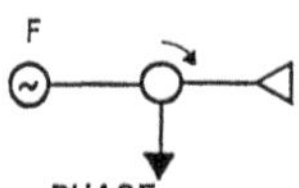

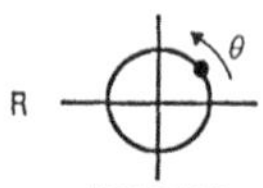

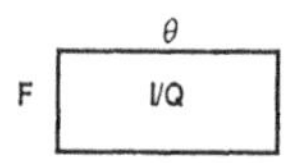

PHASE ASSUMPTION	IMAGING ALGORITHM	RESTRICTION	RESULTS
LINEAR VS F, θ $\neq f(R)$	2D FFT ON MEASURED DATA $M=2N^2 \log N$	SMALL Θ, B FAR - FIELD	UNFOCUSED IMAGE R - DEPENDENT DISTORTION
LINEAR VS F SINUSOIDAL VS θ $\neq f(R)$	2-D FT ON POLAR FORMAT DATA RESAMPLED TO RECT. GRID $M=2N^2 \log N$ + INTERP	FAR - FIELD	FOCUSED IMAGE R - DEPENDENT DISTORTION
ARBITRARY KNOWN VARIATION VS F, θ, R	CONJUGATE - PHASE MATCHED FILTER $M= N^4$	—	FOCUSED IMAGE FOR ANY R

REFERENCES

1. Scudder, H.J., "Introduction to Computer Aided Tomography," *Proc. IEEE,* Vol. 66, No. 6, June 1978, pp. 628–637.
2. Kak, A.C., "Computerized Tomography with X-Ray, Emission, and Ultra-sound Sources," *Proc. IEEE,* Vol. 67, No. 9, September 1979, pp. 1245–1271.
3. Shepp, L.A., and B.F. Logan, "The Fourier Reconstruction of a Head Section," *IEEE Trans. on Nuclear Science,* Vol. NS-21, June 1974, pp. 21–43.
4. Cho, Z.H., "General View on 3-D Image Reconstruction and Computerized Transverse Axial Tomography," *IEEE Trans. on Nuclear Science,* Vol. NS-21, June 1974, pp. 44–71.
5. Brown, W.M., and L.J. Porcello, "An Introduction to Synthetic Aperture Radar," *IEEE Spectrum,* September 1969, p. 52.
6. Chen, C.C., and H.C. Andrews, "Multi-Frequency Imaging of Radar Turntable Data," *IEEE Trans. on Aerospace and Electronic Systems,* Vol. AES-16, No. 1, January 1980, pp. 15–22.
7. Brown, W.M., and R. Fredricks, "Range-Doppler Imaging with Motion Through Resolution Cells," *IEEE Trans. Aerospace and Electronic Systems,* Vol. AES-5, January 1969, pp. 98–102.
8. Walker, J.L., "Range-Doppler Imaging of Rotating Objects," *IEEE Trans. on Aerospace and Electronic Systems,* Vol. AES-16, No. 1, January 1980, pp. 23–52.
9. Bracewell, R., *The Fourier Transform and Its Applications,* New York: McGraw-Hill, Co., 1965, pp. 244–250.
10. Harris, F.J. "On the Use of Windows for Harmonic Analyses with the Discrete Fourier Transform," *Proc. IEEE,* Vol. 66, No. 1, January 1978, pp. 51–83.
11. Simpson, R.G., H.H. Barrett, J.A. Suback, and H.D. Fisher, "Digital Processing of Annual Coded-Aperture Imagery," *Optical Engineering,* Vol. 14, No. 5, September–October 1975, pp. 490–494.
12. Ermert, H., and R. Karg, "Multi-Frequency Acoustical Holography," *IEEE Trans. on Sonics and Ultrasonics,* Vol. 26, No. 4, July 1979, pp. 279–286.
13. Kell, R.R., "On the Derivation of Bistatic RCS from Monostatic Measurements," *Proc. IEEE,* Vol. 53, August 1965, pp. 983–988.
14. Gabel, R.A., and R.A. Roberts, *Signals and Linear Systems,* New York: John Wiley and Sons, 1987.
15. Schwartz, M., and L. Shaw, *Signal Processing,* New York: McGraw-Hill, 1975.
16. Lathi, B.P., *Signals, Systems and Communication,* New York: John Wiley and Sons, 1965.
17. Munson, D.C., and J.L. Sanz, "Image Reconstruction from Frequency-Offset Fourier Data," *Proc. IEEE,* Vol. 72, No. 6, June 1984, pp. 661–669.

Chapter 6
Tomographic Imaging

Tomographic imaging [1,2], best known for its use in medical applications, has been applied to radio astronomy, geophysics, electron microscopy [3], and recently to microwave imaging [4–7]. Tomography involves the reconstruction of images from projections of the object's spatial distribution which are obtained at regular increments of viewing angle. Each projection is a sequence of line integrals of the object's distribution obtained along closely-spaced lines. The distinguishing feature of tomography is the reconstruction of two-dimensional images from measurements of one-dimensional projections obtained from a range of viewing angles. The reconstruction algorithm ascribes to the synthesized image the same spatial resolution available in the one-dimensional projections.

Figure 6.1 shows a parallel-beam x-ray tomographic system in which a collimated beam, typically one millimeter in diameter, is scanned across the object while a detector is simultaneously scanned on the opposite side of the object. Measurement of the detected radiation along each line constitutes a projection of the total attenuation in the object plane for a particular viewing angle. The scanned beam provides fine lateral resolution, but no resolution in the axial direction. In theory, a complete collection of projections measured over all possible angles provides sufficient data to uniquely reconstruct a cross-sectional image of the object with a high degree of two-dimensional spatial resolution. Because the radiation is a line source scanned in a direction normal to the rotational axis, the process is termed *transverse axial tomography*. The term "tomography," derived from the Greek root "*tomos,*" meaning a section or slice, signifies the process of imaging a section or slice of an object.

The tomographic process is directly applicable to radar imaging if range and cross-range profiles of the object's reflectivity distribution are interpreted as projections. Because such projections are readily available from radars using frequency or angular diversity, the tomographic technique offers a method of reconstructing two-dimensional images.

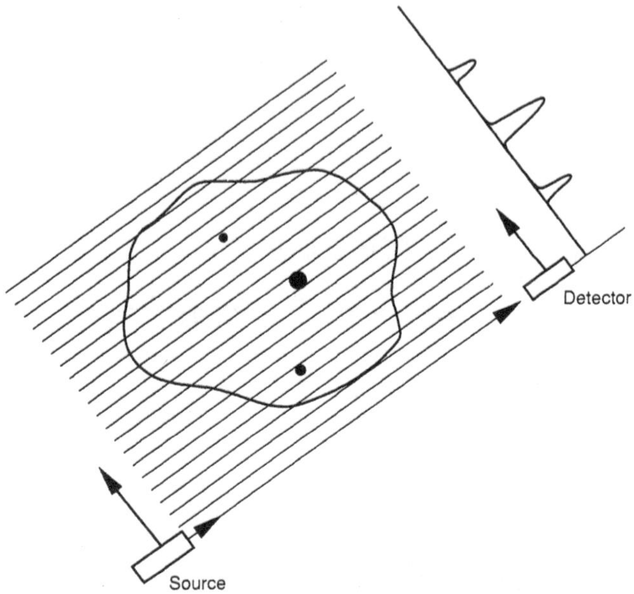

Figure 6.1 Parallel beam x-ray tomographic system.

The following sections review two implementations of tomographic imaging termed Fourier reconstruction and filtered back projection methods, explore similarities and differences between x-ray and microwave applications, and develop a tomographic viewpoint consistent with radar imaging.

6.1 FOURIER RECONSTRUCTION METHOD

Tomographic imaging using Fourier reconstruction is based on Fourier transform theory, which provides an elegant and complete mathematical approach. The key to the application of this method is the projection theorem which relates the one-dimensional Fourier transform of a projection of the spatial distribution of the object to a central section or slice of the two-dimensional Fourier transform of the object distribution. The theorem is derived by considering the geometry for data collection shown in Figure 6.2. The function to be imaged, $g(x,y)$, is scanned by a series of parallel rays oriented at an angle $\theta + \pi/2$. The output of the scanned detector provides a measure of the total object attenuation, and thus constitutes a projection of $g(x,y)$ onto the u axis.

A projection corresponding to the angle θ is a one-dimensional function of the single variable, u, represented by

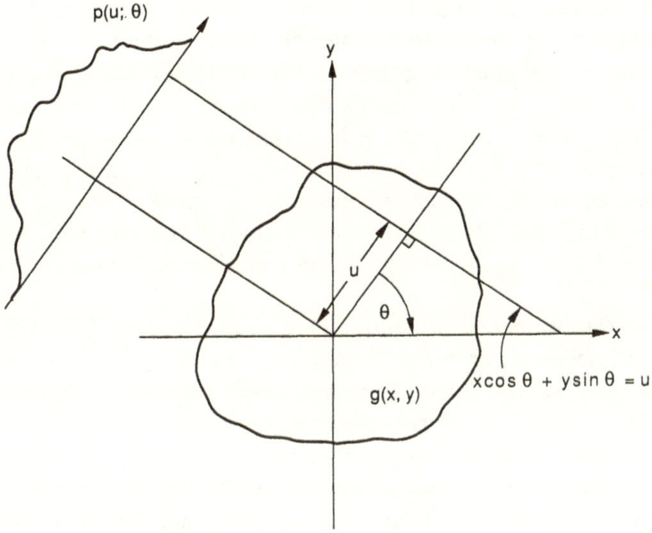

Figure 6.2 Geometry of data collection for tomographic imaging.

$$p(u;\theta) = \int_{-\infty}^{+\infty} g(u\cos\theta - v\sin\theta,\ u\sin\theta + v\cos\theta)\,\mathrm{d}v \qquad (6.1)$$

where

$$x = u\cos\theta - v\sin\theta$$

$$y = u\sin\theta + v\cos\theta$$

The notation $p(u;\theta)$ indicates that the projection is a function of the variable u, unique for each θ. A one-dimensional Fourier transform of $p(u;\theta)$ with respect to the variable u is

$$P(\omega;\theta) = \int_{-\infty}^{+\infty} p(u;\theta)\exp(-ju\omega)\mathrm{d}u \qquad (6.2)$$

A derivation of the projection theorem, developed in Appendix B, leads to

$$P(\omega;\theta) = G(\omega,\theta) \qquad (6.3)$$

Equation (6.3), termed the *projection theorem,* states that the one-dimensional Fourier transform of a projection with orientation θ is a section of the two-dimensional Fourier transform of $g(x,y)$ passing through the origin and oriented at the same angle. We see from this relation that, if projections are known for a number of different orientations, the Fourier transform of the unknown function can be determined for polar cross sections corresponding to the same angles. The various projections, therefore, provide samples of the image transform on spokes of the Fourier transform plane, as shown in Figure 6.3. In theory, if an infinite number of projections are available, a two-dimensional inverse Fourier transform provides an exact reconstruction of $g(x,y)$.

If a sufficient number of projections are available, and if the Fourier transform is relatively well behaved, an approximation to $G(\omega,\theta)$ can provide a good estimate of $g(x,y)$ by an inverse transform. The reconstruction process, therefore, consists of three steps: (1) Fourier transform each projection; (2) interpolate to fill in missing data due to the finite number of projections; and (3) inverse two-dimensional Fourier transform the results. The process is known as the *Fourier reconstruction method* because the measured projections are used to obtain samples of the spatial (Fourier) spectrum, which is subsequently inverted to estimate the image.

Practical image reconstructions using computer processing necessarily deal with discrete data from samples along each projection and angular increment. The collection of each projection processed by FFTs provides discrete samples of the spatial

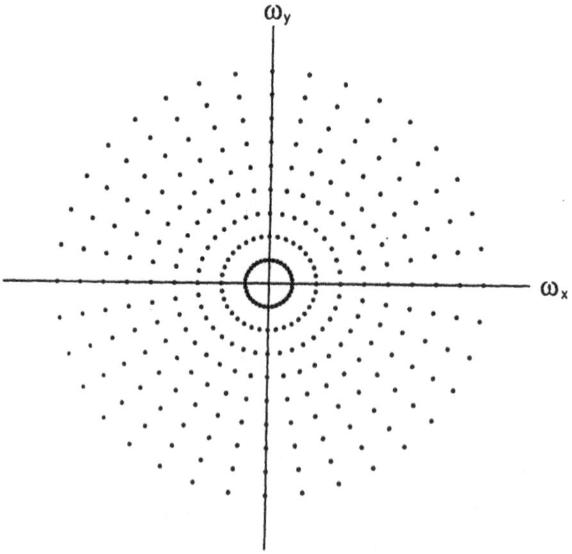

Figure 6.3 Radial samples of Fourier transform space.

spectrum of the image disposed on the polar raster shown in Figure 6.3. If FFTs are utilized for the two-dimensional inverse transform operation required in step 3, the polar raster must be converted to a Cartesian raster. Some method of interpolation is therefore required to convert polar raster samples of the spatial frequency function to Cartesian raster samples. The discrete nature of the data and the interpolation process define the basic limitations to image reconstruction. This situation is identical to the processing of polar-formatted ISAR data necessary to produce focused images, discussed in Chapter 5. The interpolation process for tomographic imaging, however, is more stringent because it involves data samples that fill a complete circular region rather than an annular region. As in the case of focused ISAR processing, the limitations due to interpolation errors can be avoided at the cost of increased computations by performing the equivalent of a two-dimensional inverse transform on samples of the transform on the polar raster. A method that utilizes two one-dimensional FFTs and a Hankel transform, the latter accounting for the majority of the computational burden, is outlined in [8–10].

6.2 FILTERED BACK-PROJECTION METHOD

An estimate of image, termed the *summation image,* can be reconstructed by the simple procedure of summing back-projections from all possible angles. Given a projection $p(u;\theta)$ for a specific angle, the back projection process assigns the value of the projection to all points in the x,y space which correspond to a given u. The summation image is formed by the superposition of back projections from all available angles. Figure 6.4 shows examples of summation images obtained from projections of a point object for various angular increments over a range of 180°. The summation images, normalized by the number of projections, show sizable artifacts for coarse angular increments. As the number of projections increases, the quality of the summation image improves and radially symmetric two-dimensional resolution is obtained. The lower plot of Figure 6.4 represents an approximation to the *point-spread function* (PSF) intrinsic to the process.

The preceding notions can be mathematically formulated as follows. The summation image is expressed by

$$\hat{g}(x,y) = \int_0^{\pi} p(u;\theta)\, d\theta \tag{6.4}$$

where

$$p(u;\theta) = \int_{-\infty}^{+\infty} g(x,y)\, dv$$

is the projection.

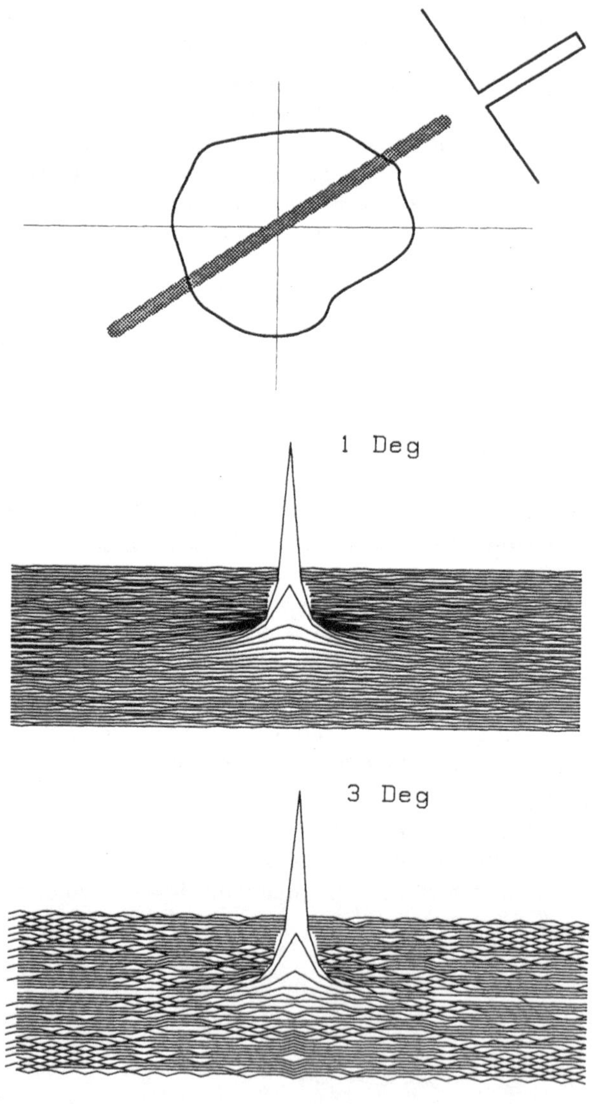

Figure 6.4 Summation images formed by summed back projections.

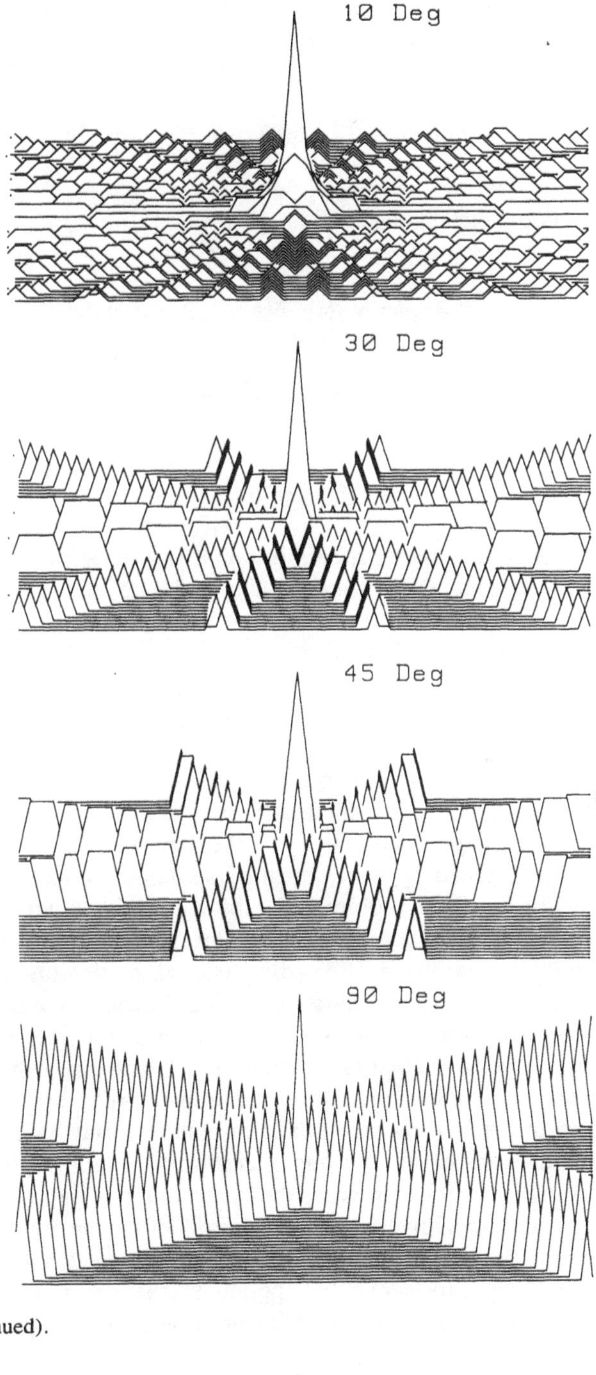

Figure 6.4 (continued).

The PSF of the summation image process, $h(x,y)$, can be found by letting $g(x,y) = \delta(x,y)$ in which case $p(u) = \delta(u)$; then,

$$h(x,y) = \int_{-\infty}^{+\infty} \delta(x \cos\theta + y \sin\theta)\, d\theta \tag{6.5}$$

Expressed in polar coordinates, the PSF is

$$h_P(r,\phi) = \int_{-\infty}^{+\infty} \delta[r \cos(\theta - \phi)]\, d\phi \tag{6.6}$$

By using the relation $\delta(ax) = (1/|\alpha|)\delta(x)$, and noting that r is not involved in the integration, (6.6) reduces to

$$h_P(r,\phi) = 1/r \tag{6.7}$$

The PSF of the summation image process is radially symmetric with a $(1/r)$ form, confirmed by the lower plot of Figure 6.4. The estimate of the true image $g(r,\phi)$ given by the summation image is, therefore, the convolution of the true object distribution with the $(1/r)$ blurring function:

$$\hat{g}(r,\phi) = g(r,\phi) \underset{2}{\star} (1/r) \tag{6.8}$$

where $\underset{2}{\star}$ denotes two-dimensional convolution.

Evidently, the image formed by the summation of back projections produces a blurred version of the image. The fidelity of the estimate is improved if the effect of the blurring can be reduced by some form of deblurring operation or filter.

The properties of the filter required to deblur the image can be established by considering the spectral transfer function which corresponds to the $(1/r)$ spatial blurring function. A radially symmetric, two-dimensional spatial function of the form $(1/r)$ is shown [11] to have a radially symmetric Fourier transform of the form $(1/\rho)$, where ρ is the spatial frequency coordinate. This means that the summation image is characterized by a low-pass response, which attenuates the high spatial frequencies. From a spectral viewpoint, therefore, the transfer function of the deblurring filter must compensate the low-pass response to restore a flat spectrum. Figure 6.5 shows the transfer functions of the various operations; the upper plot is the transfer function of the summation image process, the central plot is the transfer function of the deblurring filter, and the lower plot is the product of the two upper plots and shows the transfer function of the band-limited *deblurred* system.

The "deblurring" operation can be applied to the spectral data prior to the inverse transform which produces the final image; however, this procedure requires

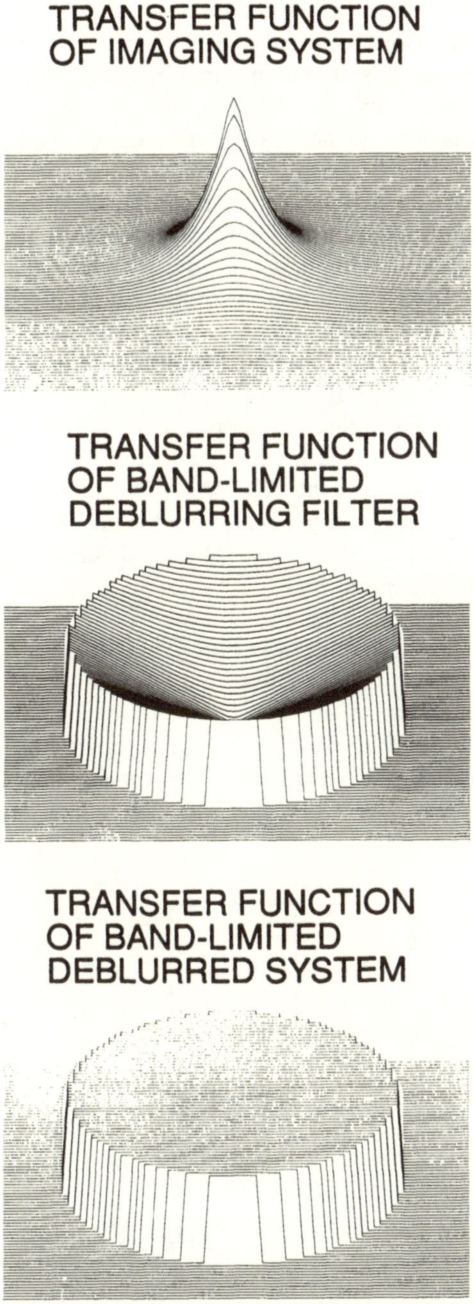

Figure 6.5 Transfer functions of imaging system, deblurring filter, and of deblurred system.

that all the data be available prior to processing the image. An equivalent alternative consists of implementing the deblurring filter operation as a convolution directly on the measured projections. This allows the final image to be progressively synthesized as the projections become available. The procedure is termed *filtered back projection* to indicate that the image is formed by superimposing back projections which have been filtered to remove the blurring.

Characteristics of the deblurring filter can be described either in the spectral or spatial domains; multiplication in the former domain being equivalent to convolution in the latter domain. Although both the transfer function and PSF are two-dimensional functions linked by a Fourier transform relation, the filtering operation can be done as a one-dimensional convolution directly on the projections. The form of the weighting function to be convolved with the measured projections is readily derived as follows using the projection theorem.

The upper left plot in Figure 6.6 shows a central section of the required two-dimensional transfer function which can be decomposed into the difference of two simpler functions at the upper right. The lower plots are Fourier transforms of the corresponding functions on the upper plots; the lower left plot shows the filter function necessary to remove the image blur. The ideal function is given by the sum of two analytic functions expressed by:

$$h(u) = 2U\,\mathrm{sinc}(2uU) - U\,[\mathrm{sinc}(uU)]^2 \qquad (6.9)$$

where $\mathrm{sinc}(x) = (\sin \pi x)/\pi x$.

Figure 6.7 shows the PSF of the process, computed by superimposing filtered back projections in one-degree increments. The reconstructed image will consist of the true object distribution convolved with the PSF, which features a high degree of

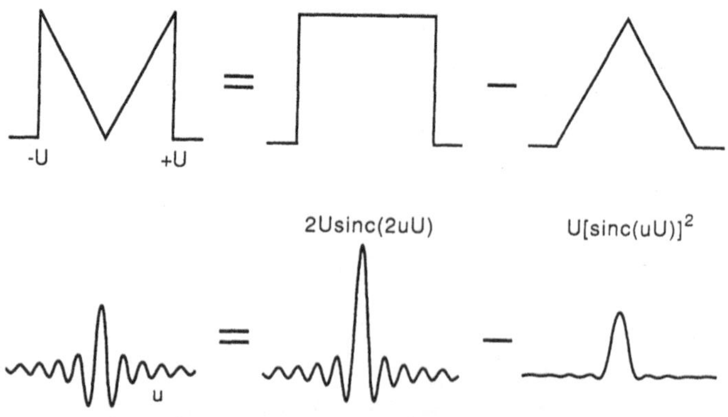

Figure 6.6 Characteristics of deblurring filter.

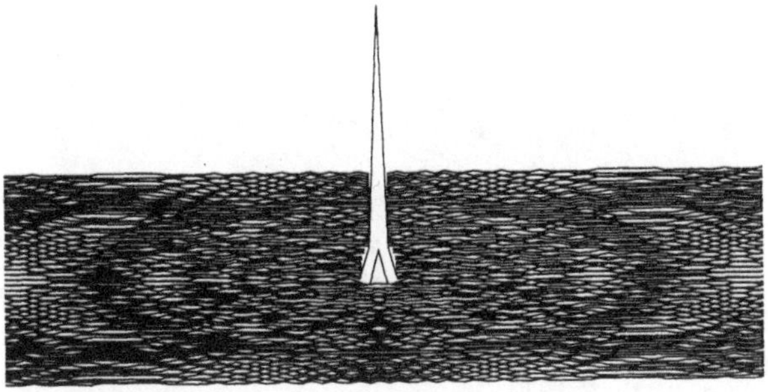

Figure 6.7 Point-spread function of filtered back projection process.

radially symmetric spatial resolution. Optimizing application of the technique in practical cases involves consideration of the discrete nature of the data and of the exact form of the deblurring filter.

Both the projections and the image are sampled at discrete increments; consequently, the value of the projection corresponding to an image pixel may fall between available samples of the projection, as shown in Figure 6.8. The projections must, therefore, be interpolated to better estimate the back projected quantity.

The spatial response of the deblurring filter, shown in Figure 6.7, is characterized by extended oscillations on each side of the main lobe. A consequence of the abrupt truncation in the corresponding transfer function, this response produces artifacts in the reconstructed image. The artifacts can be controlled at the cost of slightly degraded resolution by gradually tapering the cut-off region of the spatial frequency transfer function, which alters the form of the deblurring filter response. A number of filter responses optimized to meet resolution, artifacts, or noise criteria are used in practical systems.

In summary, the computational procedures for synthesizing an image by the filtered back projection method consist of the following steps:

1. Obtain a set of one-dimensional projections.
2. Convolve the projections with the impulse response of the deblurring filter.
3. Interpolate the filtered projections.
4. Back project the interpolated, filtered projections.
5. Superimpose the back projections.

The applicability of the tomographic imaging algorithm to microwave backscattering was tested by using experimental data. A coherent RCS pattern of a scale-model boat was obtained using 94 GHz CW irradiation. Data contained in contiguous angular sectors were processed to obtain the series of cross-range (doppler) responses shown in Figure 6.9. The magnitudes of the cross-range responses were treated as projections and processed by the filtered back projection algorithm used for x-ray

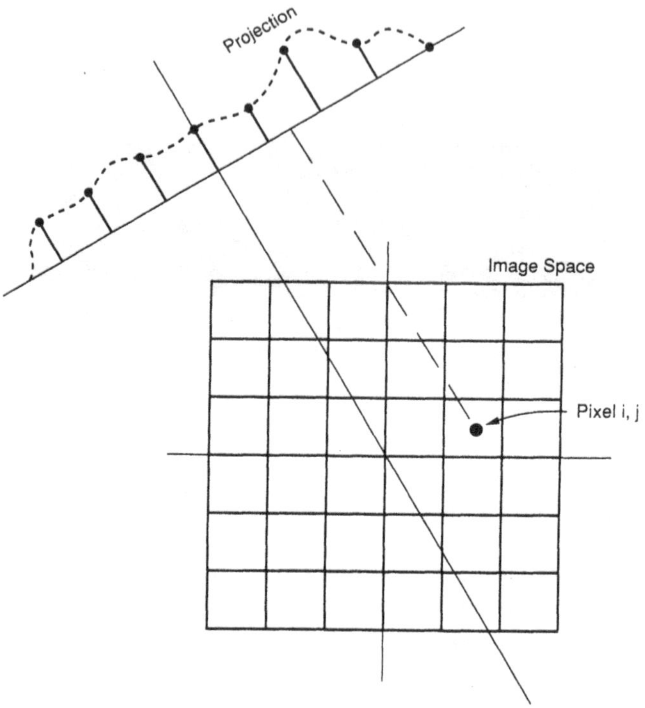

Figure 6.8 Back projection of discrete projection onto discrete image space.

imaging. The two-dimensional image, shown in Figure 6.10, is relatively primitive, but demonstrates that two-dimensional spatial resolution can be obtained from results of CW measurements.

6.3 APPLICATION TO RADAR IMAGING

Because RCS instrumentation radars can readily obtain projections in the form of range or cross-range profiles of the object scattering properties, the use of tomographic techniques to produce two-dimensional images is a logical alternative. For properly utilizing the tomographic approach, similarities and differences between x-ray and radar applications must be recognized.

Tomographic imaging applied to x-ray and radar cases differ in the following respects:

1. Projections in x-ray tomography consist of line integrals which are scalar sums of the object density; projections in microwave tomography consist of line integrals, which are phasor sums of the complex object reflectivity.

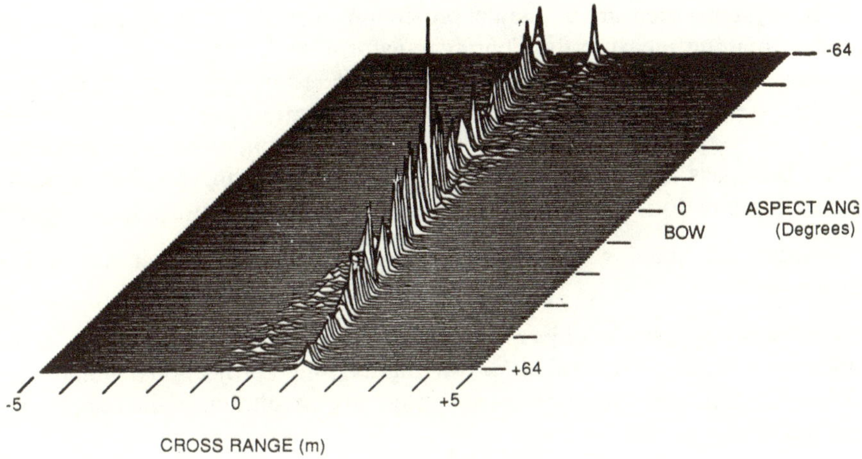

Figure 6.9 Cross-range responses obtained from CW RCS pattern.

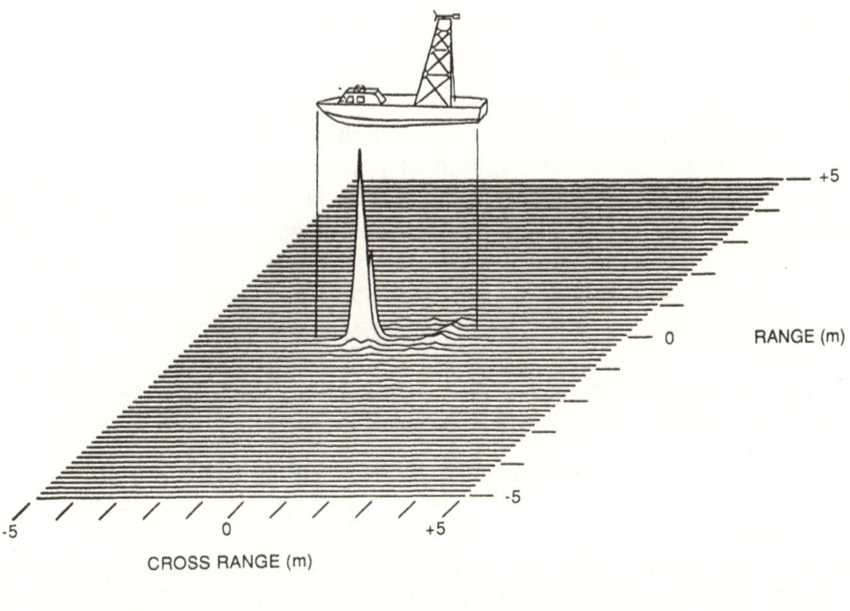

Figure 6.10 Tomographic image obtained from filtered, back projected cross-range responses.

2. X-ray systems measure the object projections directly; radar systems using synthetic imaging measure the Fourier transform of the projections.
3. Projections in x-ray systems are low-pass filtered by spatial averaging over the finite extent of the detector; the Fourier transform of each projection defines a low-pass region of the spatial spectrum. Projections in radar systems are filtered by a bandpass filter determined by the radar bandwidth; the Fourier transform of each projection, therefore, defines a bandpass region of the spatial spectrum.

Tomographic imaging applied to x-ray and radar cases are similar in the following respects:

1. Windowing applied to object's frequency responses measured by RCS instrumentation radars is equivalent to filtering the range projections; windowing the measured frequency response is equivalent to convolution of the range response with a filter function.
2. Two-dimensional ISAR imaging described in preceding chapters is equivalent to superposition of many range responses, each weighted by phase factors to account for the phasor nature of the projections; the phase factors are contained in the exponential kernel of the Fourier transform used for cross-range imaging.
3. The interpolation necessary in the x-ray back projection process is implemented in the radar case by the exponential kernel in the Fourier transform, which compensates for propagation delay within each range cell. Linear interpolation results from applying the Fourier transform directly to the data; a more proper phase compensation, which accounts for migration of object points in rotation, is provided by the image focusing process.

Tomographic implementation of image reconstruction, with proper attention in back projection to the interpolation between complex-valued samples of radar projections, offers an alternative formulation of two-dimensional imaging of rotating objects. The tomographic process uses range profiles obtained either directly or synthetically from the measured data to form the final image by superimposing modified range profiles back projected over the image space. The interpolation required to determine the proper value used in the back projection process requires that the center of rotation of the object be identified in each projection to within a small fraction of a wavelength.

The back projection process can readily incorporate the following practical features to enhance the fidelity of the reconstructed image:

1. The magnitude of each range profile can be modified to account for the $(1/r)^2$ spreading loss when spherical wave propagation is involved in near-field applications.
2. Each filtered range profile can be back projected along straight parallel lines in far-field applications; this procedure can be easily modified for back projections along circular arcs in near-field applications.

3. Because the back projection geometry can be generalized, the process is adaptable to arbitrary sampling grid structures of the image.

The preceding differences and similarities are developed in greater detail in [5,12]. These features indicate that ISAR imaging of rotating objects can be viewed as a special case of filtered back projection imaging used in tomography. Similar observations in [13] indicate that spotlight SAR processing is analogous to tomographic imaging.

To demonstrate tomographic radar imaging by the filtered back projection method, simulated data for an object consisting of four ideal point scatterers were processed to form an image [12]. An unfocused image was processed from the same data by conventional wideband ISAR methods for comparison. Data from 64 range profiles corresponding to a 9° angular sector were used for the images with Hann windows applied in both directions. Figure 6.11 shows the intensity of the unfocused image obtained by conventional methods using ISAR processing of the range responses. The image, similar to that shown in Figure 4.30, exhibits blurring for object points offset from the center. Figure 6.12 shows the intensity of the image reconstructed from the same data using the filtered back projections with linear phase interpolation. The resolution obtained by the tomographic method is virtually uniform over the

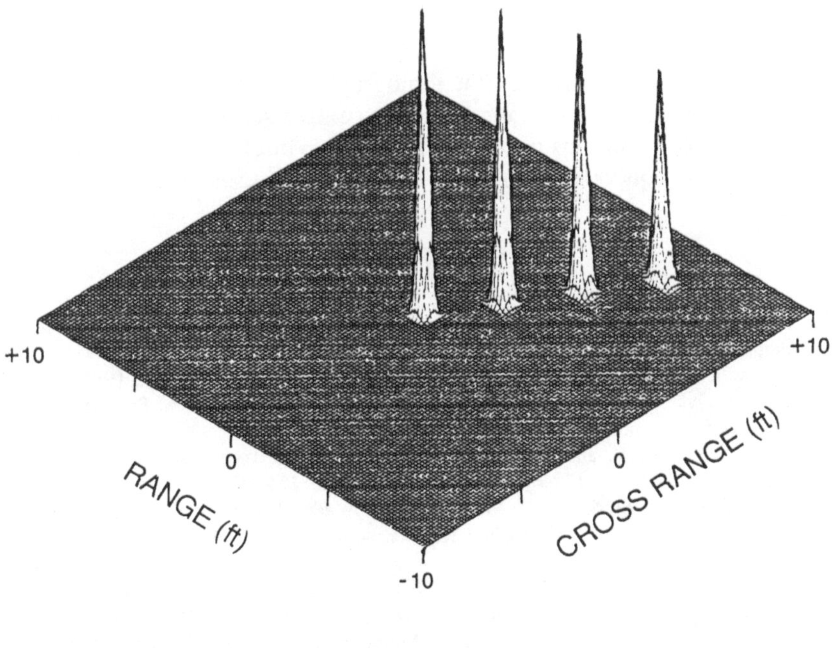

Figure 6.11 Intensity of unfocused ISAR image.

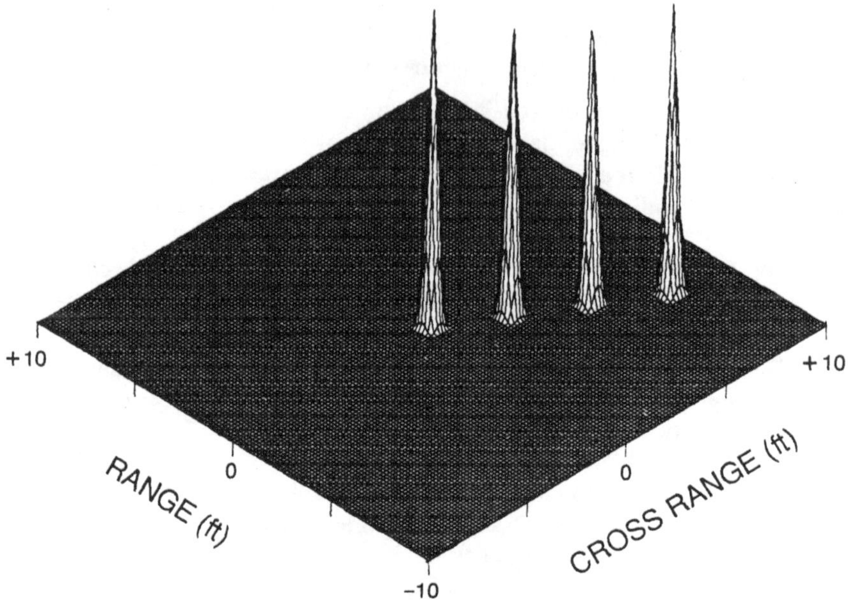

Figure 6.12 Intensity of tomographic image.

object region. The results indicate that tomographic reconstruction methods overcome the limitations imposed by range-doppler (unfocused ISAR) processing methods by accounting for variations in phase with rotation angle. The quality of resulting images, therefore, is equivalent to those obtained by focused imaging.

REFERENCES

1. Hounsfield, G.N., "A Method and Apparatus for Examination of a Body by Radiation such as X or Gamma Radiation," The Patent Office, London, Patent Specification 1283915, 1972.
2. Lee, H., and G. Wade, eds., *Imaging Technology,* IEEE Press, New York, 1985.
3. Herman, G.T., ed., *Applied Topics in Physics, Image Reconstruction from Projections,* Vol. 32, Springer-Verlag, Berlin, 1979.
4. Mensa, D.L., S. Halevy, and G. Wade, "Coherent Doppler Tomography for Microwave Imaging," *Proc. IEEE,* Vol. 71, No. 2, February 1983, pp. 254–261.
5. Halevy, S., G.R. Heidbreder, D.L. Mensa, and G. Wade, "Large Synthetic Aperture Imaging of Rotating Objects Using Tomographic Reconstruction," *Conf. Proc., Military Microwaves,* London, October 1984.
6. Farhat, N.H., C.L. Werner, and T.H. Chu, "Prospect for Three-dimensional Projective and Tomographic Imaging Radar Networks," *Radio Science,* Vol. 19, No. 5, 1984, pp. 1347–1355.
7. Feig, E., and F.A. Grünbaum, "Tomographic Methods in Range-Doppler Radar," *Inverse Problems,* Vol. 2, 1986, pp. 185–195.

8. Mersereau, R.M., and A.V. Oppenheim, "Digital Reconstruction of Multidimensional Signals from their Projections," *Proc. IEEE,* Vol. 62, No. 10, October 1974, pp. 1319–1338.
9. Halevy, S., "Two-Dimensional Discrete Fourier Transformation in Polar Coordinates for CDT Images," Annual Research Report, 1981–1982, Department of Electrical and Computer Engineering, University of California, Santa Barbara, p. 17.
10. Hansen, E.W., "Theory of Circular Harmonic Image Reconstruction," *J. Opt. Soc. Amer.,* Vol. 71, No. 3, March 1981, pp. 304–308.
11. Bracewell, R.N., *The Fourier Transform and Its Applications,* McGraw-Hill, New York, 1965.
12. Halevy, S., Microwave Imaging of Large Rotating Targets Using Tomographic Reconstruction Algorithms, Ph.D. Dissertation, University of California, Santa Barbara, CA, May 1985.
13. Munson, D.C., J.D. O'Brien, and W.K. Jenkins, "A Tomographic Formulation of Spotlight-Mode Synthetic Aperture Radar," *Proc. IEEE,* Vol. 71, No. 8, August 1983, pp. 917–925.

Chapter 7
Maximum Entropy Methods

The spatial resolution of images is fundamentally established by the signal bandwidth and angular extent of the viewing aspect. Practical limitations imposed by instrumentation or by the object itself, however, often restrict the available resolution. This precludes the ability to characterize contributions of closely-spaced scattering features; in such cases, to increase the resolution available from the frequency- and angle-limited data is desirable.

A number of techniques have been developed in recent years to increase the spectral resolution of short-time-duration signals that are beyond the limit imposed by Fourier transform methods. These methods, developed by workers in geophysics, and known as *maximum entropy* (ME) or *autoregressive* (AR) spectral estimation techniques, have been shown to provide considerable enhancement in spectral resolution. Fundamentally, such methods adopt a totally different philosophical viewpoint. While Fourier transform methods assume that the unmeasured data are either zero or periodic repetitions of the measured data, AR and ME methods remove this implied restriction and attempt to extrapolate the data outside the measurement window; by virtue of this extension, the spectral resolution is increased.

The preceding objectives are applicable to radar imaging which is obtained by processing data intrinsically limited in bandwidth and angle. On this basis, techniques used to increase spectral resolution from time-limited signals appear directly applicable to increasing the spatial resolution obtained from frequency- and angle-limited radar data. The objective of this chapter is to summarize the conceptual nature of these methods interpreted in the context of radar imaging, and to provide examples of results derived from experimental data. The following sections outline the viewpoints of maximum entropy and autoregressive spectral estimation. The treatment is limited to developing conceptual notions and summarizing general approaches. A more complete and rigorous treatment of the topic is referred to a number of published works [1–10].

7.1 MAXIMUM ENTROPY METHOD

The concept of entropy, arising in information theory [11], provides a quantitative measure of the randomness in a process. This measure provides a framework for implementing the first principle of data reduction; namely,

> The result of any transformation imposed on experimental data shall incorporate and be consistent with all relevant data and be maximally noncommittal with regard to unavailable data [12].

Violations of this apparently reasonable precept are in common practice. For example, when experimental data are processed by Fourier transforms, some of the available data are altered by tapered windows which reduce information, and unmeasured data are arbitrarily presumed to be either zero or periodic repetitions of the measured data. To gain insight into the concept of entropy applied to imaging, consider forming a simple image by placing N distinguishable particles in K pixels and representing the value of the pixel by the number of contained particles. The number of distinguishable arrangements is:

$$M = K^N \tag{7.1}$$

Examination of Table 7.1, which enumerates the 64 distinguishable arrangements for placing 6 particles in 2 pixels, indicates that the configuration consisting of equal numbers of particles in both pixels is the most prevalent.

The total number of possible arrangements of a given number of particles in a given number of pixels is termed the *multiplicity*. The multiplicity associated with placing N particles in two pixels with M is one and $N - M$ in the other is given by

$$C_N^{M,N-M} = \frac{N!}{(N - M)!\, M!} \tag{7.2}$$

The 2^{20} multiplicities for a 20-particle, 2-pixel system are summarized in Table 7.2. The table entries indicate that nearly 50% of all the possible arrangements comprise the three most uniform distributions. The fundamental conclusion to be drawn from the examples of Tables 7.1 and 7.2 is that, in the absence of any *a priori* information, the most likely occurrence is a uniformly-distributed image.

For an arbitrary number of pixels, M, the multiplicity is

$$C_N^{M_1 \ldots M_K} = \frac{N!}{M_1!\, M_2! \ldots M_K!} \tag{7.3}$$

Equation (7.3) can be simplified by use of the Stirling approximation:

Table 7.1
Distinguishable Arrangements for Six Particles in Two Pixels

Distributions of Particles in Pixels	*A*	*B*
0-6	—	1-2-3-4-5-6
1-5	1	2-3-4-5-6
1-5	2	1-3-4-5-6
1-5	3	1-2-4-5-6
1-5	4	1-2-3-5-6
1-5	5	1-2-3-4-6
1-5	6	1-2-3-4-5
2-4	1-2	3-4-5-6
2-4	1-3	2-4-5-6
2-4	1-4	2-3-5-6
2-4	1-5	2-3-4-6
2-4	1-6	2-3-4-5
2-4	2-3	1-4-5-6
2-4	2-4	1-3-5-6
2-4	2-5	1-3-4-6
2-4	2-6	1-3-4-5
2-4	3-4	1-2-5-6
2-4	3-5	1-2-4-6
2-4	3-6	1-2-4-5
2-4	4-5	1-2-3-6
2-4	4-6	1-2-3-5
2-4	5-6	1-2-3-4
3-3	1-2-3	4-5-6
3-3	1-2-4	3-5-6
3-3	1-2-5	3-4-6
3-3	1-2-6	3-4-5
3-3	1-3-4	2-5-6
3-3	1-3-5	2-4-6
3-3	1-3-6	2-4-5
3-3	1-4-5	2-3-6
3-3	1-4-6	2-3-5
3-3	1-5-6	2-3-4
3-3	2-3-4	1-5-6
3-3	2-3-5	1-4-6
3-3	2-3-6	1-4-5
3-3	2-4-5	1-2-6
3-3	2-4-6	1-3-5
3-3	2-5-6	1-3-4
3-3	3-4-5	1-2-6
3-3	3-4-6	1-2-5
3-3	3-5-6	1-2-4
3-3	4-5-6	1-2-3
4-2 ⋮ 5-1 ⋮ 0-6	These are mirror images of the first 22 distributions.	

Table 7.2
Multiplicities for Placing Twenty Particles in Two Pixels

Particles in Pixel A	*Particles in Pixel B*	*No. of Multiplicities*	*Fraction of Total Multiplicities*
20	0	1	0.000001
19	1	20	0.00002
18	2	190	0.00018
17	3	1,140	0.00109
16	4	4,845	0.00462
15	5	15,504	0.01479
14	6	38,760	0.03696
13	7	77,520	0.07393
12	8	125,970	0.12014
11	9	167,960	0.16018
10	10	184,756	0.17620
9	11	167,960	0.16018
8	12	125,970	0.12014
7	13	77,520	0.07393
6	14	38,760	0.03696
5	15	15,504	0.01479
4	16	4,845	0.00462
3	17	1,140	0.00109
2	18	190	0.00018
1	19	20	0.00002
0	20		0.00001
		1,048,576 = 2^{20}	1.00002

$$\lim_{N\to\infty}(\ln N!) = (N + 0.5)\ln(N) - N + 0.5\ln(2\pi) \tag{7.4}$$

Application of (7.4) to (7.3) can [13] result in

$$H = \frac{1}{N}\ln C_N^{M_1 \ldots M_K} = -\sum_{i=1}^{K}\left(\frac{Mi}{N}\right)\ln\left(\frac{Mi}{N}\right) \tag{7.5}$$

Interpreted in terms of an image, N represents the total number of particles comprising the image and M_i represents the number of particles in the ith pixel; the term (M_i/N), therefore, represents the relative distribution of the particles into individual pixels, and can be considered a probability density.

The quantity H represented by (7.5) is known as the *configurational entropy*. Maximizing the configurational entropy of a process, therefore, is equivalent to maximizing the multiplicity. Given a large number of possible image configurations and no *a priori* information, the configuration which maximizes the entropy should be considered the most likely; this procedure implements a maximally noncommittal treatment of unavailable information. When information about the image is available from experimental data, all possible images which are consistent with the available data should be considered, and the one with maximum entropy selected as the best estimate. The first principle of data reduction is thus implemented by the two procedures.

The processes described in preceding chapters generally obtain images by Fourier transforming the data measured in the spatial frequency domain, $G(x)$, to obtain the spatial distribution $g(x)$. $G(X)$ and $g(x)$ are a Fourier transform pair, expressed in one dimension for simplicity; multidimensional formulations follow directly. The maximum entropy imaging process is mathematically formulated as a constrained maximization; namely,

Maximize:

$$H = -\sum_i \frac{g(x_i)}{\Sigma g(x_i)} \ln \frac{g(x_i)}{\sum_i g(x_i)} \tag{7.6}$$

Subject to

$$g(x) = \sum_n G(X_n) \exp(j2\pi X_n x) \tag{7.7}$$

where x_i is the value of the ith image pixel and X_n is the measured value for the nth spectral component.

Implementation of this technique starts with an initial estimate consisting of a uniform image and proceeds with an iterative search for admissible (consistent with measured data) image configurations which maximize the entropy. This structure inherently facilitates the inclusion of any available *a priori* information. Because practical application to two-dimensional images involves large numbers of possible image configurations, nonlinear vectorial search algorithms must be used to achieve convergence with reasonable computational resources. Examples of two-dimensional applications are presented in [14–16].

7.2 AUTOREGRESSIVE MODEL METHODS

Autoregressive methods are based on linear prediction models in which a time series is represented by a linear combination of past values and samples of a white noise process. Figure 7.1 shows an autoregressive model incorporating an Mth order linear

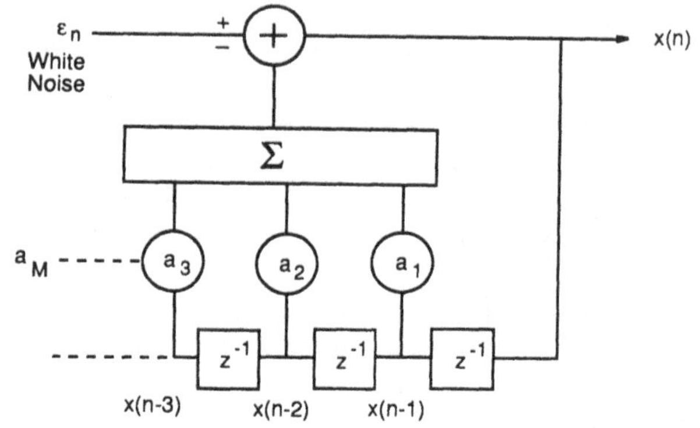

Figure 7.1 Autoregressive *M*th order linear predictor model.

predictor. The blocks labeled z^{-1} indicate unit delays, and the coefficients a_m are amplitude weights assigned to each delayed value; $x(n)$ is the signal, $\Sigma_m a_m x_{n-m}$ is the estimate of the signal, and ε_n is the unknown noise input. The order of the filter indicates the number of past values involved in the estimate. An *M*th-order linear predictor is characterized by the difference equations:

$$x(n) = -\sum_{m=1}^{M} a_m x_{n-m} + \varepsilon_n \tag{7.8}$$

If expressed in matrix notation, (7.8) can be written as

$$\varepsilon_n = \mathbf{X}_n^T \mathbf{A} \tag{7.9}$$

where

$$\mathbf{A}^T = [1, a_1, a_2, \ldots, a_M]$$

$$\mathbf{X}_n^T = [x_n, x_{n-1}, \ldots, x_{n-M}]$$

The prediction filter is optimized by forcing ε_n to be uncorrelated with all x_m for $m < n$; this results in a least-square estimation error and causes ε_n to be a white noise sequence. The preceding statements are formalized by the following relations applied to (7.9):

$$E\{x_n^* \varepsilon_n\} = E\{x_n^* \mathbf{X}_n^T \mathbf{A}\} = [p_M, 0, \ldots, 0] = \mathbf{P}_M^T \tag{7.10}$$

$p_M = E\{\varepsilon_n \varepsilon_n^*\}$ defines the mean-square value of the noise sequence;

$\mathbf{\Phi} = E\{\mathbf{X}_n^* \mathbf{X}_n^T\}$ defines the autocovariance matrix.

The final relation can be concisely summarized in matrix notation by

$$\mathbf{\Phi A} = \mathbf{P}_M$$

$$\begin{bmatrix} R(0) & R(1) & \dots & R(M) \\ R^*(1) & R(0) & \dots & R(M-1) \\ \vdots & \ddots & & \vdots \\ R^*(M) & & \dots & R(0) \end{bmatrix} \tag{7.11}$$

$R(M) = E\{x_n^* x_{n+m}\}$ is the Mth lag of the autocovariance function, and (7.11) is known as the Yule-Walker equations. If the autocovariance of the input process is known, the filter coefficients which describe the poles of the system can be derived exactly by a matrix inversion. The spectrum is then estimated from the known poles by

$$S(f) = p_M \bigg/ \left[1 + \sum_{m=1}^{M} a_m \exp(-j2\pi f_m \Delta t) \right] \tag{7.12}$$

where Δt is the time interval.

Because in most practical applications the autocovariance function is not known and must be derived from the data, an alternative formulation is used, which directly derives the autoregressive model parameters from the data. Known as the least-square method, it is heuristically described as follows.

A sequence of N values processed by an Mth order filter allows $2(N - M)$ measures of the estimation error, obtained by running the filter over the available data in the forward and backward directions. A solution for the M filter coefficients can be obtained by minimizing the total mean square error for the forward and backward direction. The computation for the coefficients involves the inversion of an $(M + 1)$ by $(M + 1)$ matrix. The resulting coefficients define the poles of a linear filter which best model the available data, and the spectrum is then estimated from the known poles. The ability to derive the filter coefficients directly from the data without the intermediate estimation of the autocovariance constitutes the major advantage of the least-square method. The disadvantages are:

1. A matrix inversion is required.
2. A different solution is required for each filter order.
3. The filters may be unstable, leading to diverging extensions.

The preceding limitations are overcome by an approach known as the *Burg method* [17]. As in the least-square method, the Burg algorithm minimizes the sum of the forward and backward linear prediction error energies. In addition, the autoregressive parameters are constrained to satisfy the Levinson recursion, which produces a stable filter. The forward linear prediction error $f_{M,k}$ is defined as

$$f_{M,k} = x_{k+M} + \sum_{i=1}^{M} a_{M,i} x_{k+M-i} = \sum_{i=0}^{M} a_{M,i} x_{k+M-i} \quad (7.13)$$

$$\text{for } 1 \leq k < N - M$$

and the backward linear prediction error $b_{M,k}$ is defined as

$$b_{M,k} = \sum_{i=0}^{M} a^*_{M,i} x_{k+i} \quad \text{for } 1 \leq k \leq N - M \quad (7.14)$$

To determine the filter estimates, the sum of the forward and backward prediction error energies e_M is minimized.

$$e_M = \sum_{k=1}^{N-M} |f_{M,k}|^2 + \sum_{k=1}^{N-M} |b_{M,k}|^2 \quad (7.15)$$

The filter parameters are constrained to satisfy the Levinson recursion:

$$a_{M,k} = a_{M-1,k} + a_{M,M} a^*_{M-1,M-k}, \quad k = 1, \ldots, M-1 \quad (7.16)$$

The Levinson recursion provides a recursive relationship between the forward and backward linear prediction errors, expressed by

$$f_{M,k} = f_{M-1,k+1} + a_{M,M} b_{M-1,k} \quad 1 \leq k \leq N - M \quad (7.17)$$

$$b_{M,k} = b_{M-1,k} + a^*_{M,M} f_{M-1,k+1} \quad 1 \leq k < N - M \quad (7.18)$$

Substitution of (7.17) and (7.18) into (7.15) yields the sum of the total error and AR parameters. If we set to zero the derivative of the error with respect to $a_{M;M}$, the result is

$$a_{M,M} = \frac{2 \sum_{k=1}^{N-M} b^*_{M-1,k} f_{M-1,k+1}}{\sum_{k=1}^{N-M} \left(|b_{M-1,k}|^2 + |f_{M-1,k+1}|^2 \right)} \quad (7.19)$$

The Levinson recursion utilized by the Burg algorithm allows the iterative computation of coefficients for successively higher model orders. The iteration starts with a first-order filter and proceeds to higher orders without a matrix inversion. In addition, the recursion constrains the filter poles to fall within the unit circle, thus assuring stability. The Burg algorithm is highly efficient and widely documented. An application for enhancing range resolution in measured RCS data is reported in [18].

When the coefficients are established, we can use the autoregressive model to describe the process. Autoregressive methods can be directly applied to one-dimensional RCS imaging; the known time series corresponds to measured samples of the angle or frequency responses of the object, and the spectrum estimated from the autoregressive coefficients corresponds to the enhanced range or cross-range profiles of the object.

The estimate of the signal spectrum based on the autoregressive model is shown in [19] to be equivalent to a maximum entropy estimate. This connection indicates that the relatively abstract maximization of entropy can be implemented by the efficient least-squares process, fitting an all-pole model to the available data. An alternative to basing the estimate of range or cross-range profiles on the autoregressive model is to use the model for extending the measured data, followed by conventional Fourier transform processing. This approach retains all the measured data and uses the model derived from the measured data to extrapolate outside the measurement window. Although the resulting process cannot be shown rigorously to have maximum entropy, the procedure intuitively satisfies the first principle of data reduction; namely, all the measured data are used without alteration, and the extensions are based on a maximum entropy model.

The applicability of autoregressive modeling to RCS imaging is illustrated in the following results obtained from experimental data [20]. Five point scatterers separated in range by 10 cm were irradiated with a 10 GHz signal using a bandwidth of 3.175 GHz. Figure 7.2 shows 128 complex samples of the measured frequency response and a range profile obtained by conventional FFT processing using a Hann window. The range profile, interpolated to 1024 samples by zero-padding, shows individual scatterers not clearly resolved by the available 10 cm resolution. Autoregressive models of orders 16, 32, and 64 applied to the measured data using the Burg algorithm yield filter poles shown in relation to the unit circle in Figure 7.3. As the model order increases, the poles approach the unit circle. Because the proximity to the unit circle determines the sharpness of the peaks and the rate of decay of the extensions, the higher order models provide better representations of peaks in the estimated range profiles. The preceding assertions are supported by Figures 7.4 and 7.5. Figure 7.4 shows the maximum entropy estimates of the range profiles derived from the autoregressive model. The scatterers are not resolved by the lower order model, but are well resolved by the higher order model. Figure 7.5 shows the

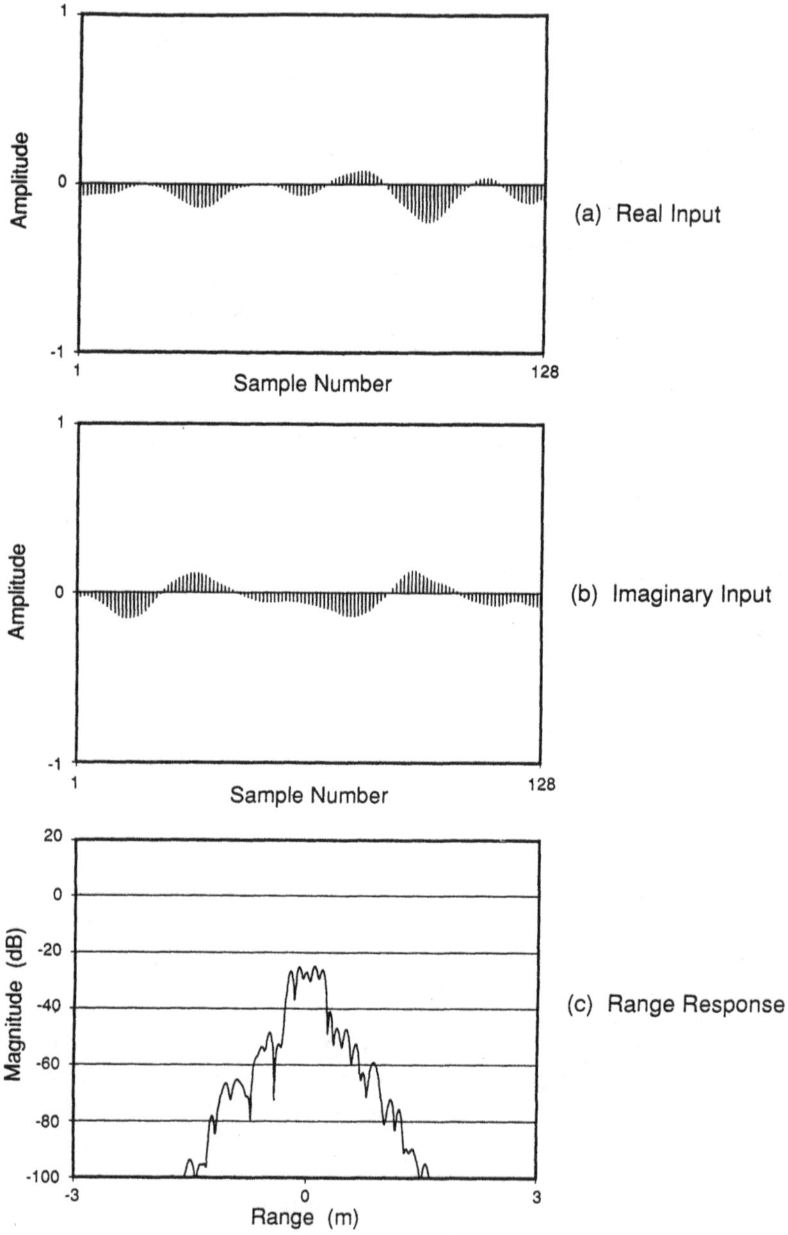

Figure 7.2 Responses of experimental data from five point scatterers.

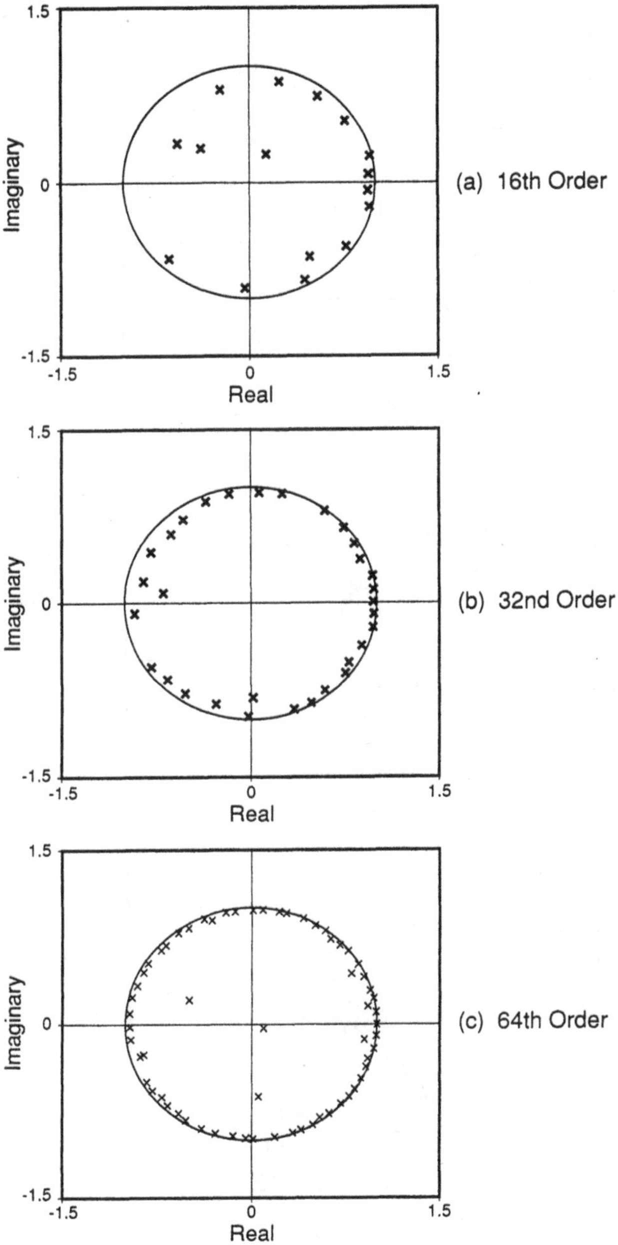

Figure 7.3 Pole locations of autoregressive models for five point scatterers.

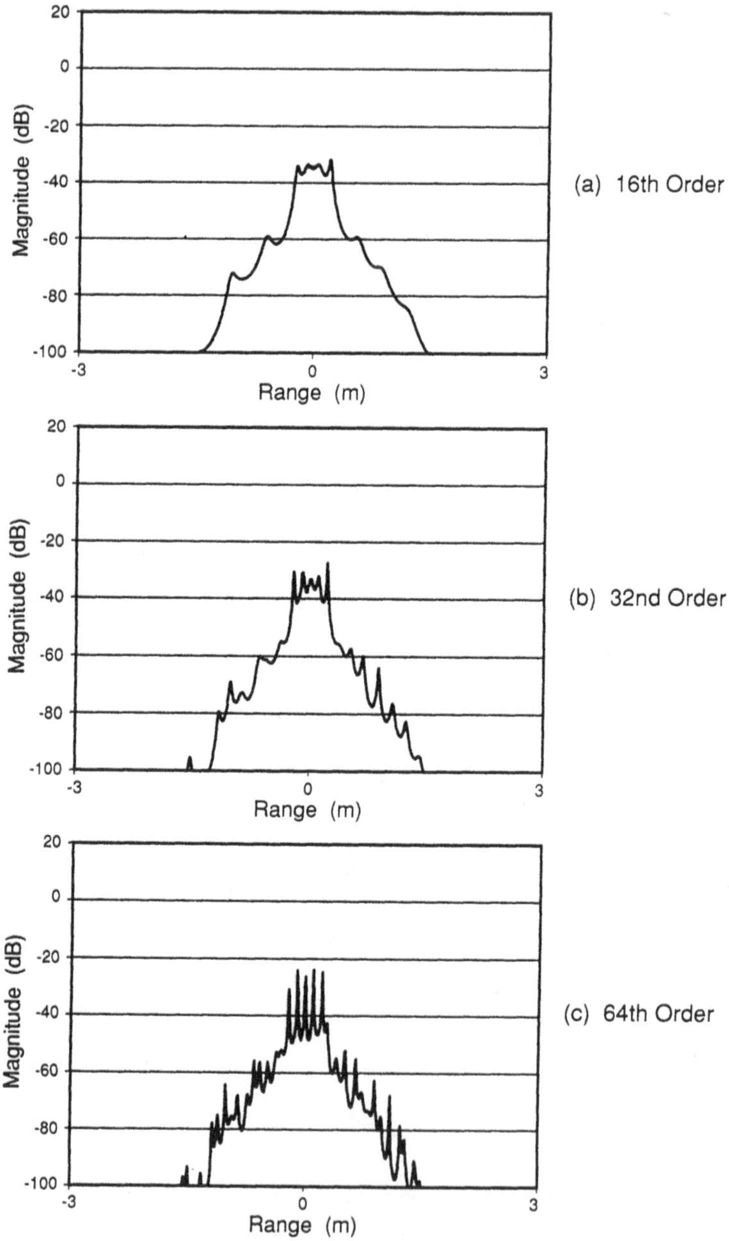

Figure 7.4 Maximum entropy estimates of range profiles for five point scatterers.

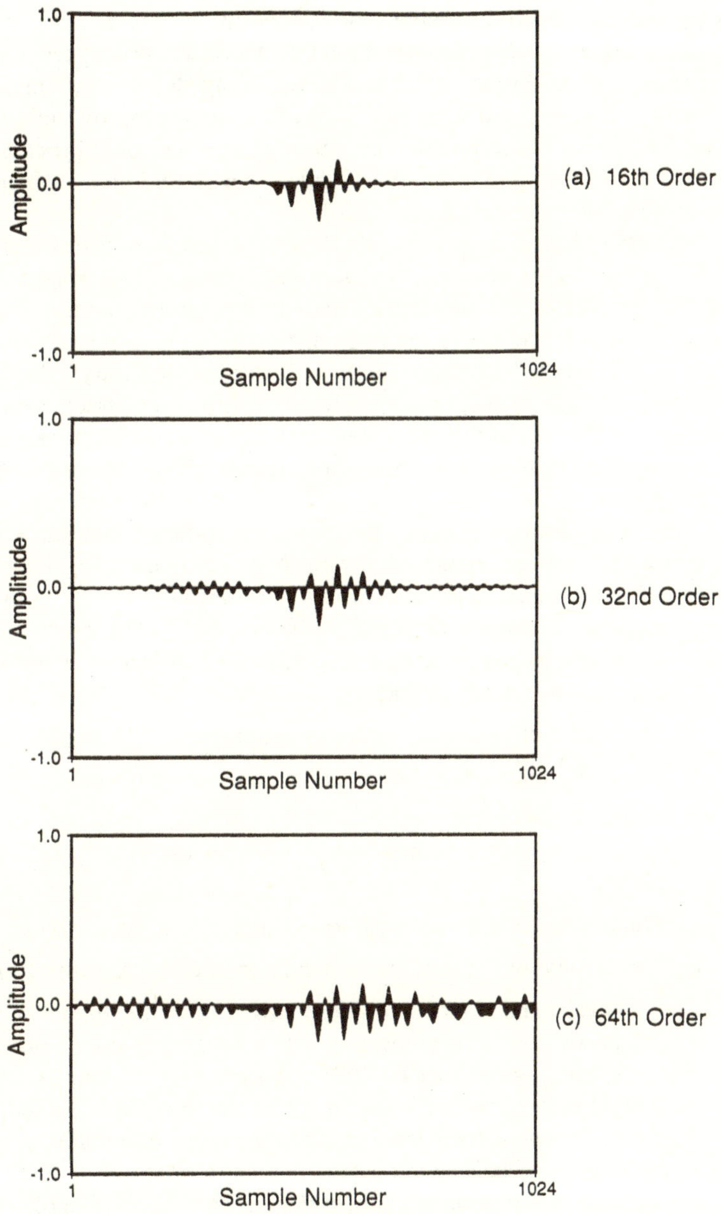

Figure 7.5 Data from five point scatterers by extended autoregressive models.

original 128 real samples (plotted in the center of the figure) extended to 1024 samples by using the autoregressive model. The imaginary samples, not shown, exhibit a similar behavior. The extensions based on the lower order model are highly damped, while those for the higher order model are more persistent; consequently, the magnitudes of the range peaks are not faithfully represented by the lower order models. Figure 7.6 shows range profiles obtained by conventional processing applied to the extended data. The results for the higher order model show a marked improvement in resolution over the original results.

Results obtained by processing data obtained from a small drone aircraft under similar conditions are shown in Figures 7.7 through 7.11. Figure 7.7 shows the 128 complex samples of the measured frequency response for the aircraft viewed from the frontal aspect, and the range response obtained by conventional FFT processing. Figures 7.8 through 7.11 show the pole locations, the maximum entropy estimates, extensions, and range responses obtained from the extended data for model orders of 16, 32, and 64. As in the preceding example, the results show increased resolution over conventional processing, but some degree of uncertainty in the magnitudes of spectral peaks.

Although limited in scope, the examples indicate that autoregressive modeling can be used to increase range resolution over methods using Fourier transform processing. Similar increases in cross-range resolution result from the analogous treatment of angular responses of complex objects [21]. The practical utility of autoregressive methods to experimental RCS data is subject to a number of limitations discussed in the literature, including:

a. The selection of model order is arbitrary;

b. Underestimating the model order can omit individual peaks;

c. Overestimating the model order can produce spurious peaks or "peak splitting," in which a proper single peak is incorrectly split into two adjacent peaks;

d. The enhancement in resolution is dependent on the data;

e. The achievable resolution depends on the signal-to-noise ratio.

The analysis and control of these limitations are topics of current research. Because the performance of autoregressive modeling is not as predictable as Fourier transform processing, guidelines for RCS applications will likely be derived empirically.

Although autoregressive modeling can be rigorously linked to maximum entropy methods in one-dimensional applications, no equivalent link is available for two-dimensional applications. One-dimensional modeling, coupled with the property that two-dimensional Fourier transform data in rectangular coordinates are separable, can be applied in the *ad hoc* procedure represented in Figure 7.12. The central array represents data from which a two-dimensional image is obtained by conventional

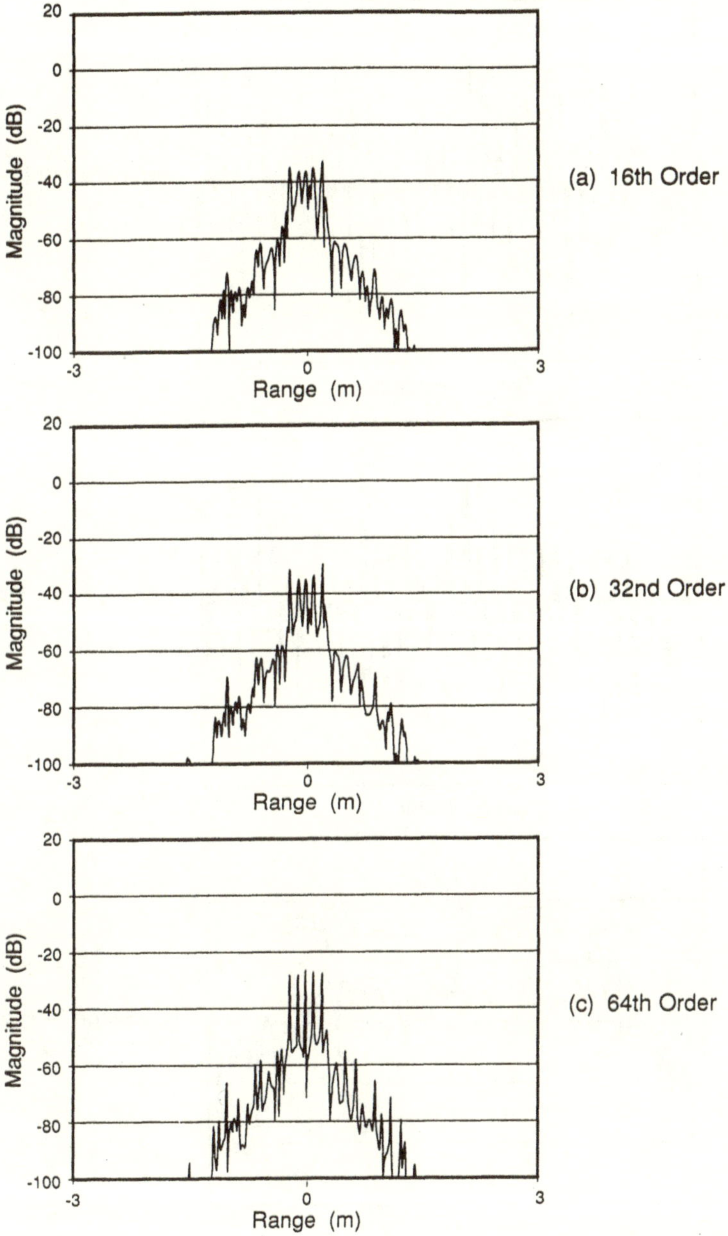

Figure 7.6 Range profiles obtained from extended data for five point scatterers.

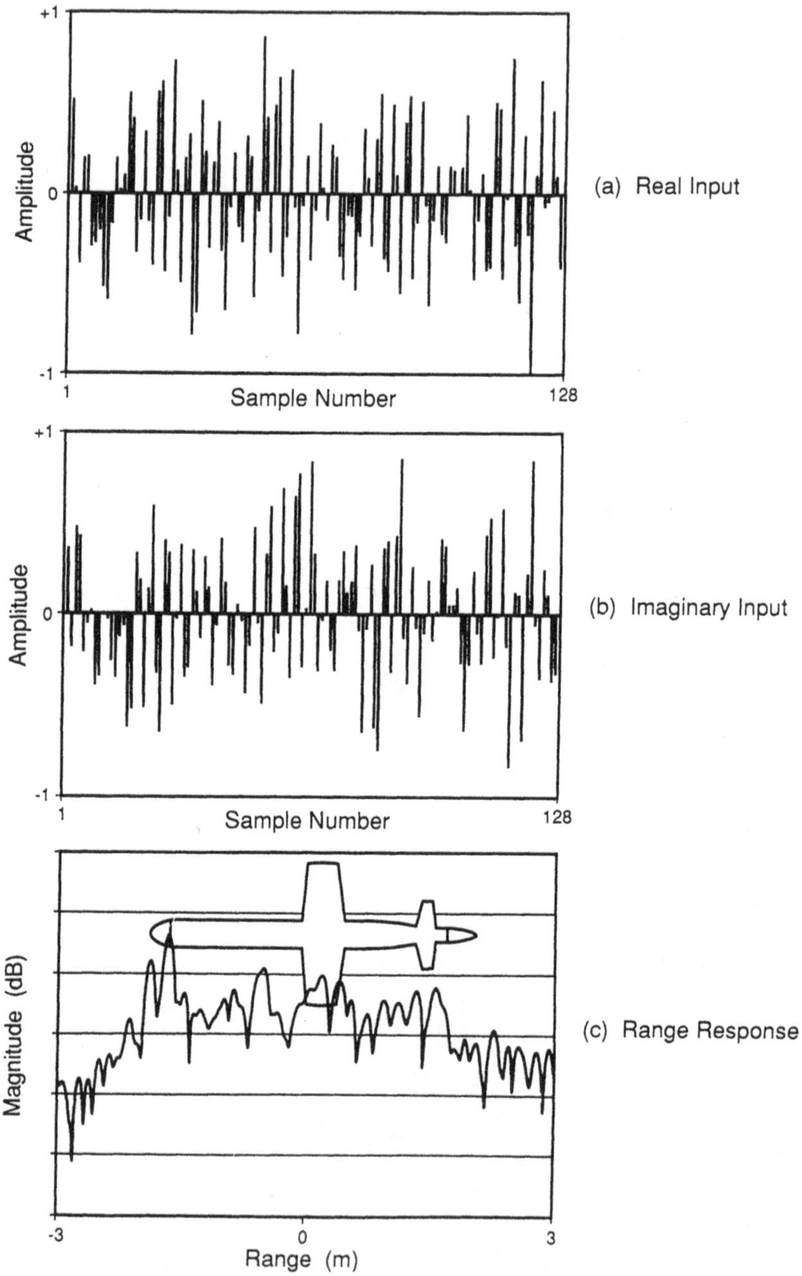

Figure 7.7 Responses of experimental data from drone aircraft.

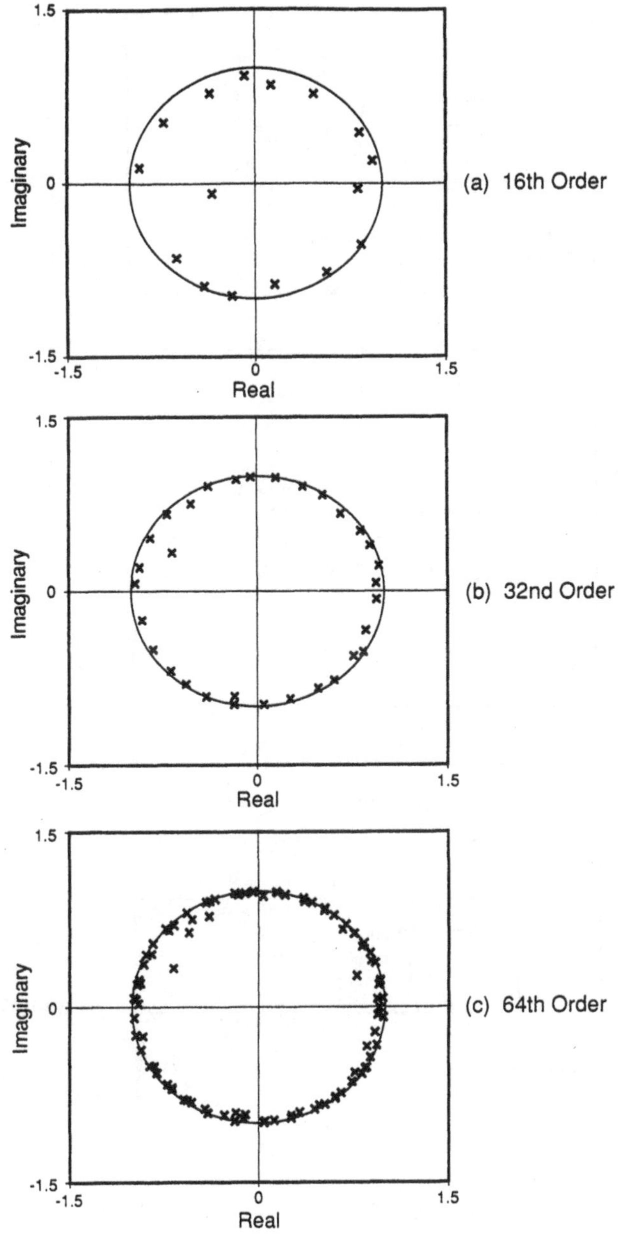

Figure 7.8 Pole locations of autoregressive models for drone aircraft.

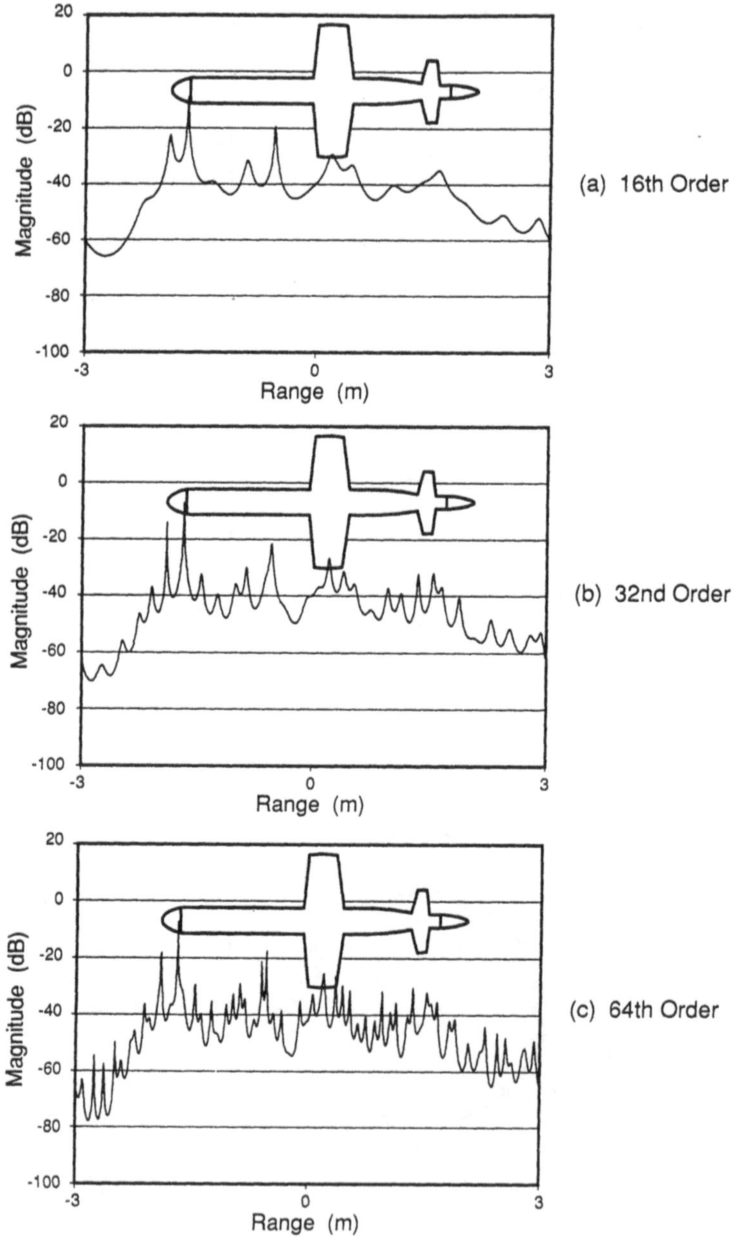

Figure 7.9 Maximum entropy estimates of range profiles for drone aircraft.

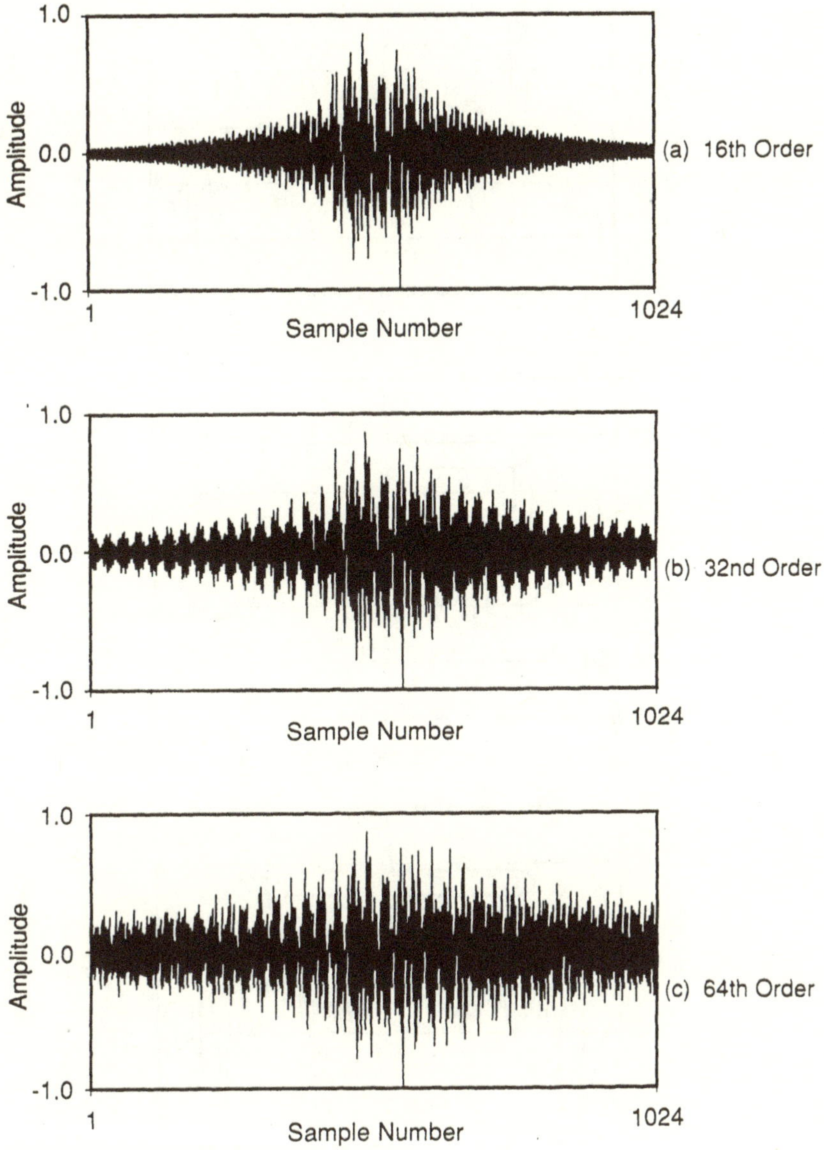

Figure 7.10 Data from drone aircraft by extended autoregressive models.

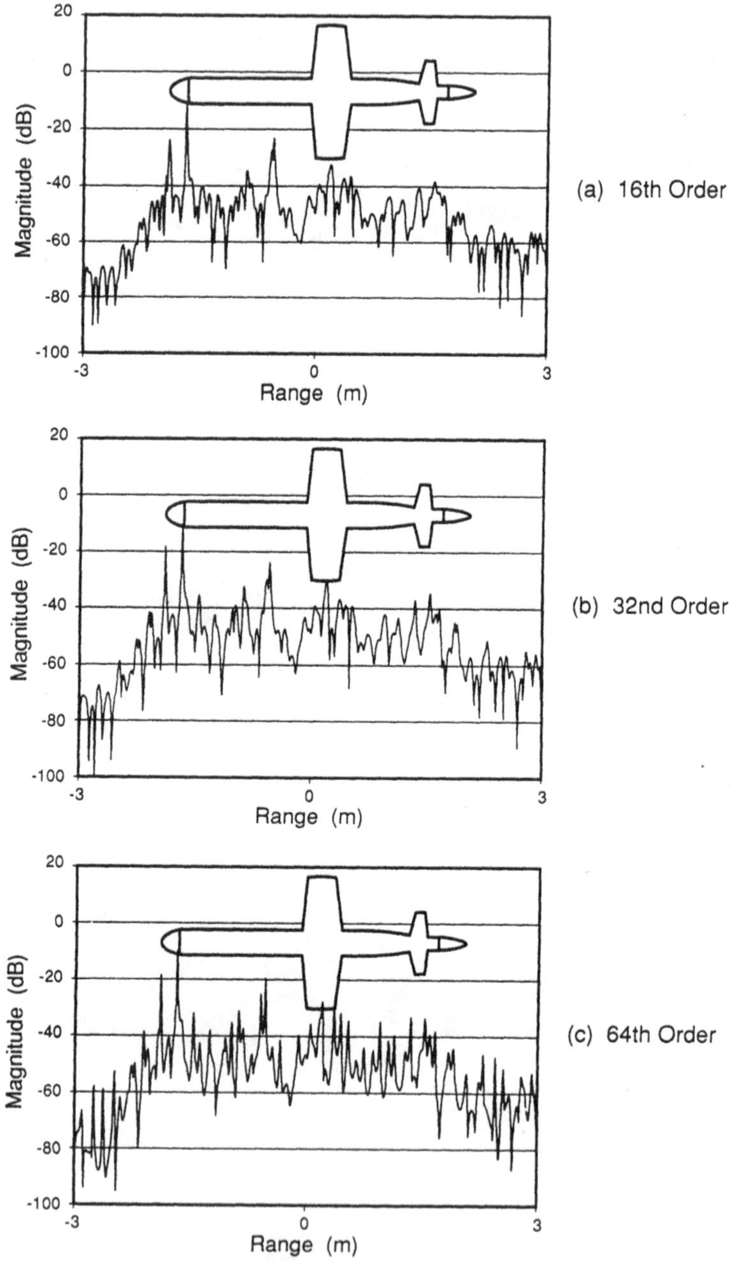

Figure 7.11 Range profiles obtained from extended data for drone aircraft.

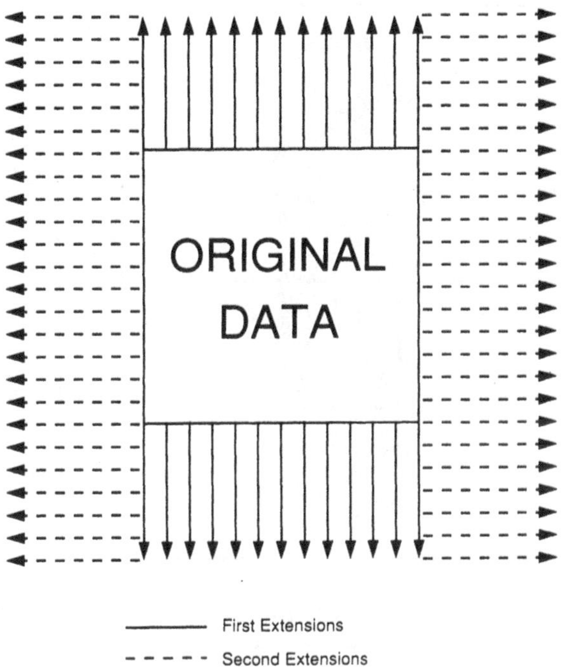

Figure 7.12 Two-dimensional extension of data array by autoregressive modeling.

processing. One-dimensional autoregressive modeling can be used to extend the data in the vertical direction by processing columns of the array, followed by similar processing on rows of the array. The data in the available array are extrapolated by this method, leading to increased resolution. The data occupying the corners of the extended array constitute "extensions of extensions." Although the procedure has no theoretical justification, it appears intuitively acceptable because the measured data are retained intact, and the extensions are tapered by two-dimensional windows prior to a Fourier transform. This method was applied to the set of data obtained from the five point scatterers. Figure 7.13 shows images obtained from conventional processing applied to the original data, and to the data array extended by factors of 2, 4, and 8 in each direction. The images derived from the extended data show the expected increase in resolution in both range and cross range. Figure 7.14 shows images obtained from conventional processing applied to the drone aircraft data, and to the data extended by a factor of 2 in each direction. These results show a marked improvement in resolution.

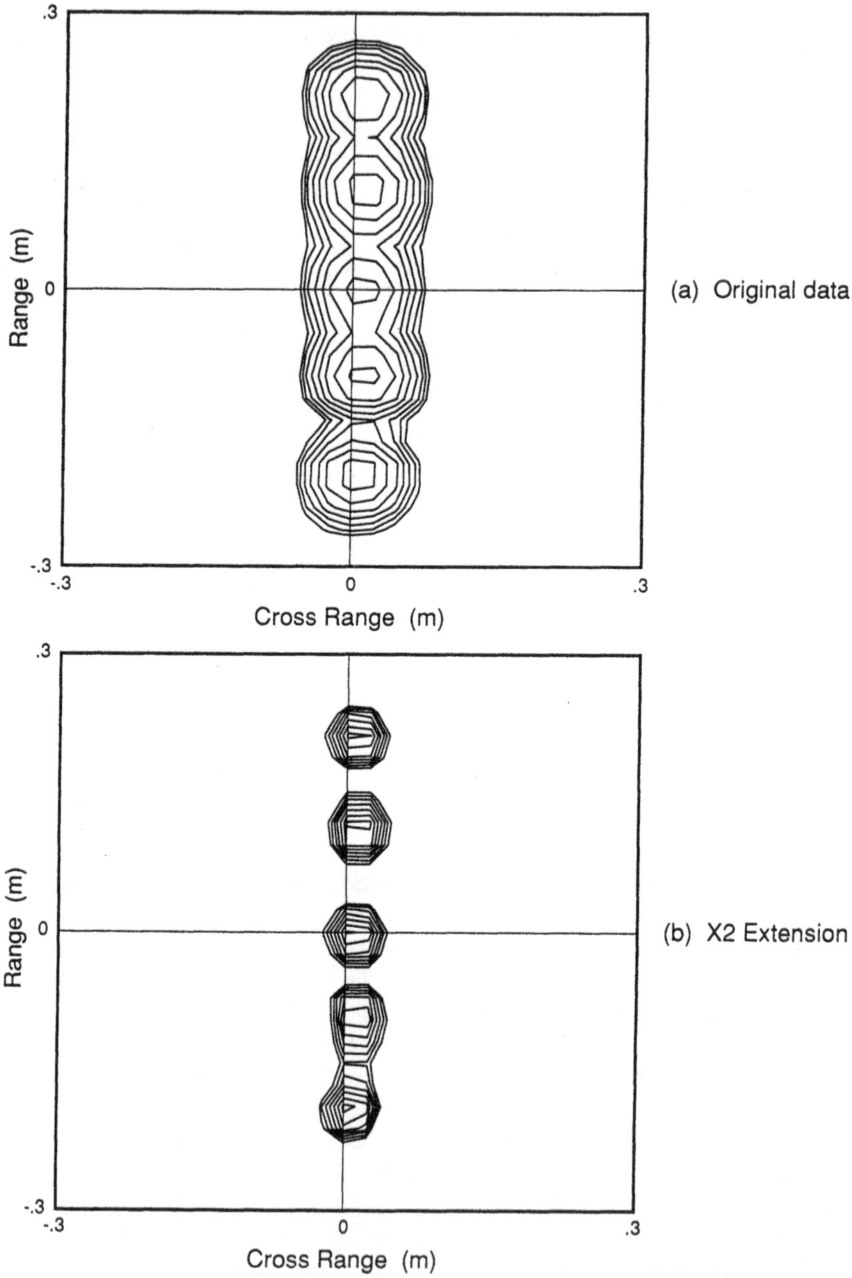

Figure 7.13 Images obtained from extended data for five point scatterers.

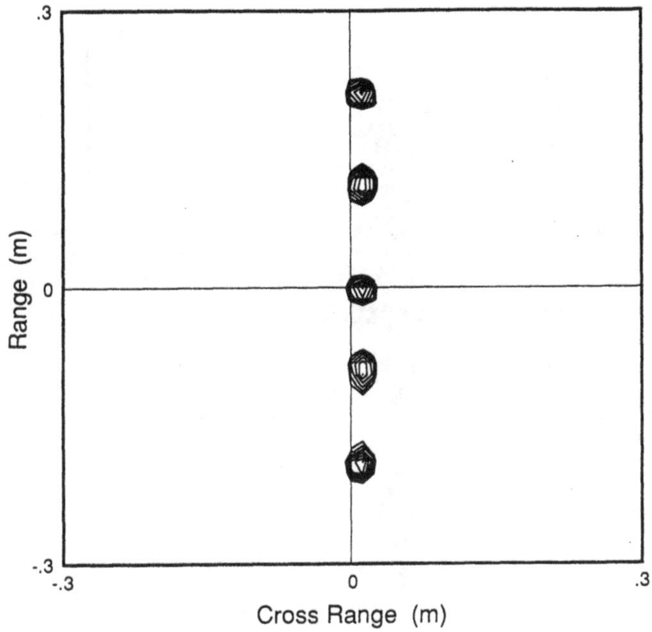

(a) X4 Extension

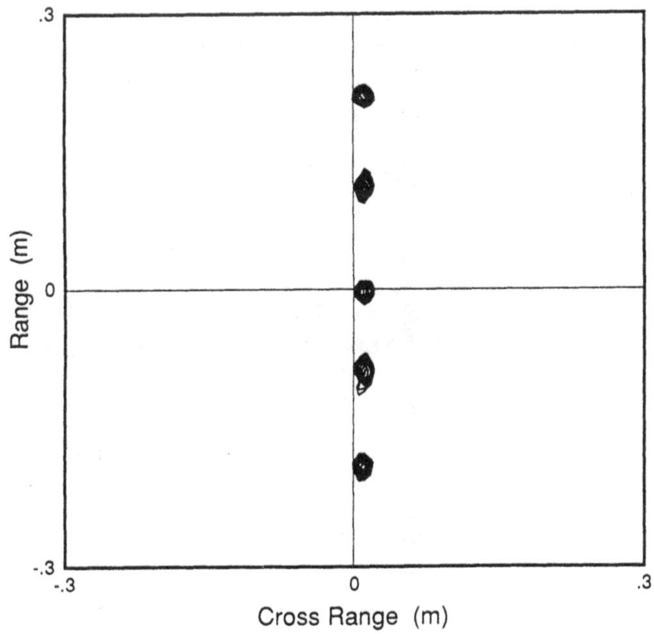

(b) X8 Extension

Figure 7.13 (continued).

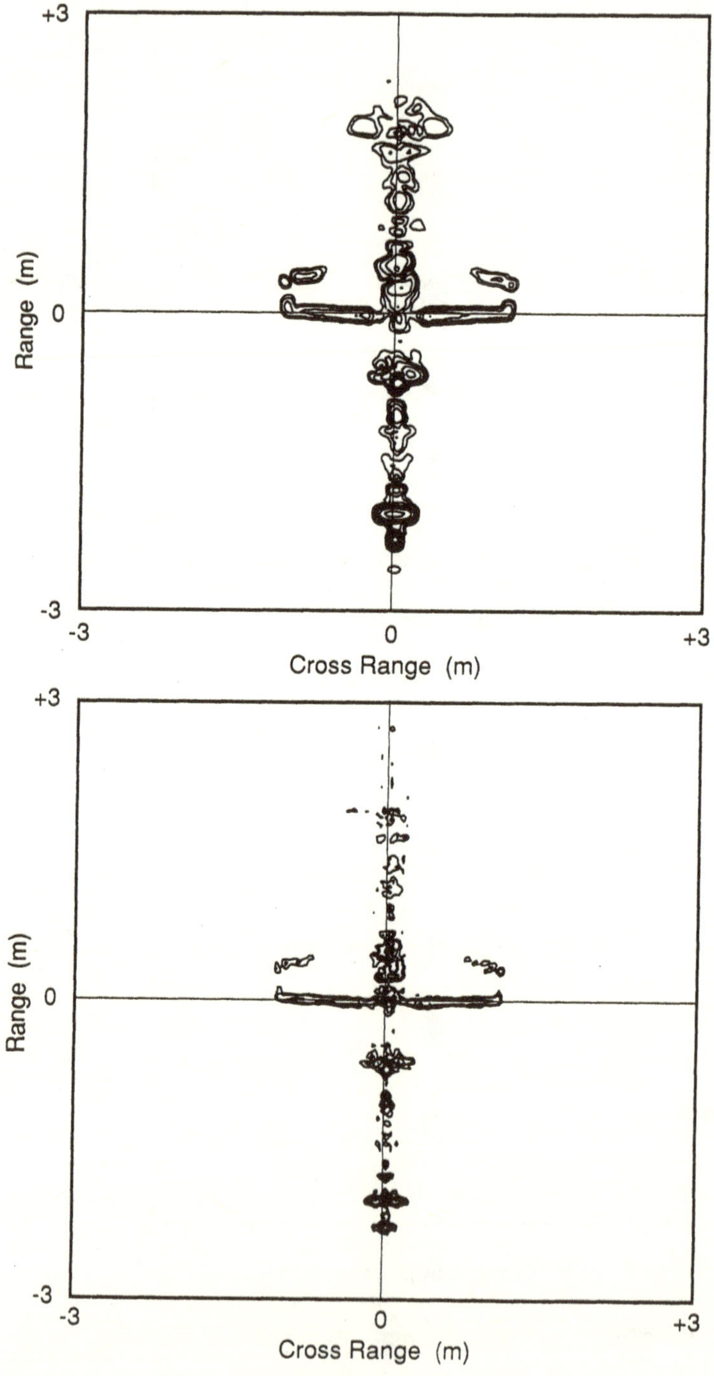

Figure 7.14 Images obtained from extended data of drone aircraft.

REFERENCES

1. Childers, D.G., ed., *Modern Spectrum Analysis,* IEEE Press, New York, 1978.
2. Kesler, S., ed., *Modern Spectrum Analysis II,* IEEE Press, 1986.
3. Special Issue on Spectral Estimation, *Proc. IEEE,* Vol. 70, No. 9, September 1982.
4. Marple, L., "A New Autoregressive Spectrum Analysis Algorithm," *IEEE Trans. Acoustics, Speech, and Signal Processing,* Vol. ASSP-28, No. 4, August 1980.
5. Kay, S., and S.L. Marple, "Spectrum Analysis—A Modern Perspective," *Proc. IEEE,* Vol. 69, No. 11, November 1981.
6. Jaynes, E.T., "On the Rationale of Maximum Entropy Methods," *Proc. IEEE,* Vol. 70, No. 9, September 1982.
7. Gull, S.F., and G.J. Daniell, "Image Reconstruction from Incomplete and Noisy Data," *Nature,* vol. 272, April 1978.
8. Burg, J.P., "A New Analysis Technique for Time Series Data," NATO Advanced Study Institute on Signal Processing with Emphasis on Underwater Acoustics, August 12–23, 1968.
9. Gull, S.F., and J. Skilling, "The Entropy of an Image" in Smith, C.R., and W.T. Grandy, eds., *Maximum-Entropy and Bayesian Methods in Inverse Problems,* D. Reidel, Hingham, MA, 1985.
10. Skilling, J., and R.K. Bryan, "Maximum Entropy Image Reconstruction: General Algorithm," *Monthly Notices, Royal Astronomical Society,* 211, 1984, pg. 111–124.
11. Shannon, C.E., "A Mathematical Theory of Communication," *Bell System Technical J.,* Vol. 27, 1948, pp. 379–423, 623–656.
12. Ables, J.G., "Maximum Entropy Spectral Analysis," *Astron. Astrophys.,* Suppl. 15, 1964, pp. 383–393.
13. Goldman, S., *Information Theory,* Prentice-Hall, Englewood Cliffs, NJ, 1953, pp. 318–320.
14. Wernecke, S.J., and L.R. D'Addario, "Maximum Entropy Image Reconstruction," *IEEE Trans. Computers,* Vol. C-26, April 1977, pp. 351–364.
15. Lim, J.S., and N.A. Malik, "A New Algorithmn for Two-Dimensional Maximum Entropy Power Spectrum Estimation," *IEEE Trans. on Acoustics, Speech, and Signal Processing,* Vol. ASSP-29, No. 3, June 1981.
16. Heidbreder, G., and F. Van Roekeghem, "Maximum Entropy Reconstruction of Radar Images of Rotating Targets," *IASTED International Symposium on Signal Processing and Its Applications,* Brisbane, Australia, August 1987.
17. Burg, J.P., "Maximum Entropy Spectral Analysis," *Proc. 37th Meeting of the Society of Exploration Geophysicists,* 1967.
18. Walton, E.K., "Comparison of Fourier and Maximum Entropy Techniques for High-resolution Scattering Studies," *Radio Science,* Vol. 22, No. 3, May–June 1987, pp. 350–356.
19. Van Den Bos, A., "Alternative Interpretation of Maximum Entropy Spectral Analysis," *IEEE Trans. Information Theory,* Vol. IT-17, July 1971, 493–494.
20. Bati, A.C., "Application of Maximum Entropy Methods to Radar Imaging," Master of Science Thesis, California State University, Northridge, CA, June 1990.
21. Chen, Y., and R.S. Berkowitz, "Characteristics of Autoregressive Spectral Estimation in Range-Doppler Imaging," *Proc. IEEE National Radar Conf.,* 1988, pp. 118–122.

Appendix A
Rectification of Curved Two-Dimensional Radar Image Fields

Two-dimensional images of rotating objects irradiated by microwave signals can be generated by methods described in the text. Each image pixel represents the magnitude of the signal scattered from a two-dimensional cell defined by its range and cross-range coordinates. For an infinite observation distance, the loci of constant range and cross range describe two sets of parallel, mutually orthogonal lines. When the observation distance is finite, the lines of constant range and cross range respectively describe a set of radial lines and concentric circular arcs disposed about the radar antenna. This results in a curved image field as shown in Figure A.1.

If the two-dimensional array of image cells is plotted on a rectangular coordinate system, a distortion of the image results. This appendix derives the formulas to correctly locate image points based on the values of the measured range and apparent cross-range. As illustrated in Figure A.2, consider a point scatterer on a two-dimensional target field $P(x, y)$, rotating about a center of rotation, O, at a radius, ρ. An antenna, A, is located at a distance, R, from the center of rotation. The line OA defines the y-axis of a Cartesian coordinate system, the origin of which is located at O. The radius vector OP to the point P subtends an angle, θ, with the x-axis of a θ-centered coordinate system. At any instant, the point P is at a distance r from the antenna A.

For each angle θ, the radar imaging system located at A measures the range r. By processing signals recorded over a finite angular sector, the system measures a "doppler" quantity, which is proportional to $dr/d\theta$. For each angle, therefore, the system outputs two quantities proportional to range and doppler and denoted by r and s, respectively. Given the two measured quantities r and s, in conjunction with the measurement distance R, we seek a transformation to determine the x and y coordinates of the object point in a Cartesian coordinate system centered on O.

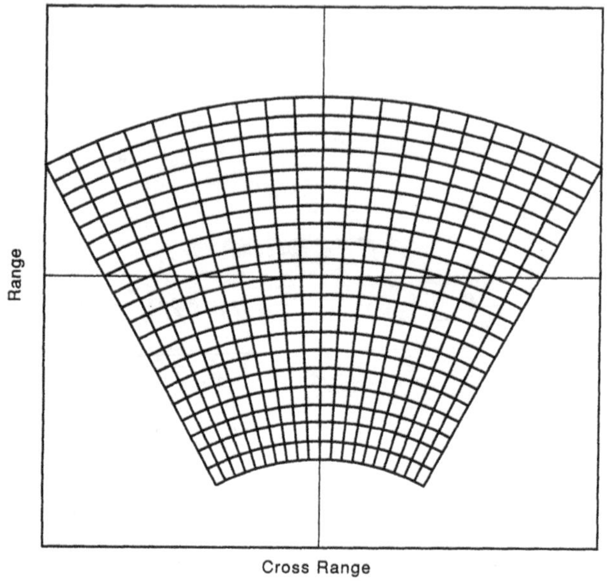

Figure A.1 Curved image field resulting from short observation distance.

From geometric considerations of Figure A.2, the following relations can be stated:

$$x = \rho \cos\theta \tag{A.1}$$

$$y = (r^2 - x^2)^{1/2} - R \tag{A.2}$$

$$r = (R^2 + 2R\rho \sin\theta + \rho^2)^{1/2} \tag{A.3}$$

The measured doppler quantity is proportional to $dr/d\theta$, obtained by differentiating (A.3):

$$s \simeq \frac{dr}{d\theta} = \frac{R\rho \cos\theta}{(R^2 + 2R\rho \sin\theta + \rho^2)^{1/2}} \tag{A.4}$$

Note that, in the limit of infinite R, (A.4) properly reduces to

$$\lim_{R\to\infty} (dr/d\theta) = \rho \cos\theta = x \tag{A.5}$$

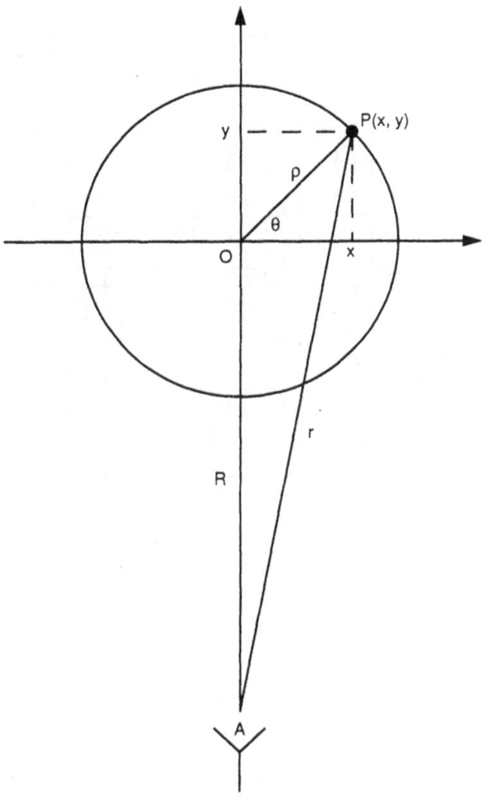

Figure A.2 Geometry of radar observing a rotating point.

which is the true cross range for an infinite observation distance. We combine (A.1), (A.3), and (A.4) to yield:

$$s = R\rho \cos\theta / r = x(R/r) \tag{A.6}$$

The necessary transformation relations are obtained from (A.2) and (A.6).

$$x = s(r/R) \tag{A.7}$$

$$y = (r^2 - x^2)^{1/2} - R \tag{A.8}$$

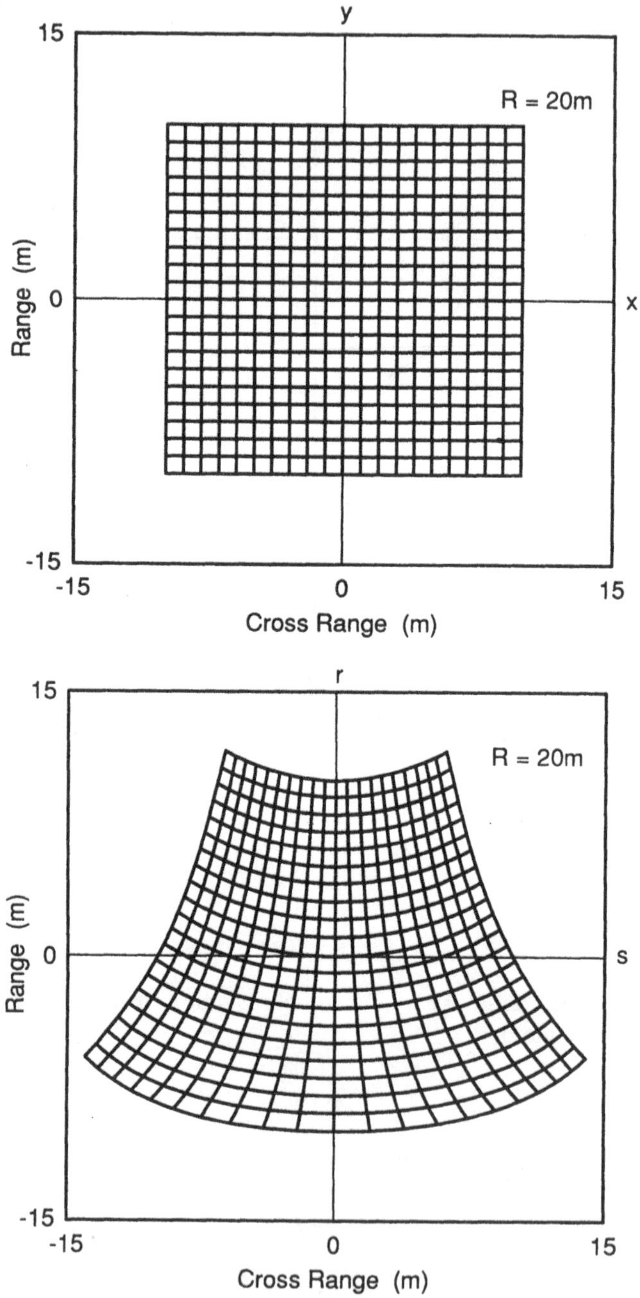

Figure A.3 Mapping of objects coordinates into perceived radar coordinates.

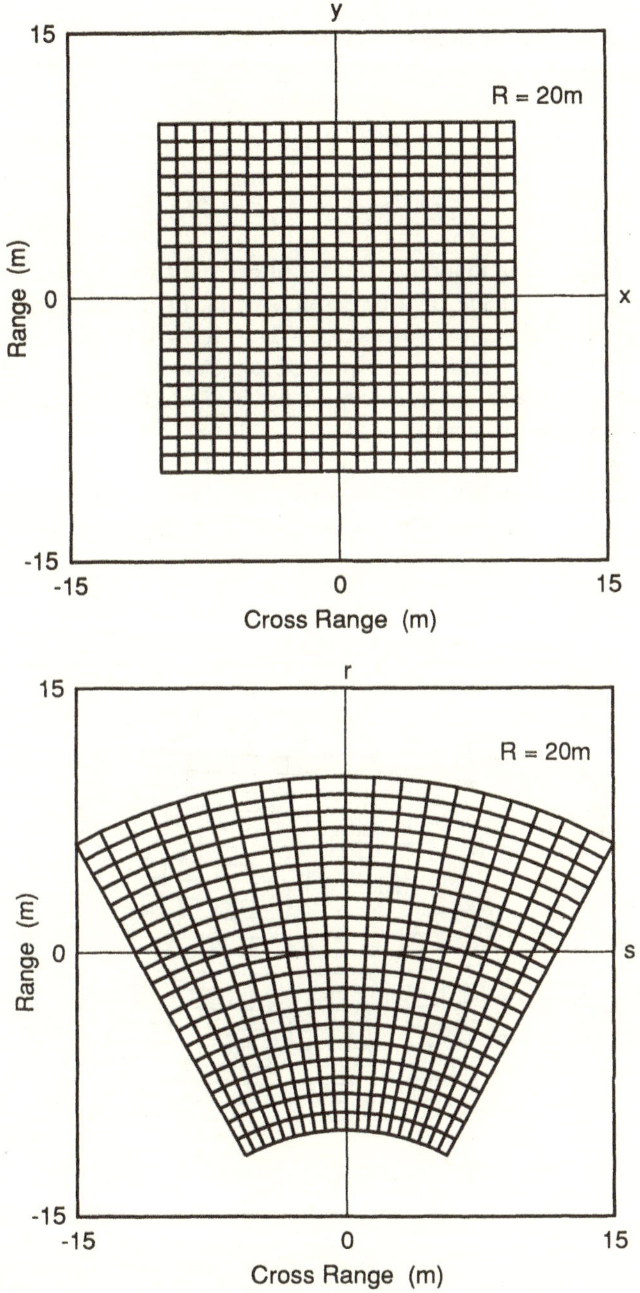

Figure A.4 Mapping of perceived radar coordinates into object coordinates.

Equations (A.7) and (A.8) are sufficient to transform the measured and known quantities r, s, and R into the required quantities x and y. The corresponding reciprocal transformations are

$$s = x(R/r) \tag{A.9}$$

$$r = (x^2 + y^2 + R^2 + 2Ry)^{1/2} \tag{A.10}$$

Figure A.3 shows the mapping of lines of constant x and y into lines of constant r and s. Therefore, the image of a target consisting of a rectangular grid observed from a relatively short distance is distorted when the range and doppler coordinates perceived by the radar, r and s, are plotted on Cartesian coordinates. When the perceived target coordinates are transformed by (A.7) and (A.8), the resulting image is rectified into a Cartesian grid.

Figure A.4 shows the x and y coordinates of lines which map into lines of constant r and s. As expected, these sets of lines consist of a polar raster centered on the location of the antenna. Figures A.5 and A.6 illustrate the distortion which

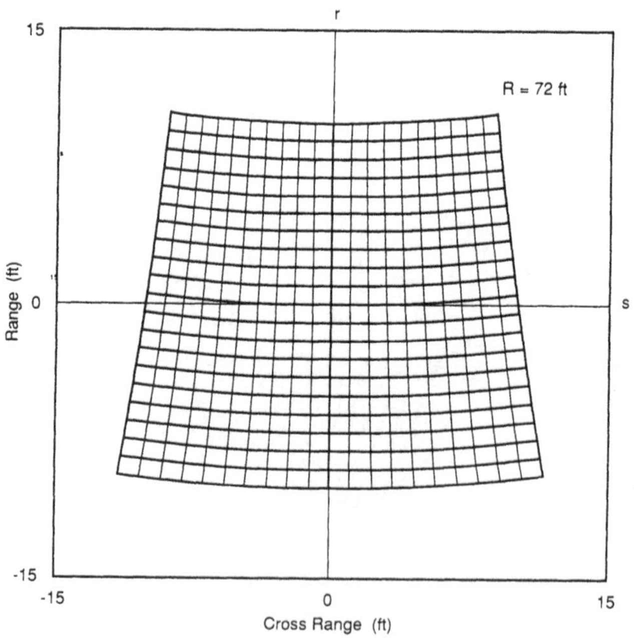

Figure A.5 Radar map of orthogonal grid target for a radar distance of 72 ft.

typically occurs when a 20-ft square target field is mapped from a measurement range of 72 ft.

Figure A.5 represents the radar map of a target consisting of lines that are parallel and orthogonally disposed in the object space. Figure A.6 is the spatial distribution of target points that map into lines of constant range and cross range as perceived by the radar system.

Figures A.7 and A.8 show two-dimensional images of an aircraft obtained by processing measurements obtained from a distance of 72 ft for viewing angles of 0° and 12.8°. The images, shown with an outline of the aircraft superimposed, were processed without the geometric correction for the curved field. The scattering from the two wing tips is shown displaced from the correct location as a consequence of the geometric distortion.

Figures A.9 and A.10 show the same images as Figures A.7 and A.8, replotted with the geometric correction applied. The scattering from the two wing tips appears properly registered, indicating that the correction technique is effective in rectifying the geometric distortion induced by measuring large targets at short ranges.

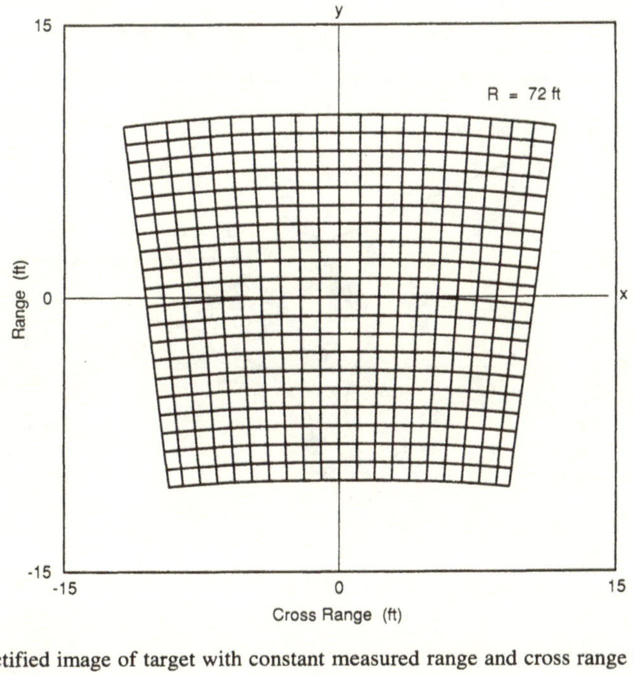

Figure A.6 Rectified image of target with constant measured range and cross range for a radar distance of 72 ft.

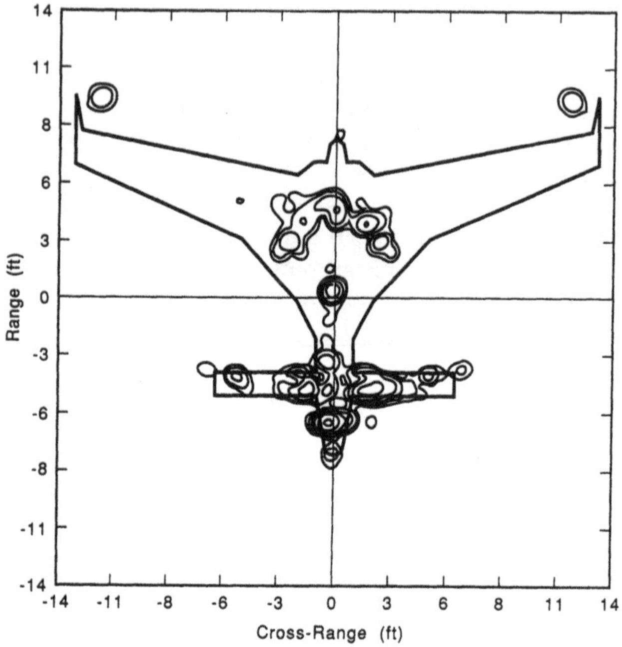

Figure A.7 Two-dimensional radar image of aircraft viewed at 0° without geometric correction.

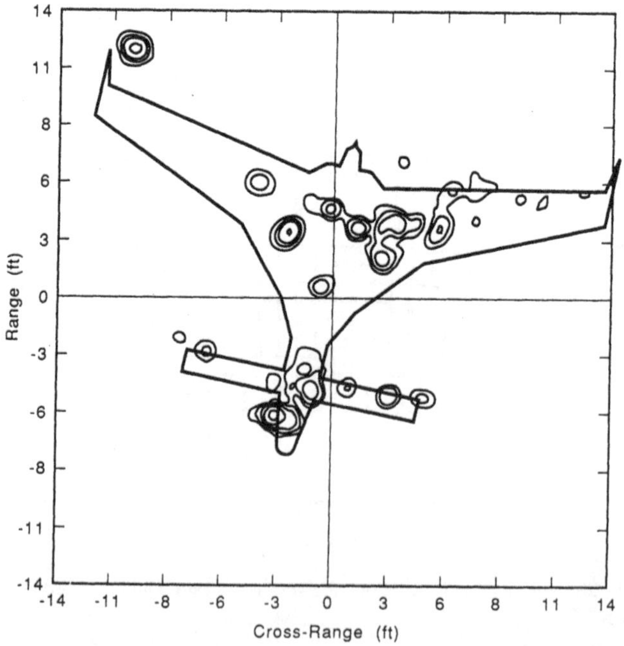

Figure A.8 Two-dimensional radar image of aircraft viewed at 12.8° without geometric correction.

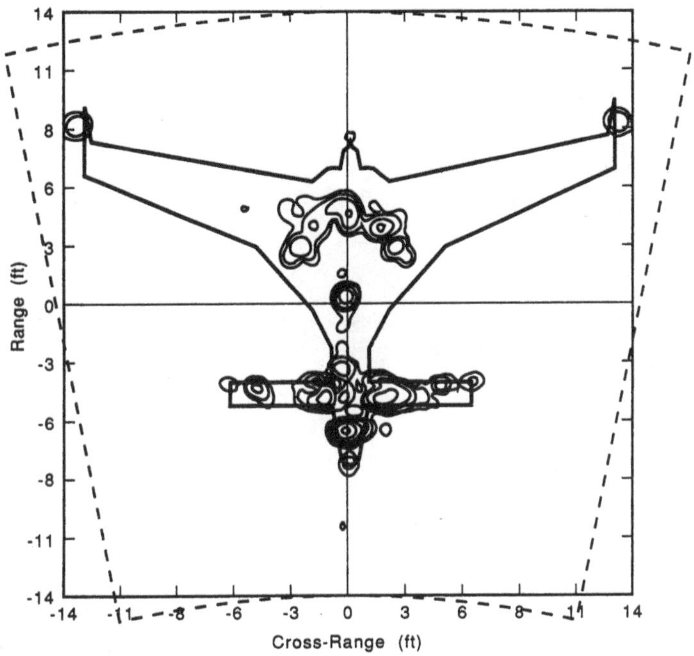

Figure A.9 Two-dimensional radar image of aircraft viewed at 0° with geometric correction applied.

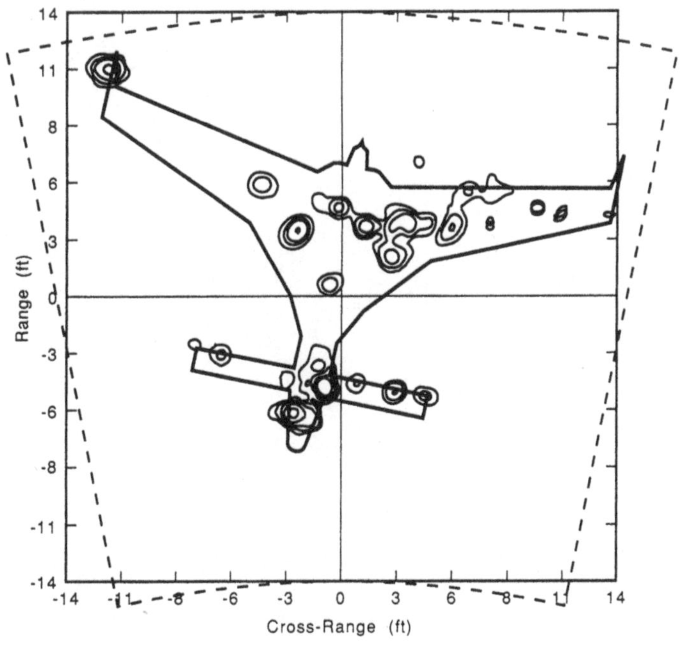

Figure A.10 Two-dimensional radar image of aircraft viewed at 12.8° with geometric correction applied.

Appendix B
Application of Tapered Windows to Annular-Shaped Spectra

Processing a two-dimensional record of signals reflected from a rotating object as a function of frequency and rotation angle provides a two-dimensional spatial image of the object's reflectivity density. When mapped in a polar format, the recorded data represent a spatial spectrum related to the object image by a two-dimensional Fourier transform. Because in practical measurements the range of irradiating frequencies is restricted to a limited bandwidth, the available data represent a sample of the spatial spectrum over an annular region. The point-spread function of the imaging process, given by the Fourier transform of the annular spectrum, consists of a central peak surrounded by circular sidelobes, which generally increase as the width of the spectral annulus decreases. In view of practical limitations, we therefore seek the best image reconstruction from narrow annular spectra. In an analogous situation, spectral estimates obtained from time-limited signals are improved by the use of tapered windows which have been the subject of considerable research [1]. The mechanism for reducing the spectral sidelobes is intuitively acceptable: discontinuities in the time function represent high frequencies in the spectrum and tapering to reduce discontinuities therefore reduces high frequency sidelobes. As a result, smoothly tapered time windows reduce spectral sidelobes.

Indiscriminate applications of tapered windows to two-dimensional functions may not produce the desired reduction in the sidelobes of the transforms, as shown in the example of Figure 4.24. In this case, a cosinusoidal taper applied to an annular spectrum increased the sidelobe level rather than decreasing it. On first consideration, this appears contrary to the intuitive notion that smooth tapers in one domain decrease sidelobes in the other. This apparent paradox is resolved in the analysis which follows.

Because the spectra considered are radially symmetric functions, the corresponding spatial functions, being related by a Fourier transform, are also radially symmetric [2]. This fact allows each of the functions to be characterized by a single variable expressing the variations as a function of radial distance. Although the functions may be expressed by a single variable, they are nevertheless two-dimensional functions, and must be treated as such when transformed. The Fourier transform can be reduced to a one-dimensional integral by use of the projection theorem which has been widely applied in the analysis of tomographic imaging [3–5]. The theorem states that a central slice of the two-dimensional spectrum at an angle θ can be obtained by a one-dimensional transform of the projection of the spatial function onto an axis oriented at the same angle θ. The theorem can be proved by the following arguments.

Consider a two-dimensional function $f(x, y)$ expressed in terms of coordinates u, v rotated about x, y by the angle θ as shown in Figure B.1. The function $f(x, y)$ can be expressed in terms of u and v by the rotational transformation $u = x \cos\theta + y \sin\theta$, $v = -x \sin\theta + y \cos\theta$:

$$f(x,y) = f(u \cos\theta - v \sin\theta, u \sin\theta + v \cos\theta) \tag{B.1}$$

The projection of $f(x,y)$ onto the u axis corresponding to the angle θ is a one-dimensional function of the single variable u:

$$p(u;\theta) = \int f(u \cos\theta - v \sin\theta, u \sin\theta + v \cos\theta)\, dv \tag{B.2}$$

The notation $p(u;\theta)$ indicates that the projection is a function of the variable u and is unique for each θ. The one-dimensional Fourier transform of $p(u;\theta)$ with respect to the variable u is

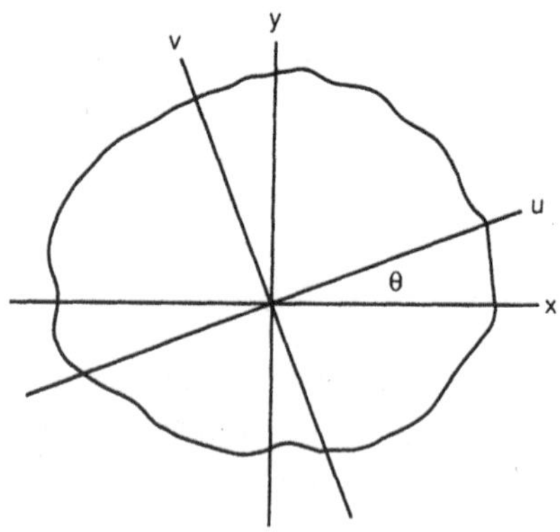

Figure B.1 Rotated coordinate system.

$$P(\omega;\theta) = \int p(u;\theta)e^{-j\omega u}\,du \tag{B.3}$$

The Fourier transform of the function $f(x,y)$, expressed in polar coordinates, is

$$F(\omega,\theta) = \int\int f(x,y)e^{-j(\omega x \cos\theta + \omega y \sin\theta)}dx\,dy \tag{B.4}$$

where ω and θ are the radial and angular coordinates in the spatial frequency plane. Equation (B.4), expressed in terms of the coordinate rotation relations, can be rewritten as

$$F(\omega,\theta) = \int\int f(u\cos\theta - v\sin\theta,\, u\sin\theta + v\cos\theta)e^{-ju\omega}\,du\,dv \tag{B.5}$$

$$F(\omega,\theta) = \int p(u;\theta)e^{-ju\omega}\,du \tag{B.6}$$

Equation (B.6) is thus equal to (B.3), yielding:

$$F(\omega,\theta) = P(\omega;\theta) \tag{B.7}$$

Equation (B.7), termed the projection theorem, states that the one-dimensional Fourier transform of the projection with orientation θ is a section of the two-dimensional Fourier transform of $f(x,y)$ passing through the origin and subtending an angle θ with the ω_x axis.

In our application, we use the above theorem by computing the projection of the annular spectrum and then performing a one-dimensional Fourier transform on the result. Because both spatial and spectral functions are radially symmetric, the result completely characterizes the point-spread function. The recognition that the radial behavior of the spatial function is determined by the Fourier transform of the projection of the spectral function is significant. All the features of one-dimensional tapered windows, which link smooth tapers to sidelobe reduction, are pertinent to the two-dimensional case if they are applied to the one-dimensional projection of the spectral function. Thus, smoothing the projection of the annular spectrum determines the sidelobe structure of the point-spread function in the radial direction.

The upper plot of Figure B.2 shows the cross section of an annular spectrum with unit amplitude for radial values between 32 and 64 units, and zero elsewhere. The projection of the annular spectrum is shown on the central plot, and the Fourier transform of the projection on the lower plot. The Fourier transform, representing the point-spread function corresponding to the annular spectrum, has sidelobes of magnitude 0.3 relative to the peak. The two central peaks of the projection are the

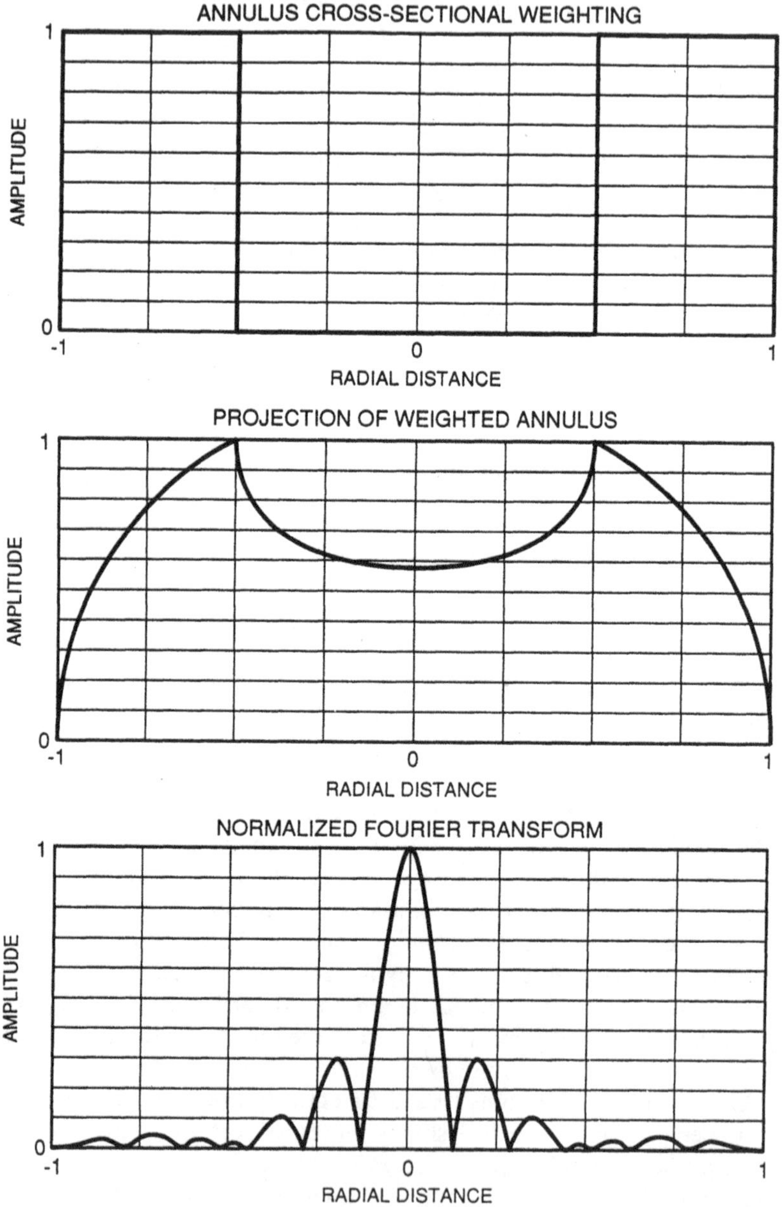

Figure B.2 Annular spectrum with uniform weighting.

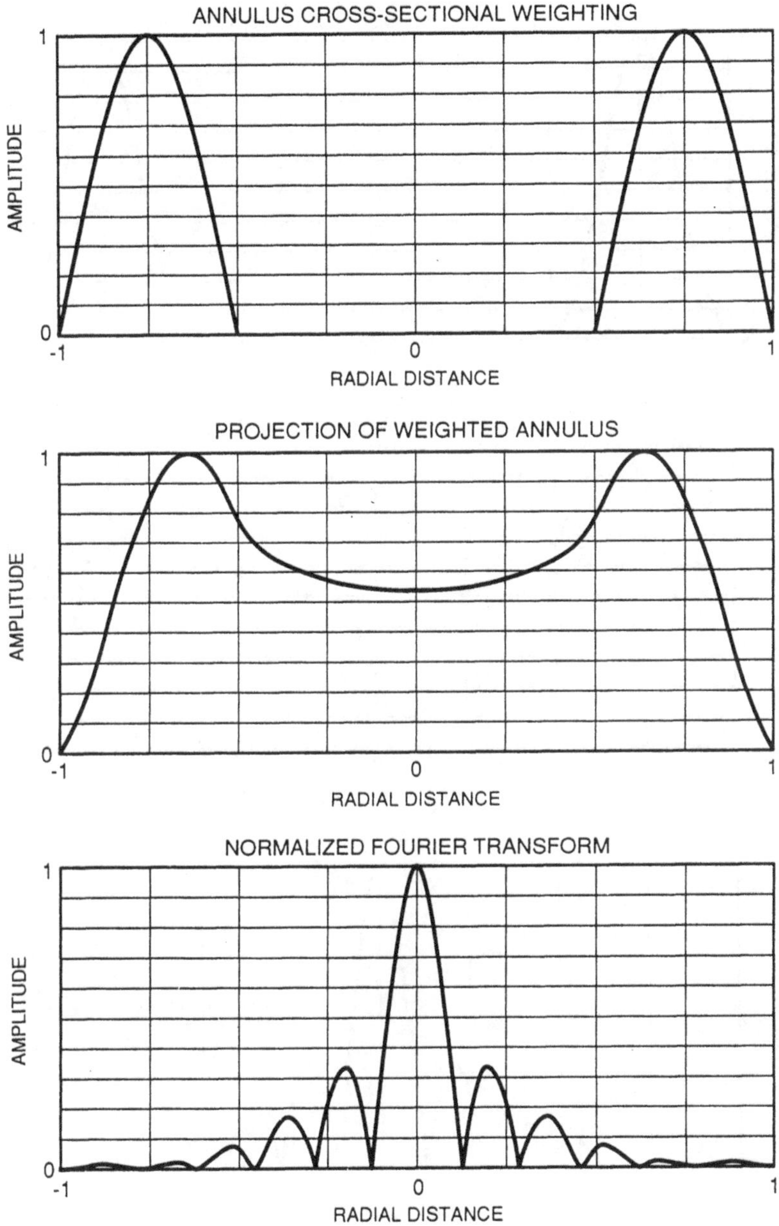

Figure B.3 Annular spectrum with cos(·) tapered weighting.

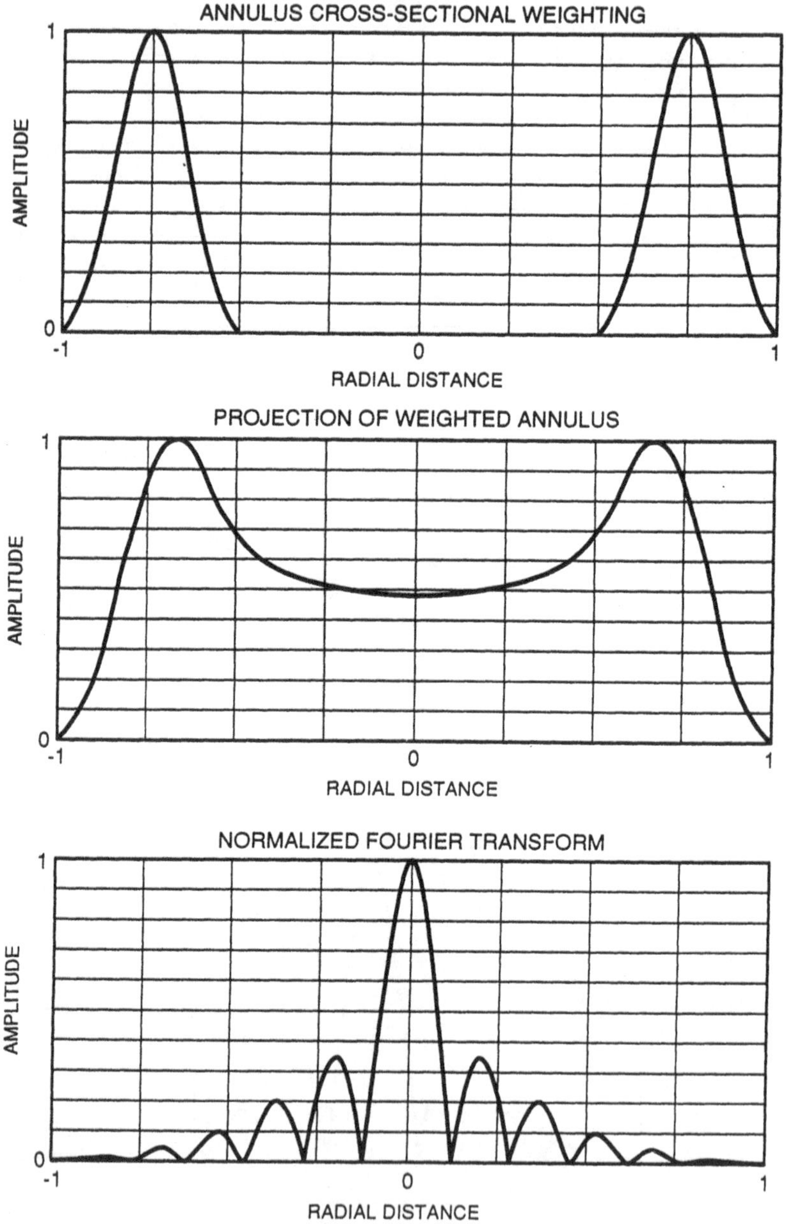

Figure B.4 Annular spectrum with $\cos^2(\cdot)$ tapered weighting.

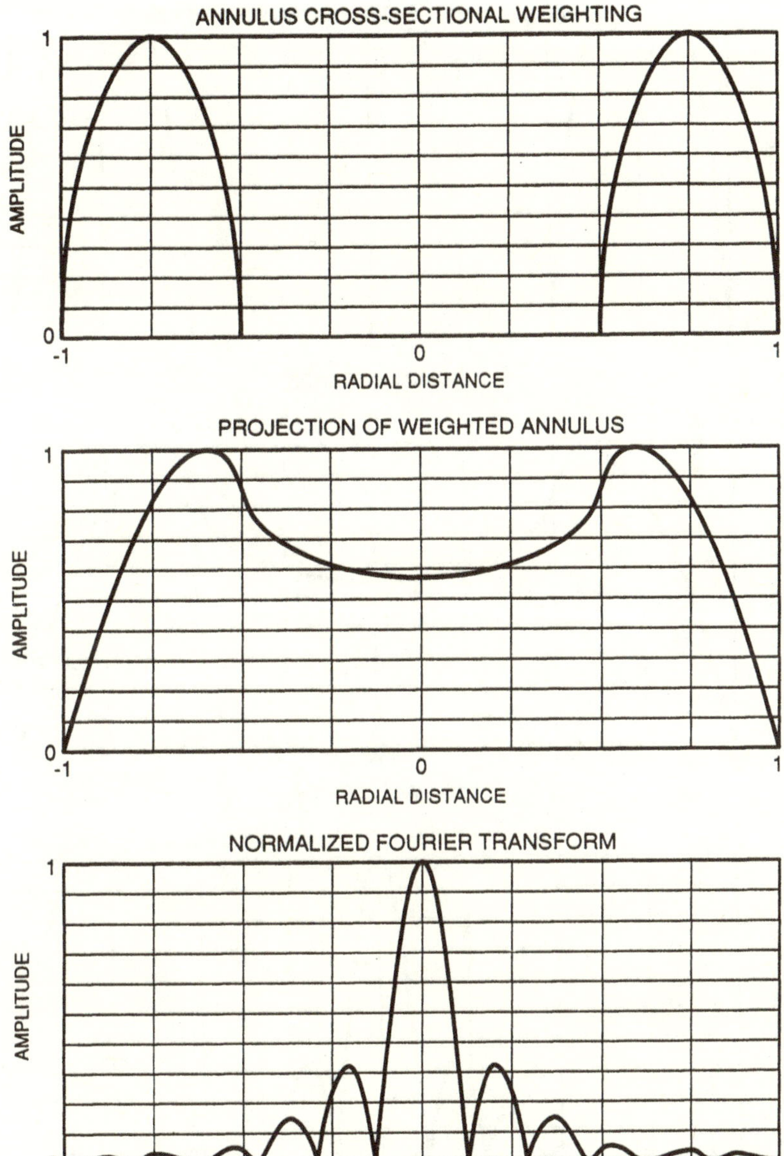

Figure B.5 Annular spectrum with $\cos^{1/2}(\cdot)$ tapered weighting.

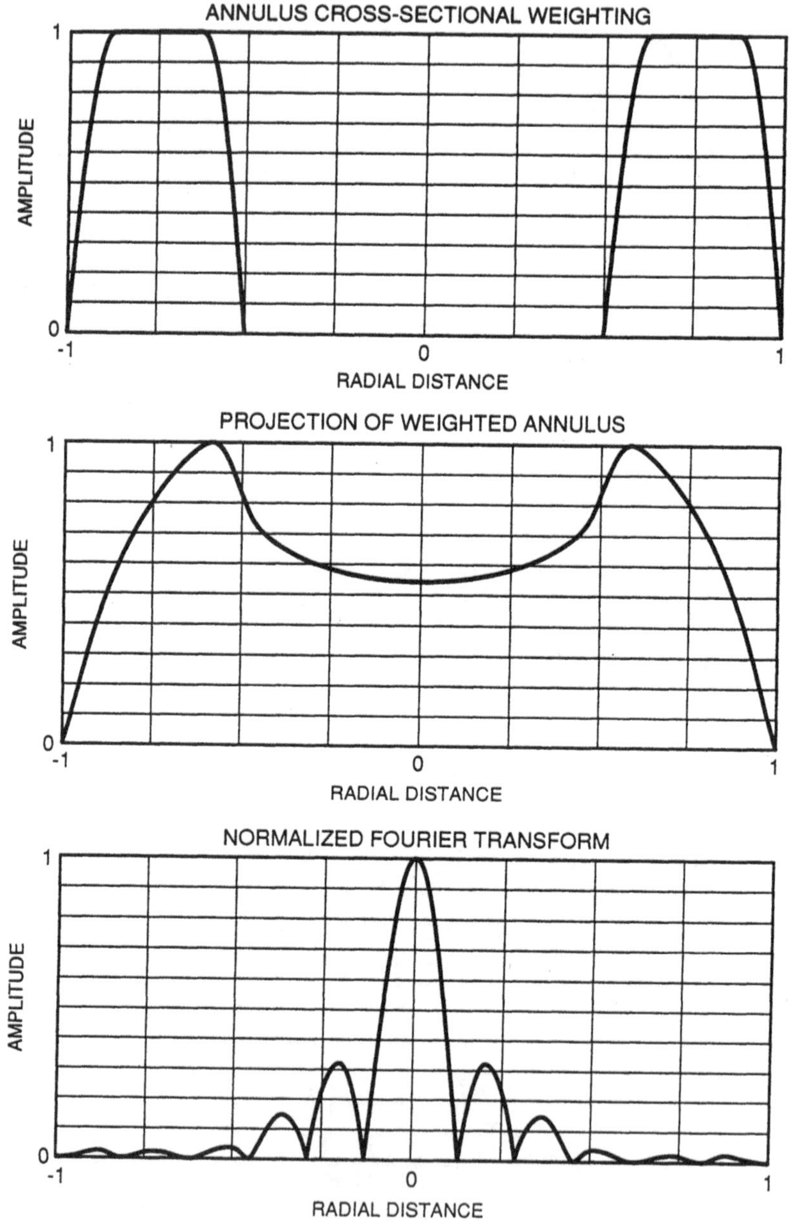

Figure B.6 Annular spectrum with cos(·) tapered edge weighting.

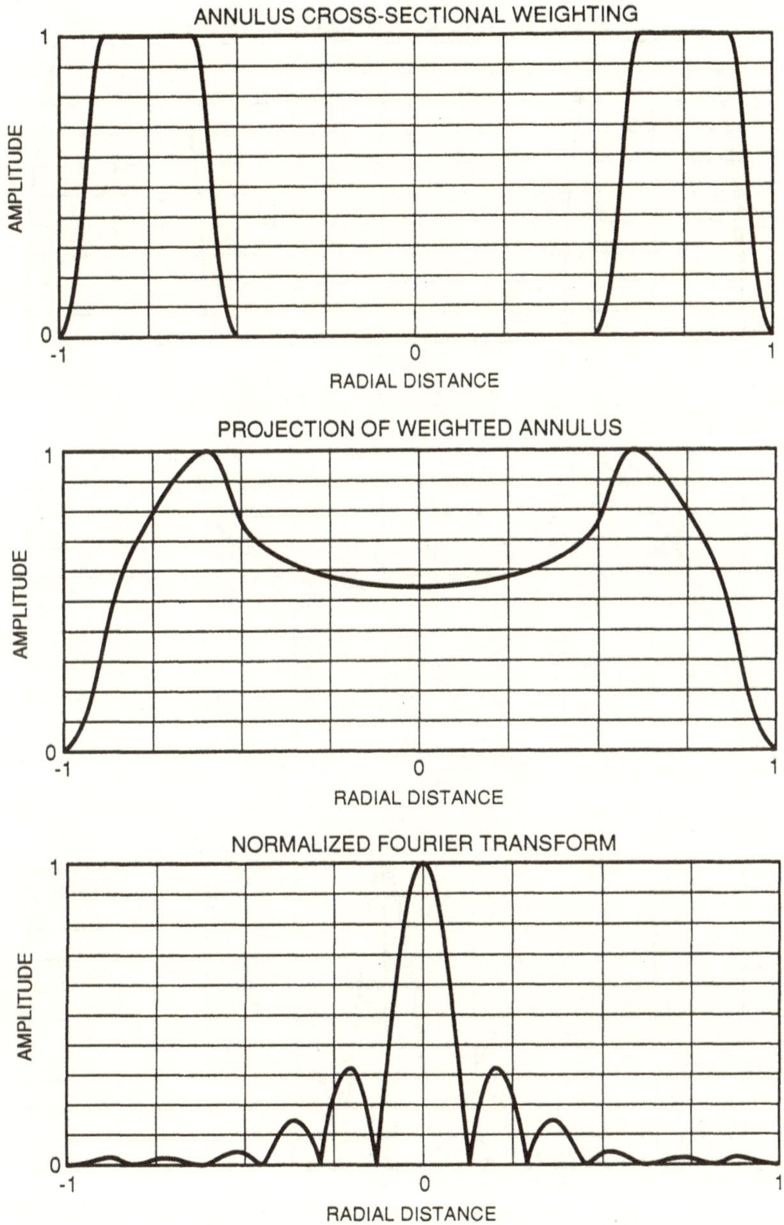

Figure B.7 Annular spectrum with $\cos^2(\cdot)$ tapered edge weighting.

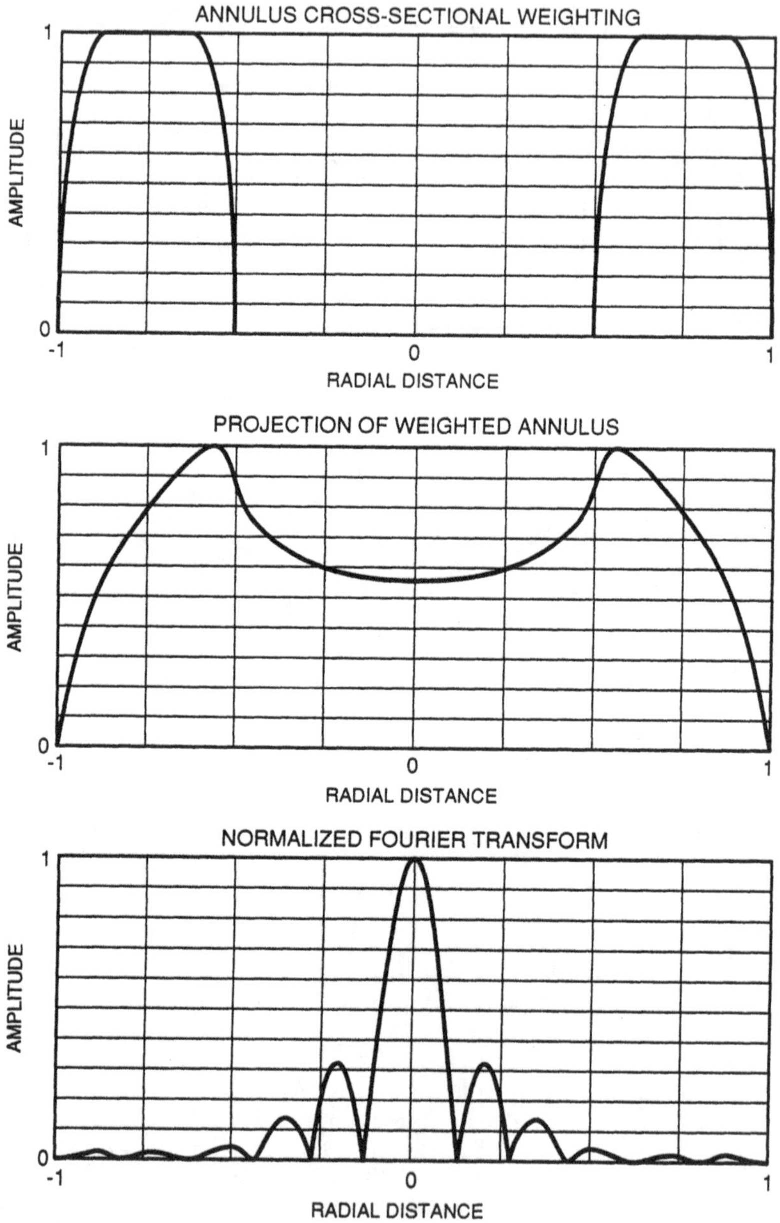

Figure B.8 Annular spectrum with $\cos^{1/2}(\cdot)$ tapered edge weighting.

cause of the relatively high sidelobes in the point-spread function. As the radial extent of the annulus decreases, the peaks become more pronounced and the sidelobes in the point-spread function increase. Figures B.3–B.5 show effects of radially tapering the annulus with functions of the form $\cos^{\alpha}(\cdot)$ for α of 1, 2, and 0.5. Figures B.6–B.8 show the effects of similar tapers applied to the inner and outer edges of the annulus. Each of the tapers results in an increase in the level of the sidelobe adjacent to the peak.

In each case, the tapering enhances the peaked behavior of the projection, thereby raising the sidelobe level of the corresponding spectrum. A number of additional tapers were tested, results for which are not shown. These included functions which provided smooth tapers radially outward and inward. For each taper tested, the sidelobes of the point-spread function were increased over those for the uniform annular spectrum. Although the results are not analytically conclusive, they support the assertion that smooth tapering of the annular spectra does not decrease the sidelobes of the point-spread function, but rather increases them.

REFERENCES

1. Harris, F.J., "On the Use of Windows for Harmonic Analyses with the Discrete Fourier Transform," *Proc. IEEE,* Vol. 66, No. 1, January 1978, pp. 51–83.
2. Bracewell, R., *The Fourier Transform and Its Applications,* McGraw-Hill, New York, 1965, pp. 244–250.
3. Bracewell, R.N., and A.C. Riddle, "Inversion of Fan-Beam Scans in Radio Astronomy," *The Astrophysical Journal,* Vol. 150, November 1967, pp. 427–434.
4. Shepp, L.A., and J.B. Kruskal, "Computerized Tomography: The New Medical X-Ray Technology," *American Mathematical Monthly,* Vol. 85, No. 6, June 1978, 420–439.
5. Mersereau, R.M., and A.V. Oppenheim, "Digital Reconstruction of Multidimensional Signals from their Projections," *Proc. IEEE,* Vol. 62, No. 10, October 1974, pp. 1319–1338.

Index

The Artech House Radar Library

David K. Barton, *Series Editor*

Active Radar Electronic Countermeasures by Edward J. Chrzanowski

Airborne Pulsed Doppler Radar by Guy V. Morris

AIRCOVER: Airborne Radar Vertical Coverage Calculation Software and User's Manual by William A. Skillman

Analog Automatic Control Loops in Radar and EW by Richard S. Hughes

Aspects of Modern Radar, by Eli Brookner, *et al.*

Aspects of Radar Signal Processing by Bernard Lewis, Frank Kretschmer, and Wesley Shelton

Detectability of Spread-Spectrum Signals by Robin A. Dillard and George M. Dillard

Electronic Homing Systems by M.V. Maksimov and G.I. Gorgonov

Electronic Intelligence: The Analysis of Radar Signals by Richard G. Wiley

Electronic Intelligence: The Interception of Radar Signals by Richard G. Wiley

EREPS: Engineer's Refractive Effects Prediction System Software and User's Manual, developed by NOSC

Handbook of Radar Measurement by David K. Barton and Harold R. Ward

High Resolution Radar by Donald R. Wehner

High Resolution Radar Imaging by Dean L. Mensa

Interference Suppression Techniques for Microwave Antennas and Transmitters by Ernest R. Freeman

Introduction to Electronic Warfare by D. Curtis Schleher

Introduction to Sensor Systems by S.A. Hovanessian

Logarithmic Amplification by Richard Smith Hughes

Modern Radar System Analysis by David K. Barton

Monopulse Principles and Techniques by Samuel M. Sherman

Monopulse Radar by A.I. Leonov and K.I. Fomichev

Multifunction Array Radar Design by Dale R. Billetter

Multisensor Data Fusion by Edward L. Waltz and James Llinas

Multiple-Target Tracking with Radar Applications by Samuel S. Blackman

Multitarget-Multisensor Tracking: Advanced Applications, Yaakov Bar-Shalom, ed.

Over-The-Horizon Radar by A.A. Kolosov, et al.

Principles and Applications of Millimeter-Wave Radar, Charles E. Brown and Nicholas C. Currie, eds.

Principles of Modern Radar Systems by Michel H. Carpentier

Pulse Train Analysis Using Personal Computers by Richard G. Wiley and Michael B. Szymanski

Radar and the Atmosophere by Alfred J. Bogush, Jr.

Radar Anti-Jamming Techniques by M.V. Maksimov, *et al.*

Radar Cross Section by Eugene F. Knott, *et al.*

Radar Detection by J.V. DiFranco and W.L. Rubin

Radar Propagation at Low Altitudes by M.L. Meeks

Radar Range-Performance Analysis by Lamont V. Blake

Radar Reflectivity Measurement: Techniques and Applications, Nicholas C. Currie, ed.

Radar Reflectivity of Land and Sea by Maurice W. Long

Radar System Design and Analysis by S.A. Hovanessian

Radar Technology, Eli Brookner, ed.

Receiving Systems Design by Stephen J. Erst

Radar Vulnerability to Jamming by Robert N. Lothes, Michael B. Szymanski, and Richard G. Wiley

RGCALC: Radar Range Detection Software and User's Manual by John E. Fielding and Gary D. Reynolds

SACALC: Signal Analysis Software and User's Guide by William T. Hardy

Secondary Surveillance Radar by Michael C. Stevens

SIGCLUT: Surface and Volumetric Clutter-to-Noise, Jammer and Target Signal-to-Noise Radar Calculation Software and User's Manual by William A. Skillman

Signal Theory and Random Processes by Harry Urkowitz

Solid-State Radar Transmitters by Edward D. Ostroff, *et al.*

Space-Based Radar Handbook, Leopold J. Cantafio, ed.

Statistical Theory of Extended Radar Targets by R.V. Ostrovityanov and F.A. Basalov

The Scattering of Electromagnetic Waves from Rough Surfaces by Petr Beckmann and Andre Spizzichino

VCCALC: Vertical Coverage Plotting Software and User's Manual by John E. Fielding and Gary D. Reynolds

www.ingramcontent.com/pod-product-compliance
Lightning Source LLC
LaVergne TN
LVHW101500200926
848483LV00001B/174

* 9 7 8 0 8 9 0 0 6 3 8 9 7 *